**Multimetallic Catalysts
in Organic Synthesis**

*Edited by
M. Shibasaki, Y. Yamamoto*

*Further Titles of Interest*

J.-E. Bäckvall (Ed.)

## Modern Oxidation Methods

2004
ISBN 3-527-30642-0


M. Beller, C. Bolm (Eds.)

## Transition Metals for Organic Synthesis

**Building Blocks and Fine Chemicals**
**Second Edition, 2 Volumes**

2004
ISBN 3-527-30613-7


A. de Meijere, F. Diederich (Eds.)

## Metal-Catalyzed Cross-Coupling Reactions

**Second Edition, 2 Volumes**

2004
ISBN 3-527-30518-1


S.-I. Murahashi (Ed.)

## Ruthenium in Organic Synthesis

2004
ISBN 3-527-30692-7

# Multimetallic Catalysts in Organic Synthesis

*Edited by*
*Masakatsu Shibasaki and Yoshinori Yamamoto*

**WILEY-VCH**

WILEY-VCH Verlag GmbH & Co. KGaA

Editors:

**Prof. Dr. Masakatsu Shibasaki**
Graduate School of Pharmaceutical Sciences
The University of Tokyo
7-3-1 Hongo, Bunkyo-ku
Japan

**Prof. Dr. Yoshinori Yamamoto**
Department of Chemistry
Tohoku University
Graduate School of Science
Aramaki-Aoba, Aobaku
Sendai 980-8578
Japan

■ All books published by Wiley-VCH are carefully produced. Nevertheless, authors, editors, and publisher do not warrant the information contained in these books, including this book, to be free of errors. Readers are advised to keep in mind that statements, data, illustrations, procedural details or other items may inadvertently be inaccurate.

**Library of Congress Card No.:** Applied for
British Library Cataloging-in-Publication Data
A catalogue record for this book is available from the British Library.

**Bibliographic information published by Die Deutsche Bibliothek**
Die Deutsche Bibliothek lists this publication in the Deutsche Nationalbibliografie; detailed bibliographic data is available in the internet at http://dnb.ddb.de.

© 2004 Wiley-VCH Verlag GmbH & Co. KGaA, Weinheim

Printed in the Federal Republic of Germany.
Printed on acid-free paper.

**Composition**   Manuela Treindl, Laaber
**Printing**   Strauss GmbH, Mörlenbach
**Bookbinding**   Litges & Dopf Buchbinderei GmbH, Heppenheim

**ISBN**   3-527-30828-8

# Contents

*Multimetallic Catalysts in Organic Synthesis.* Edited by M. Shibasaki and Y. Yamamoto
Copyright © 2004 WILEY-VCH Verlag GmbH & Co. KGaA, Weinheim
ISBN: 3-527-30828-8

# Preface

Tyrosinase monooxygenases, which catalyze the ortho-hydroxylation of phenols, are dioxygen-activating enzymes. These proteins contain bimetallic centers, but it is not known in detail how the bimetallic centers are involved in enzyme activity, although there might be cooperative effects between the two metal atoms. Chemical transformations induced by bimetallic (or multimetallic) catalyst systems often have higher reaction rates and selectivities than those induced by monometallic and mononuclear complex catalysts, indicating that the development of multimetallic catalysts is one of the most important fields in chemical science for the 21st century.

A few years ago, Dr. Elke Maase suggested that we edit a book about multimetallic catalysts. Although we were very interested, there was some hesitation, since multimetallic catalysts were still very much under development. After several discussions, we decided to edit the book, because we believed that it would be very helpful to chemists who were interested in considering new ideas.

This book is about one of the major developments in current organic chemistry: the use and application of multimetallic catalysts for the synthesis of fine chemicals and structurally more complicated organic building blocks. Multimetallic catalysts have contributed tremendously to organic synthesis over the last few decades, and it seems clear that this trend will continue in future. The aim of this book is to provide readers with an interest in multimetallic chemistry for organic synthesis an insightful, up-to-date survey of this area. We have focused on the most important multimetal catalyzed methods as well as multitransition metal catalyzed reactions with a particular emphasis on the most recent developments. The literature available up to early 2003 was carefully reviewed, and, in some cases, important reactions that were published even in late 2003 were also incorporated. Due to space restrictions, we emphasized the methodology and paid less attention to the total synthesis of natural products compared to the synthesis of structurally less complicated building blocks and fine chemicals. We believe that in most cases the latter examples more clearly illustrate the principles that govern the reactivity of multimetallic catalysts in organic synthesis.

The book consists of three sections. The first describes efficient chemical transformations using two different metals such as the Nozaki-Hiyama-Kishi reaction, where two metals exist independently to effectively promote chemical transformations. The second section discusses asymmetric catalyses promoted by

*Multimetallic Catalysts in Organic Synthesis.* Edited by M. Shibasaki and Y. Yamamoto
Copyright © 2004 WILEY-VCH Verlag GmbH & Co. KGaA, Weinheim
ISBN: 3-527-30828-8

heterobimetallic (or homobimetallic) catalysts, while the final section describes bimetallic (or multimetallic) catalyst-promoted transformations.

Finally, we sincerely hope that this book will be a valuable source of information for researchers working in academia and industry and that it will stimulate new development in this fascinating and intellectually appealing interdisciplinary area.

Masakatsu Shibasaki
Yoshinori Yamamoto

# List of Contributors

A. E. Gekhman
N. S. Kurnakov Institute of General
and Inorganic Chemistry
Russian Academy of Sciences
Leninskii pr., 31
119991 Moscow
Russia

L. S. Glebov
A. V. Topchiev Institute of
Petrochemical Synthesis
Russian Academy of Sciences
Leninskii pr., 29
119991 Moscow
Russia

Patrick M. Henry
Loyola University of Chicago
Department of Chemistry
6525 North Sheridan Road
Chicago, IL 60626
USA

Masanobu Hidai
Tokyo University of Science
Faculty of Industrial Science and
Technology
Department of Material Science
and Technology
Noda
Chiba 278-8510
Japan

Zhaomin Hou
RIKEN Institute
Organometallic Chemistry Laboratory
Hirosawa 2-1, Wako
Saitama 351-0198
Japan

Youichi Ishii
Chuo University
Faculty of Science and Engineering
Department of Applied Chemistry
Kasuga, Bunkyo-ku
Tokyo 112-8551
Japan

Shin Kamijo
Research Center for Sustainable
Materials Engineering
Institute of Multidisciplinary Research
for Advanced Materials
Tohoku University
Sendai 980-8578
Japan

Motomu Kanai
The University of Tokyo
Graduate School of Pharmaceutical
Sciences
Hongo 7-3-1, Bunkyo-ku
Tokyo 113-0033
Japan

*Multimetallic Catalysts in Organic Synthesis.* Edited by M. Shibasaki and Y. Yamamoto
Copyright © 2004 WILEY-VCH Verlag GmbH & Co. KGaA, Weinheim
ISBN: 3-527-30828-8

G. Yu. Kliger
A. V. Topchiev Institute of
Petrochemical Synthesis
Russian Academy of Sciences
Leninskii pr., 29
119991 Moscow
Russia

D. I. Kochubey
G. K. Boreskov Institute of Catalysis
S.B.
Russian Academy of Sciences
Lavrentyev pr., 5
630000 Novosibirsk
Russia

V. V. Kriventsov
G. K. Boreskov Institute of Catalysis
S.B.
Russian Academy of Sciences
Lavrentyev pr., 5
630000 Novosibirsk
Russia

V. Ya. Kugel
A. V. Topchiev Institute of
Petrochemical Synthesis
Russian Academy of Sciences
Leninskii pr., 29
119991 Moscow
Russia

Naoya Kumagai
The University of Tokyo
Graduate School of Pharmaceutical
Sciences
Hongo 7-3-1, Bunkyo-ku
Tokyo 113-0033
Japan

Yu. V. Maksimov
N. N. Semenov Institute of Chemical
Physics
Russian Academy of Sciences
Kosygin st., 4
117977 Moscow
Russia

Shigeki Matsunaga
The University of Tokyo
Graduate School of Pharmaceutical
Sciences
Hongo 7-3-1, Bunkyo-ku
Tokyo 113-0033
Japan

A. I. Mikaya
National Institute of Standards and
Technology
Gaithersburg, MD
USA

Ilya Moiseev
N. S. Kurnakov Institute of General
and Inorganic Chemistry
Russian Academy of Sciences
Lenninskey pr., 31
117907 Moscow, GSP-1

V. P. Mordovin
A. A. Baikov Institute of Metallurgy
and Material Sciences
Russian Academy of Sciences
Leninskii pr., 49
119991 Moscow
Russia

J. A. Navio
Instituto de Ciença de Materiales de
Sevilla
C/Americo Vespucio, s/n
41092 Sevilla
Spain

Takashi Ohshima
The University of Tokyo
Graduate School of Pharmaceutical
Sciences
Hongo 7-3-1, Bunkyo-ku
Tokyo 113-0033
Japan

Joseph E. Remias
Lyondell Chemical Company
Newtown Square Technical Center
3801 West Chester Pike
Newtown Square, PA 19073
USA

Ayusman Sen
Department of Chemistry
The Pennsylvania State University
University Park, PA 16802
USA

Masakatsu Shibasaki
The University of Tokyo
Graduate School of Pharmaceutical
Sciences
Hongo 7-3-1, Bunkyo-ku
Tokyo 113-0033
Japan
mshibasa@mol.f.u-tokyo.ac.jp

George G. Stanley
Department of Chemistry
Louisiana State University
Baton Rouge, LA 70803-1804
USA

M. V. Tsodikov
A. V. Topchiev Institute of
Petrochemical Synthesis
Russian Academy of Sciences
Leninskii pr., 29
119991 Moscow
Russia

Yoshinori Yamamoto
Department of Chemistry
Tohoku University, Graduate School of
Science
Aramaki-Aoba, Aobaku
Sendai 980-8578
Japan

F. A. Yandieva
A. V. Topchiev Institute of
Petrochemical Synthesis
Russian Academy of Sciences
Leninskii pr., 29
119991 Moscow
Russia

V. G. Zaikin
A. V. Topchiev Institute of
Petrochemical Synthesis
Russian Academy of Sciences
Leninskii pr., 29
119991 Moscow
Russia

# 1
# Organic Synthesis with Bimetallic Systems

*Shin Kamijo and Yoshinori Yamamoto*

## 1.1
## Introduction

The application of bimetallic systems to organic synthesis has emerged dramatically in recent years, and great progress has been made in research aimed at developing reactions promoted with catalytic amounts of activating reagents. The cross-coupling reaction is a representative example of this type of transformation. In the early stages of the investigations, most studies were focused on transition metal (TM)-catalyzed reactions using main group organometallic compounds ($R^2$–M). The organometallic compound ($R^2$–M) was used as a coupling partner of the substrate ($R^1$–X); the cross-coupling reactions can be regarded as transformations promoted by a bimetallic system (*cat.* TM/*stoichiometric* $R^2$–M) (Figure 1.1a). Not only cross-coupling reactions (Section 1.2.1), but also reactions of $\pi$-allylpalladium complexes (Section 1.2.4) and nickel-catalyzed three-component coupling (TCC) reactions (Section 1.2.5) can be classified as belonging to category **a**. The conjugate addition of organomagnesium and -lithium reagents to Michael acceptors in the presence of catalytic amounts of copper salts also belongs to this category, but such organocopper reactions are not mentioned in this chapter since many excellent reviews and monographs have been published on these topics in recent years [1]. Another characteristic feature of these cross-coupling reactions is that an enhancement of the reaction rate is often observed in the presence of an additional metal salt (MX). The coupling reaction between $R^1$–X and $R^2$–M proceeds very smoothly in the presence of catalytic amounts of TM and stoichiometric amounts of MX (Figure 1.1b). Wacker reactions (Section 1.2.2), Heck reactions (Section 1.2.3), most of the reactions involving $\pi$-allylpalladium complexes (Section 1.2.4), and Nozaki–Hiyama–Kishi (NHK) reactions (Section 1.2.6) belong to this category **b**. We will discuss the reactions promoted by a combination of catalytic and stoichiometric amounts of metals (categories **a** and **b**) in the first section.

Recent studies have revealed that a wide variety of bimetallic catalytic systems composed of a transition metal and an additional metal salt (*cat.* TM/*cat.* MX) efficiently catalyze organic transformations, such as the cross-coupling reaction

*Multimetallic Catalysts in Organic Synthesis.* Edited by M. Shibasaki and Y. Yamamoto
Copyright © 2004 WILEY-VCH Verlag GmbH & Co. KGaA, Weinheim
ISBN: 3-527-30828-8

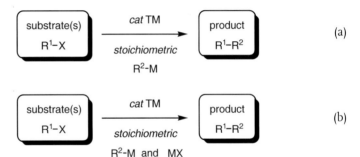

**Figure 1.1**

between $R^1$–X and $R^2$–M (Figure 1.2c), the Wacker reaction, reactions involving $\pi$-allylpalladium complexes, and so forth (category **c**). The MX catalyst often promotes these reactions by forming a reactive species in situ via transmetalation and halide abstraction. In some cases, the MX catalyst behaves as a Lewis acid and activates the substrates and intermediates through coordination. The reactions belonging to category **c** are mentioned in Sections 1.3.1 to 1.3.5. The utilization of dinuclear metal catalysts (cat. M–M) is one of the approaches to realize new catalytic transformations, although application of such catalytic systems to organic synthesis is not so popular and relatively few examples have been reported to date. The reactions belonging to this category **d** are mentioned in Sections 1.3.6.1 and 1.3.6.2 (Figure 1.2d). The unique catalytic properties of M–M catalysts originate from double activation by the two metal centers. Several combinations of two transition metals (cat. $TM^1$/cat. $TM^2$) have been applied in both one-pot and sequential reactions (Figure 1.2e). The overall transformation is achieved through the two successive reactions promoted by each transition metal catalyst. The Pauson–Khand (Section 1.3.4.4) and sequential reactions (Section 1.3.6.3) are classified as belonging to this category **e**. We will discuss the reactions promoted by a combination of catalytic amounts of two metals (categories **c**, **d**, **e**) in the second part of this chapter. The details of each reaction will be considered in each section.

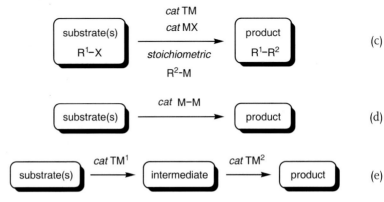

**Figure 1.2**

## 1.2
## Reactions Promoted by a Combination of Catalytic and Stoichiometric Amounts of Metals

### 1.2.1
### Transition Metal-Catalyzed Cross-Coupling Reactions

Transition metal-catalyzed cross-coupling reactions have been developed by employing various combinations of catalysts and organometallic compounds [2]. Although the organometallic compounds are exploited as a reagent, the cross-coupling reactions can be regarded as transformations promoted by a bimetallic system. A representative reaction scheme and catalytic cycle are depicted in Scheme 1.1. Generally, the reactions begin with oxidative addition of transition metal catalysts (TM) to aryl and alkenyl halides $R^1$–X (**1**) to form the intermediates **4**. Transmetalation between **4** and organometallic compounds $R^2$–M (**2**) affords the intermediates **5**, and reductive elimination of TM catalysts results in the formation of a carbon–carbon bond to furnish the coupling products $R^1$–$R^2$ (**3**). Palladium and nickel catalysts usually show excellent activities in these transformations.

$R^1$–X **1**

$+$ $\xrightarrow{cat\ TM}$ $R^1$–$R^2$

$R^2$–M **2**        **3**

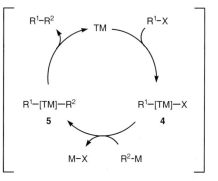

| coupling reaction | TM | M | 2 |
|---|---|---|---|
| Kumada–Tamao–Corriu | Pd, Ni | Mg | **2a** |
| Negishi | Pd, Ni | Zn | **2b** |
| Suzuki–Miyaura | Pd, Ni | B | **2c** |
| Stille | Pd | Sn | **2d** |
| Hiyama | Pd | Si | **2e** |

**Scheme 1.1**

The first successful nickel-catalyzed cross-coupling reaction between halides and Grignard reagents $R^2$–MgX (**2a**) was simultaneously achieved by two research groups and is now referred to as the Kumada–Tamao–Corriu reaction. The palladium-catalyzed cross-coupling reaction between halides and organozinc reagents $R^2$–ZnX (**2b**) was developed by Nigishi and co-workers. Some reactions were also catalyzed by a nickel complex. Organozinc reagents are most conveniently prepared in situ from organolithium, -magnesium, or -aluminum compounds with a $ZnX_2$ salt. The use of organozinc reagents extended the range of compatibility to functional groups such as ketones, esters, cyano, and amide groups, which react with Grignard reagents. Suzuki and Miyaura et al. investigated the palladium-catalyzed coupling reaction of halides with organoboron compounds $R^2$–$BR'_2$ (**2c**). The addition of base is required to activate either the boron reagent or the Pd catalyst in order to promote the coupling reaction. This transformation is widely applied due to its high tolerance of a broad range of functionalities and of water.

Coupling reactions between halides and organotin compounds $R^2$–$SnR'_3$ (**2d**) in the presence of a palladium catalyst have been extensively studied and are now referred to as the Stille coupling reaction. This transformation proceeds under essentially neutral conditions. The standard catalyst system is a combination of a Pd catalyst with a ligand such as (2-furyl)$_3$P or $AsPh_3$. Although the Stille coupling reaction is an excellent carbon–carbon bond-forming procedure, unsatisfactory results are sometimes encountered due to substrate instability. Therefore, many additives have been examined for an accelerating effect on the reaction. Gronowitz and co-workers [3] employed $Ag_2O$ as an activator in the reaction between 2-iodopyridine and a stannylthiophene (Scheme 1.2). Migita and co-workers [4] observed an acceleration of the coupling reaction between aryl bromides and α-stannylacetate by utilizing $ZnBr_2$ as an additive (Scheme 1.3). Corey and co-workers [5] applied CuCl as an activator for coupling reactions between aryl nonaflate and vinylstannanes (Scheme 1.4).

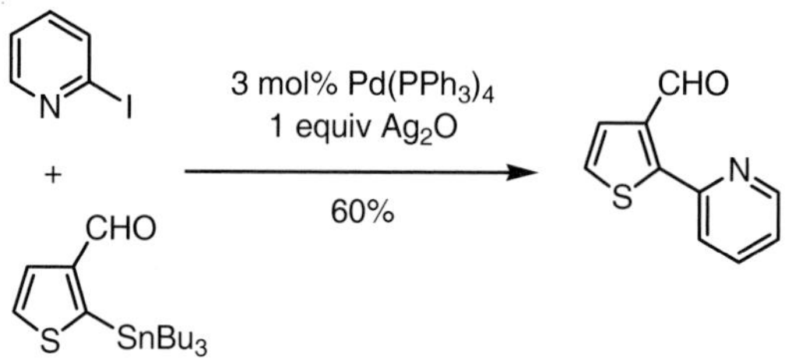
N
I
+
CHO
S
SnBu3
3 mol% Pd(PPh3)4
1 equiv Ag2O
60%
CHO
S
N

Scheme 1.2

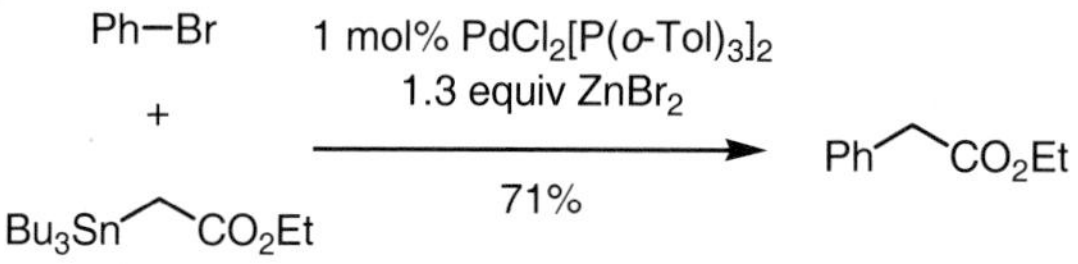
Ph—Br
+
Bu3Sn
CO2Et
1 mol% PdCl2[P(o-Tol)3]2
1.3 equiv ZnBr2
71%
Ph
CO2Et

Scheme 1.3

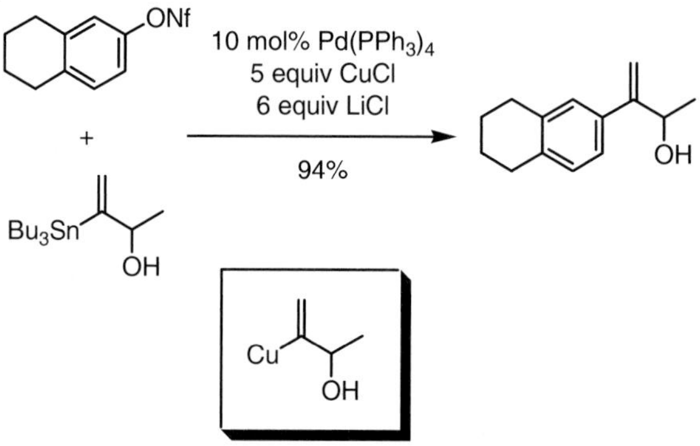
ONf
+
Bu3Sn
OH
10 mol% Pd(PPh3)4
5 equiv CuCl
6 equiv LiCl
94%
OH
Cu
OH

Scheme 1.4

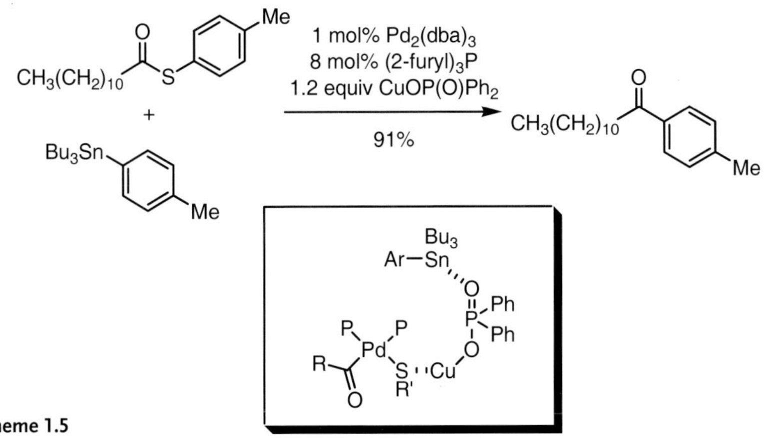

**Scheme 1.5**

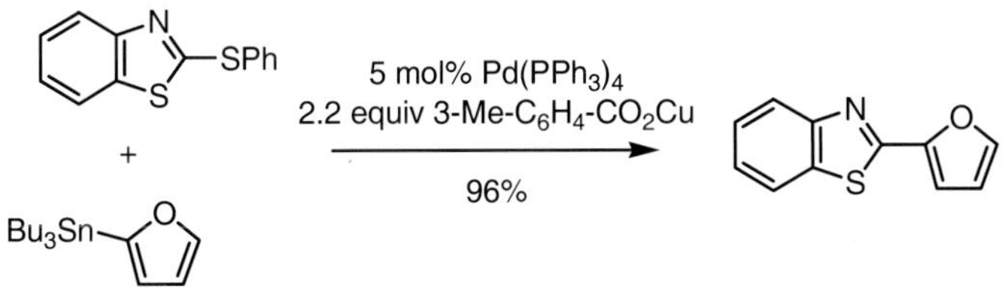

**Scheme 1.6**

These reaction activators most probably facilitate transmetalation of the organic group from the stannane to the palladium center through the formation of more reactive organometallic species such as vinylcopper, as shown in Scheme 1.4. Liebeskind and co-workers [6] have recently reported the Stille-type coupling reaction between thiol esters and organostannanes (Scheme 1.5) The reaction proceeded in the presence of a combination of Pd$_2$(dba)$_3$/(2-furyl)$_3$P as catalyst and CuOP(O)Ph$_2$ as an activator. The role of the Cu additive is activation of the acylpalladium thiolate intermediate to facilitate transmetalation of the aryl group from the stannane to the palladium atom. These authors applied a similar Pd–Cu bimetallic catalyst to the coupling reaction between a heteroaromatic thioether and an organostannane (Scheme 1.6) [7]. The same type of reaction was also reported by Guillaumet and co-workers [8]. Migita and co-workers [9] developed a method for the α-arylation of ketones based on a coupling reaction between enol acetates and bromobenzene (Scheme 1.7) The key to realizing this coupling reaction is to perform it in the presence of a combination of PdCl$_2$(o-tolyl$_3$P)$_2$ catalyst and Bu$_3$SnOMe. The enol acetates are transformed in situ to the corresponding tributyltin enolates, which react with the Pd intermediate to afford the coupling product. A similar coupling reaction utilizing silyl enol ethers and aryl bromides was investigated by Kuwajima et al. (Scheme 1.8) [10]. In this case, the addition of Bu$_3$SnF in combination with PdCl$_2$(o-tolyl$_3$P)$_2$ is essential for successful reaction.

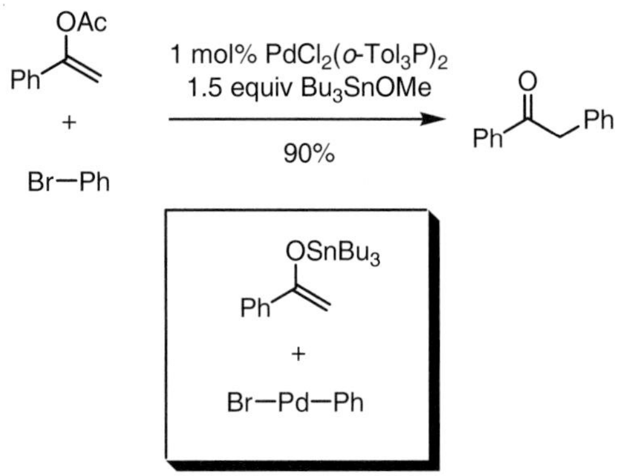

**Scheme 1.7**

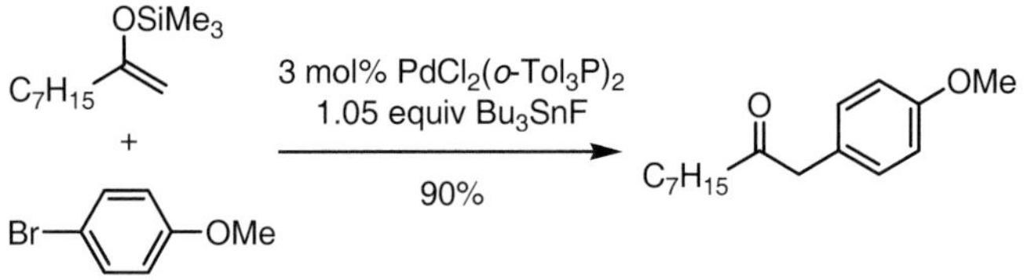

**Scheme 1.8**

Hiyama and co-workers found that a cross-coupling reaction between halides and organosilanes $R^2$–$SiR'_3$ (**2e**) could be achieved in the presence of a palladium catalyst combined with a fluoride anion source such as tetrabutylammonium fluoride (TBAF) or tris(dimethylamino)sulfur (trimethylsilyl)difluoride (TASF). The addition of a fluoride source is essential to facilitate transmetalation of the organic group through the formation of a five-coordinate silicate species. Hiyama and Mori et al. [11] applied silanols as coupling partners with aryl iodides in the cross-coupling reaction (Scheme 1.9). The reaction proceeded particularly well in the presence of $Ag_2O$ as an activator with a catalytic amount of $Pd(PPh_3)_4$. A fluoride activator such as TBAF, which is often employed for the usual Hiyama coupling reaction using a fluorosilane as a starting material, failed to give the corresponding adduct. The role of the Ag additive can be rationalized in terms of two cooperative

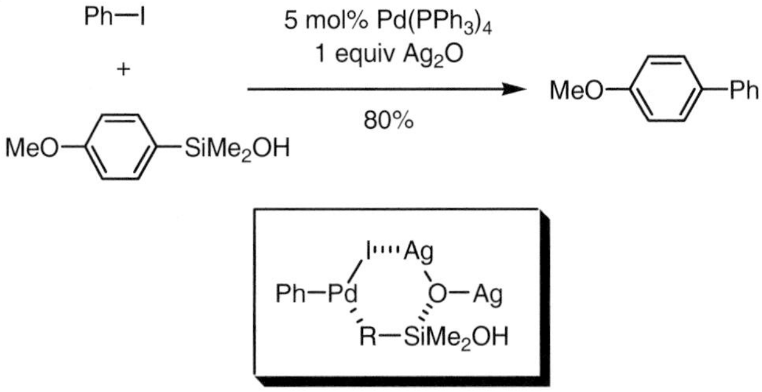

**Scheme 1.9**

functions, which may be outlined as follows. One is the formation of a penta-coordinate silicate species to facilitate transmetalation of the aryl group from the Si to the Pd atom, and the other is interaction with the iodide on the organopalladium intermediate to make transmetalation feasible. A similar reaction using alkynyl-silanols as coupling partners has been reported by Chang and co-workers [12].

## 1.2.2
## The Wacker Reaction

The synthesis of acetaldehyde **7** (R = H) by oxidation of ethylene **6** (R = H) in the presence of a catalytic amount of $PdCl_2$ and a stoichiometric amount of $CuCl_2$ is known as the Wacker process (Scheme 1.10) [13]. The reaction involves nucleophilic addition of $H_2O$ to ethylene coordinated by $PdCl_2$. The Pd catalyst activates the carbon-carbon double bond by $\pi$-coordination. The Cu additive serves to oxidize the generated $Pd^0$ species so as to regenerate the $PdCl_2$ catalyst. The role of oxygen is assumed to be oxidation of the CuCl produced to regenerate the $CuCl_2$ additive, although a stoichiometric amount of $CuCl_2$ is employed in most cases. The reaction has been extended to substituted alkenes.

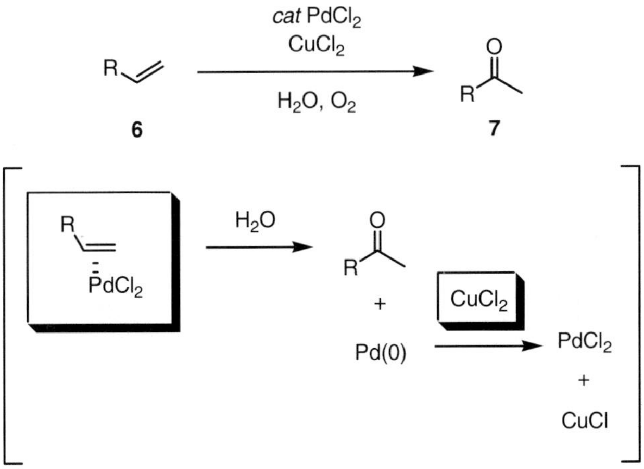

**Scheme 1.10**

Wacker-type reactions have been successfully applied for the formation of a wide variety of heterocyclic compounds. For example, various kinds of oxygen-containing heterocycles, such as tetrahydrofuran (Scheme 1.11) [14] and benzofuran derivatives (Scheme 1.12) [15], have been synthesized in the presence of a catalytic amount of

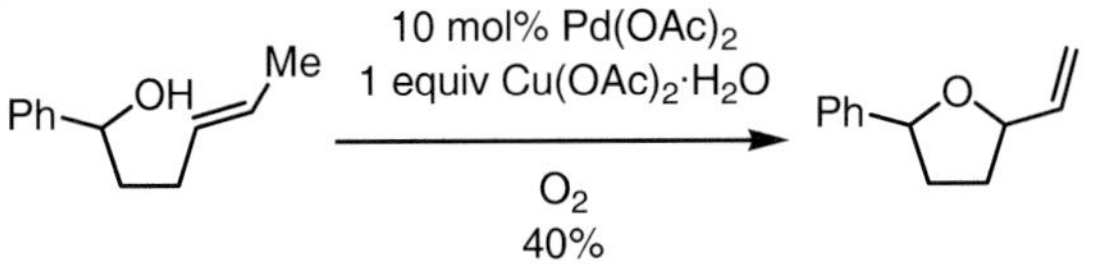

**Scheme 1.11**

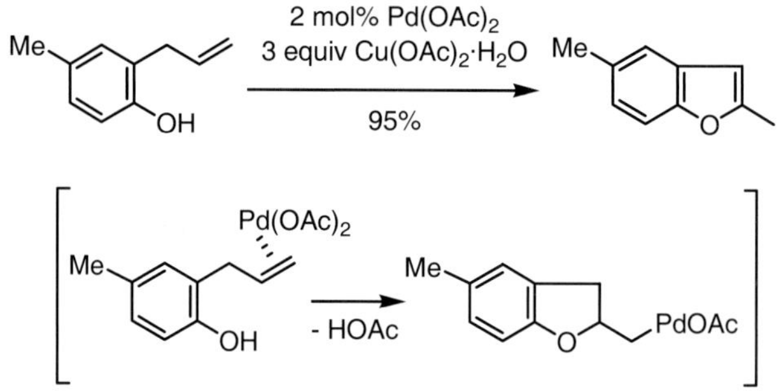

**Scheme 1.12**

a Pd$^{II}$ species and a stoichiometric amount of Cu oxidant. The reactions involve intramolecular alkoxypalladation of the alkene to form a σ-alkylpalladium intermediate, followed by β-elimination to furnish the cyclized products.

When the reaction was conducted under CO atmosphere, insertion of CO into the σ-alkylpalladium species took place to afford carbonylated products. Semmelhack et al. investigated the synthesis of pyran derivatives by using a combination of Pd(OAc)$_2$ and CuCl$_2$ oxidant under CO gas (Scheme 1.13) [16]. Tamaru and co-workers [17] succeeded in obtaining bicyclic lactone derivatives via the intramolecular alkoxycarbonylation reaction using a Pd–Cu system (Scheme 1.14). Semmelhack et al. [18] also succeeded in trapping an alkylpalladium intermediate with alkenes and the corresponding carbon chain elongated products were obtained (Scheme 1.15).

The reaction has also been applied for the synthesis of nitrogen-containing cyclic compounds. Gallagher and co-workers [19] subjected allenyl amides to a CO atmosphere in the presence of PdCl$_2$ and CuCl$_2$ and thereby obtained pyrrolidine

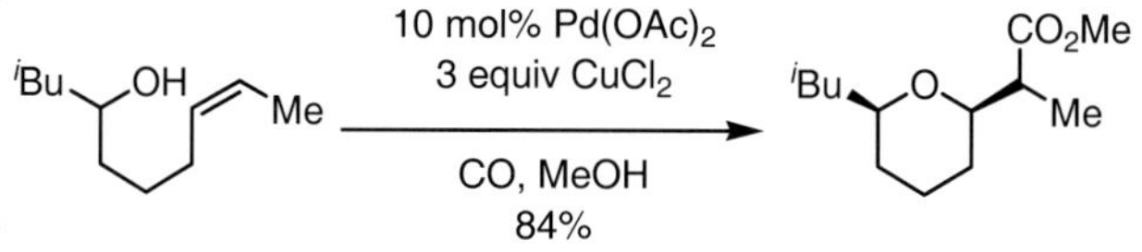

**Scheme 1.13**

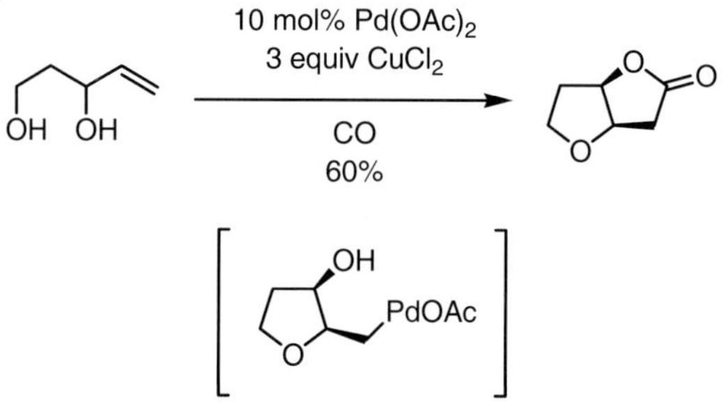

**Scheme 1.14**

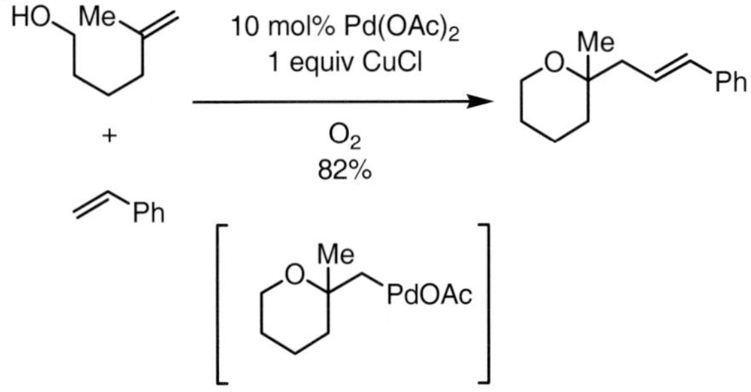

**Scheme 1.15**

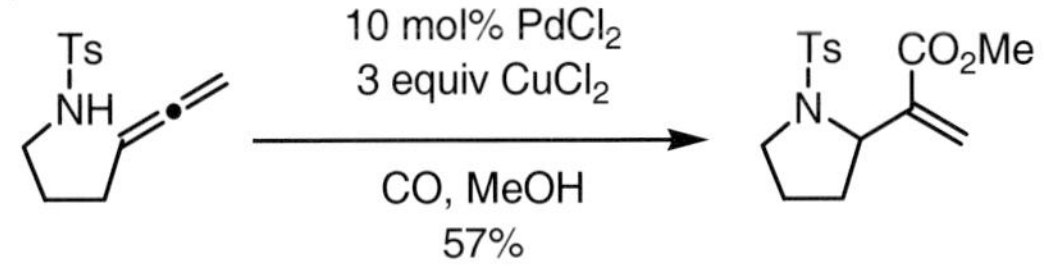

**Scheme 1.16**

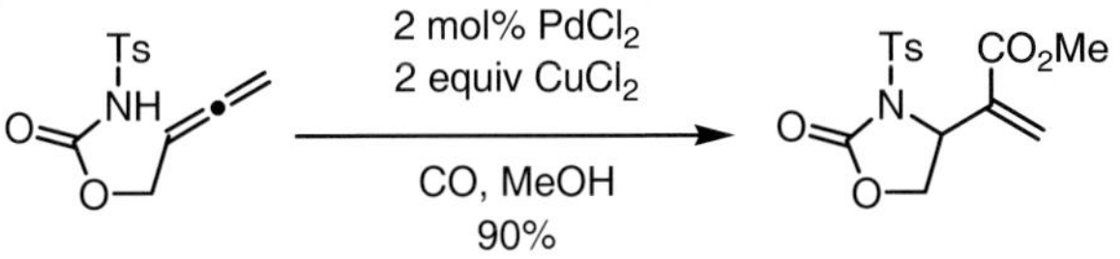

**Scheme 1.17**

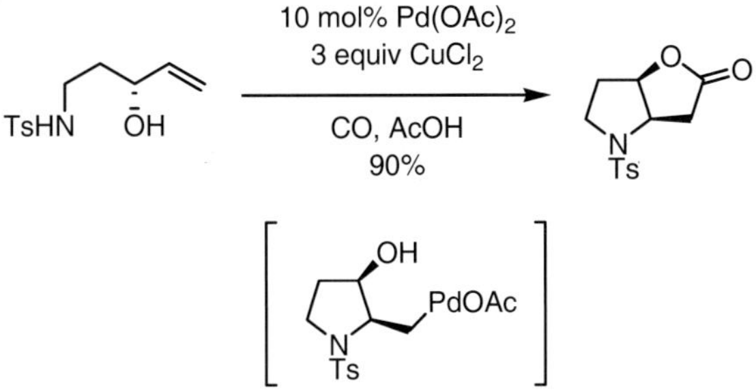

**Scheme 1.18**

derivatives (Scheme 1.16). A similar cyclization has been reported by Tamaru and co-workers [20] (Scheme 1.17). Tamaru and co-workers [21] also observed the formation of bicyclic compounds through an intramolecular aminocarbonylation reaction using a Pd–Cu system (Scheme 1.18).

Widenhoefer et al. [22] reported a similar type of cyclization reaction using a Pd–Cu bimetallic system. The reaction probably proceeds through addition of the enolate to the alkene activated by coordination of the Pd complex (Scheme 1.19).

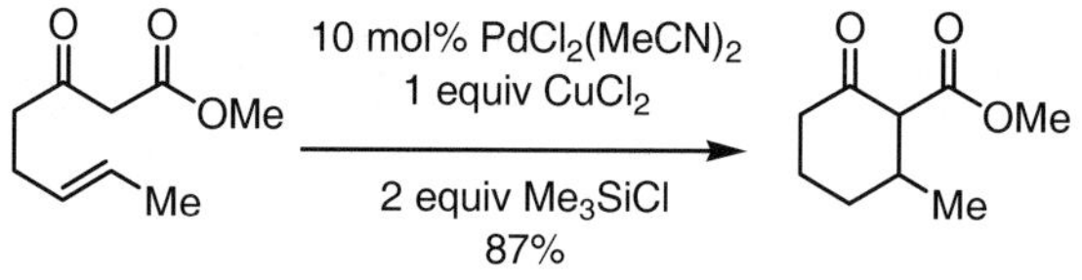

Scheme 1.19

$$\text{(diketone substrate)} \xrightarrow[\substack{2 \text{ equiv Me}_3\text{SiCl} \\ 87\%}]{\substack{10 \text{ mol\% PdCl}_2(\text{MeCN})_2 \\ 1 \text{ equiv CuCl}_2}} \text{(cyclized product)}$$

The combination a catalytic amount of $PdCl_2(MeCN)_2$ and a stoichiometric amount of $Yb(OTf)_3$ was also found to be effective by Yang and co-workers [23]. The catalytic use of $Yb(OTf)_3$ proved to be effective for some substrates, although the use of a stoichiometric amount of $Yb(OTf)_3$ was seemingly required to obtain the desired carbocycles in good yields.

## 1.2.3
## The Heck Reaction

The $Pd^0$-catalyzed coupling of an aryl or vinyl halide or triflate with an alkene is known as the Heck reaction [24]. The reaction is normally carried out using a Pd catalyst with phosphine ligands under basic conditions. Investigations aimed at selecting the optimal additive for the Heck reaction revealed that the addition of an Ag salt increased the reaction rate and led to consistently higher reactivities compared to the original catalyst systems.

A stoichiometric amount of silver additive was first employed in the coupling reaction between iodobenzenes and vinylsilane by Hallberg and co-workers [25] (Scheme 1.20). When this reaction was carried out in the presence of a combination of a catalytic amount of $Pd(OAc)_2/PPh_3$ and a stoichiometric amount of $AgNO_3$, styrylsilanes were obtained without the formation of any desilylated styrenes, which are the products under the standard conditions of the Heck reaction. The addition of the Ag salt suppressed cleavage of silyl group during the course of the reaction.

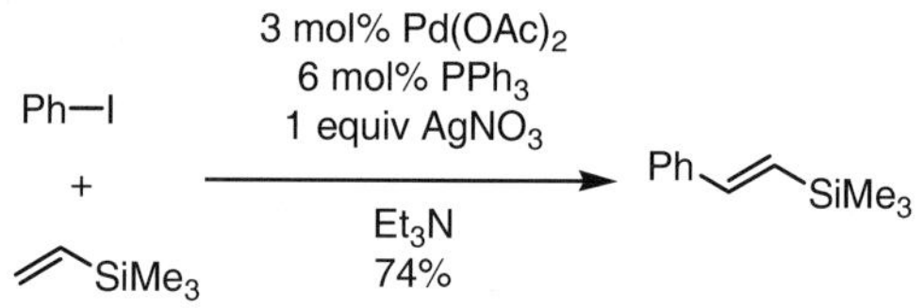

Scheme 1.20

$$\begin{array}{c} Ph-I \\ + \\ \text{(vinyl)SiMe}_3 \end{array} \xrightarrow[\substack{Et_3N \\ 74\%}]{\substack{3 \text{ mol\% Pd(OAc)}_2 \\ 6 \text{ mol\% PPh}_3 \\ 1 \text{ equiv AgNO}_3}} Ph\text{—}\text{(CH=CH)}\text{SiMe}_3$$

Overman and co-workers [26] found that the addition of a silver salt minimized alkene isomerization in the derived products (Scheme 1.21). Jeffery [27] reported that the addition of an Ag salt to suppress alkene isomerization was especially effective in reactions using allyl alcohols as coupling partners (Scheme 1.22). The formation of aldehydes could be avoided by simply adding a stoichiometric amount of AgOAc to the reaction mixture.

Acceleration of reaction rates and enhancement of enantioselectivity were observed in the presence of a Pd–Ag catalyst system, and extensive investigations on the construction of quaternary carbon centers via the intramolecular Heck reaction were carried out. Overman and co-workers first reported the intramolecular

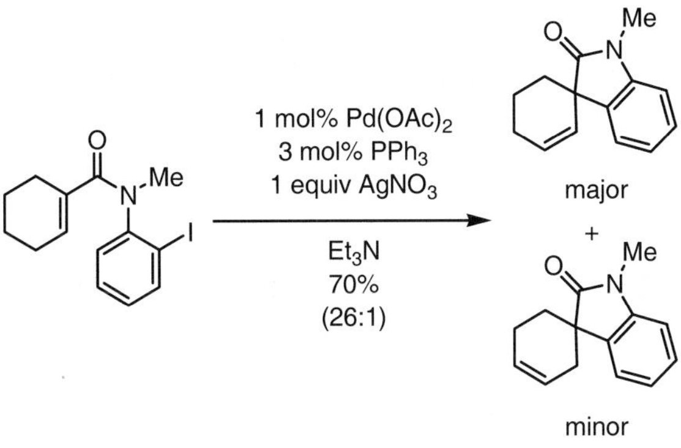

**Scheme 1.21**

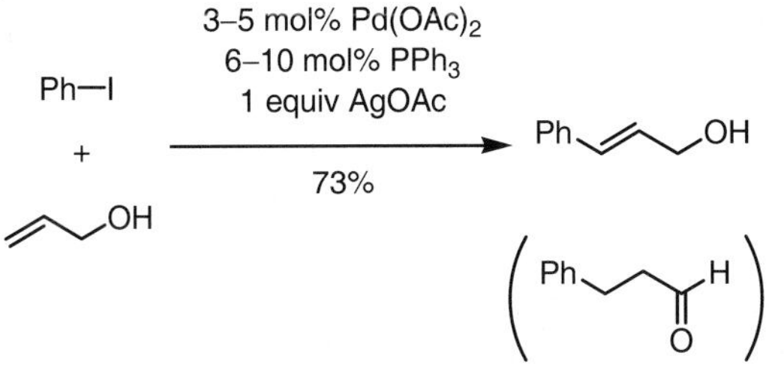

**Scheme 1.22**

Heck reaction using a combination of a catalytic amount of Pd(OAc)$_2$/PPh$_3$ complex and a stoichiometric amount of AgNO$_3$, as indicated in Scheme 1.21. Shibasaki and co-workers [28] demonstrated for the first time that an enantioselective intramolecular Heck reaction could be achieved by utilizing a catalytic amount of Pd(OAc)$_2$/(R)-BINAP with a stoichiometric amount of Ag$_2$CO$_3$ (Scheme 1.23). The Heck reaction in the presence of an Ag salt is proposed to proceed through a cationic pathway. The role of the Ag salt is to abstract the halide ion from the palladium intermediate to leave a vacant site for coordination of the alkene. Grigg and co-workers [29] reported that the addition of a Tl$^I$ salt instead of an Ag$^I$ salt led to a similar effect.

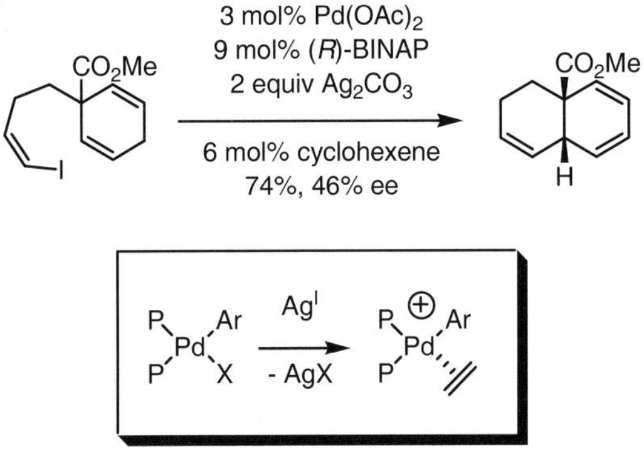

**Scheme 1.23**

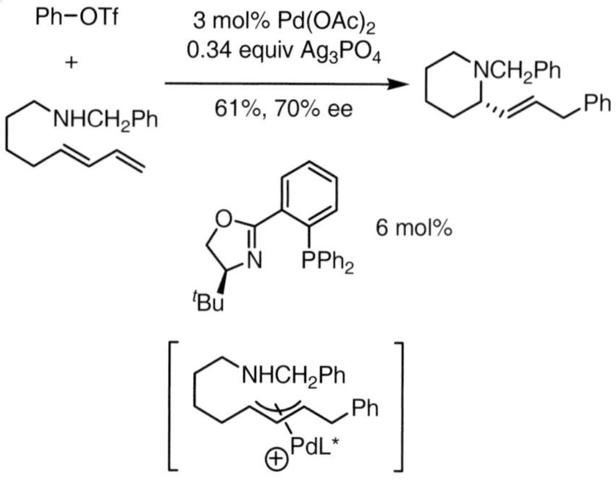

**Scheme 1.24**

A silver salt is often utilized in enantioselective tandem reactions involving the Heck reaction. For example, Helmchen et al. [30] reported the synthesis of chiral piperidine derivatives through the Heck reaction and subsequent asymmetric allylic amination of the resulting π-allylpalladium intermediate (Scheme 1.24). The addition of a stoichiometric amount of an Ag salt proved essential to achieve high enantioselectivity. Larock and co-workers [31] reported the asymmetric hetero-annulation of allenes using functionalized aryl iodides (Scheme 1.25).

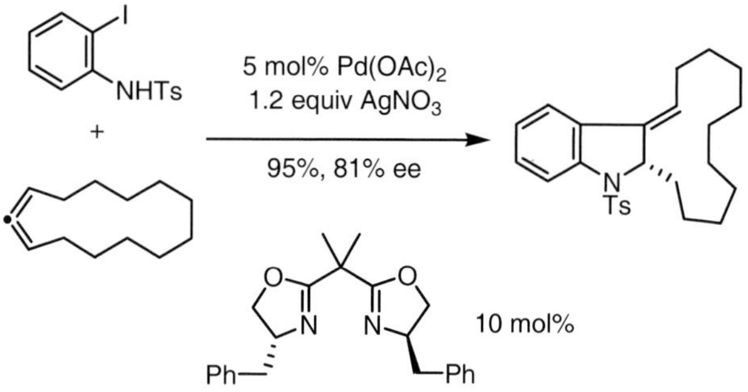

**Scheme 1.25**

## 1.2.4
### Reactions Involving π-Allylpalladium Intermediates

#### 1.2.4.1 Electrophilic Reactions
Poli and co-workers [32] applied a $Pd_2(dba)_3/PPh_3$ and $Ti(O^iPr)_4$ bimetallic catalyst system to the allylation of active methylene compounds (Scheme 1.26). They investigated the reactions between allyl acetates and various carbon nucleophiles. The Ti additive coordinates to the nucleophiles and produces titanium enolates after deprotonation. The $pK_a$ value of the active methylene compound is lowered by the coordination of the Ti additive, making it more reactive. Accordingly, a wide range of nucleophilic agents can be used as reaction partners with π-allylpalladium

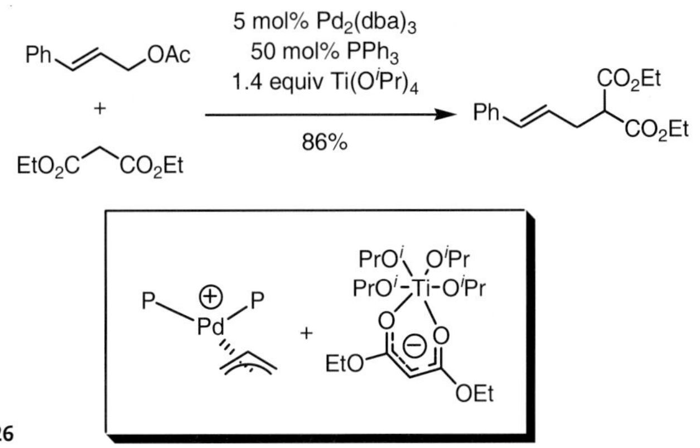

**Scheme 1.26**

complexes in the presence of the Pd–Ti catalyst system. Applying this methodology to the intramolecular reaction, these authors were able to obtain 2,3-disubstituted pyrrolidinones (Scheme 1.27).

Fuji and co-workers [33] used $Et_2Zn$ as a base in a palladium-catalyzed asymmetric allylic alkylation reaction (Scheme 1.28). They found that the use of a combination of $[(\eta^3\text{-}C_3H_5)PdCl]_2/(R)$-BINAP and a stoichiometric amount of $Et_2Zn$ led to a remarkable enhancement of the enantiomeric excess. In contrast to $Et_2Zn$, the use of bases such as KH, NaH, LiH, and BuLi resulted in low enantioselectivities.

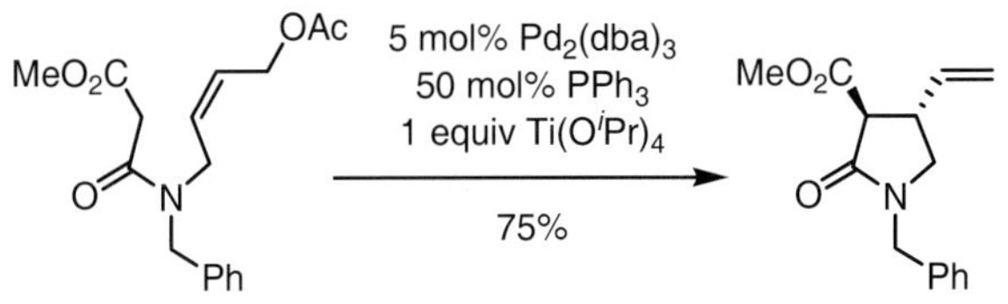

**Scheme 1.27**

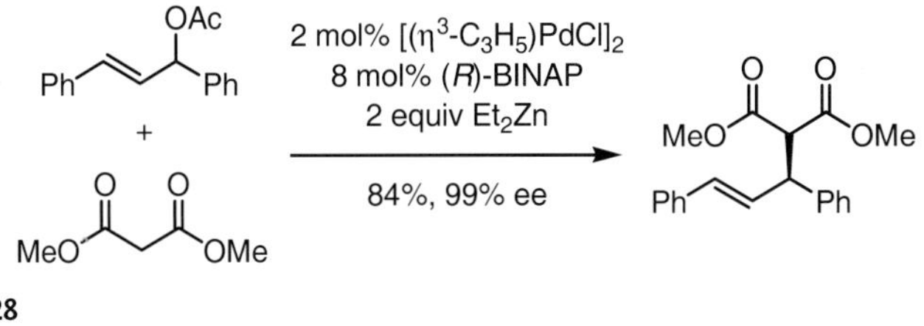

**Scheme 1.28**

### 1.2.4.2 Nucleophilic Reactions

It is well known that $\pi$-allylpalladium intermediates generally behave as allyl cation equivalents and they are widely used for the allylation of various nucleophiles. The reactivity of $\pi$-allylpalladium intermediates can be switched from the ordinary electrophilic one to nucleophilic reactivity when they are combined with a stoichiometric amount of an appropriate metal reagent.

Inanaga and co-workers [34] found that reactions of allyl acetates with ketones in the presence of a combination of catalytic Pd(PPh$_3$)$_4$ and a stoichiometric amount of SmI$_2$ produced homoallylic alcohols (Scheme 1.29). Under these conditions, however, aromatic and α,β-unsaturated carbonyl compounds could not be used because pinacol-type self-coupling products were obtained predominantly.

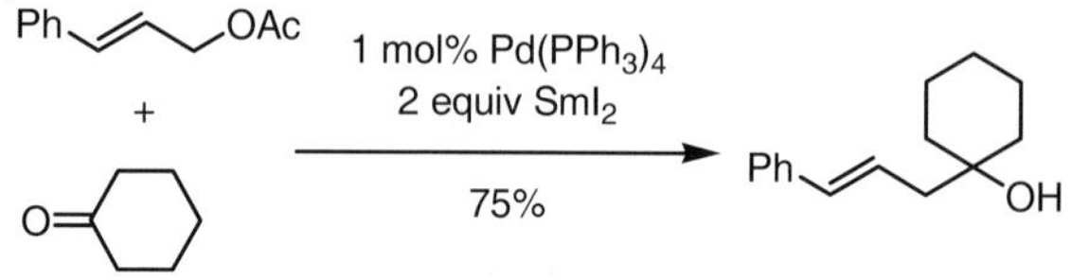

**Scheme 1.29**

Masuyama and co-workers [35] reported a similar type of reaction using Pd(PPh$_3$)$_4$ combined with a stoichiometric amount of Zn powder (Scheme 1.30). Aromatic and α,β-unsaturated aldehydes could be used with this Pd–Zn system.

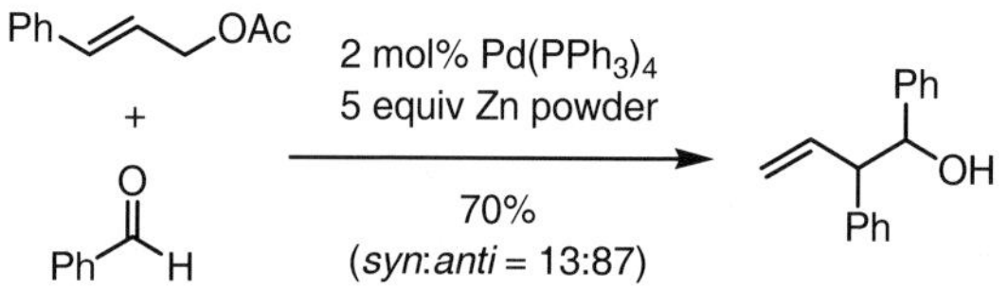

**Scheme 1.30**

Masuyama and co-workers [36] also developed a new catalyst combination, PdCl$_2$(PhCN)$_2$ and SnCl$_2$, for charge reversal of the π-allylpalladium complex derived from allyl alcohols (Scheme 1.31). An NMR study revealed that the actual allylating agent was an allyltrichlorotin species.

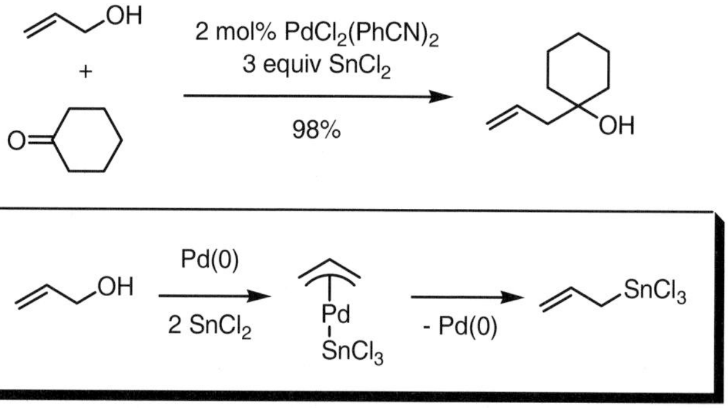

**Scheme 1.31**

Tamaru and co-workers [37] investigated the umpolung of π-allylpalladium complexes via an alkyl–allyl exchange reaction using a stoichiometric amount of Et$_2$Zn (Scheme 1.32). The reaction is proposed to involve the in situ formation of allylzinc species by transmetalation between a π-allylpalladium benzoate and Et$_2$Zn.

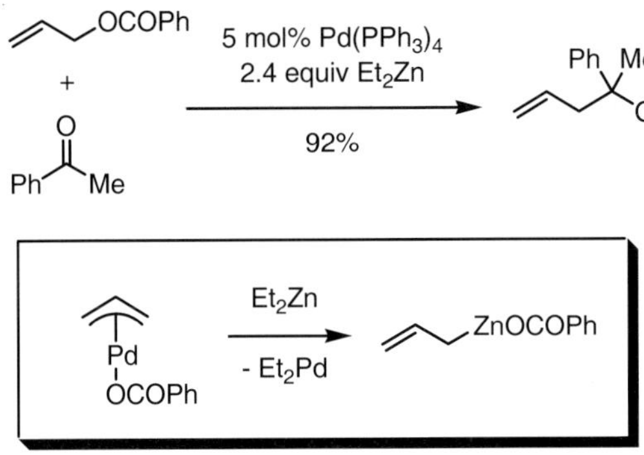

**Scheme 1.32**

A wide variety of carbonyl compounds, such as aldehydes, ketones, esters, lactones, and acid anhydrides, are applicable under these allylation conditions. The same authors also succeeded in allylating aldehydes by using a combination of Pd(PPh$_3$)$_4$ catalyst and a stoichiometric amount of Et$_3$B (Scheme 1.33) [38]. In this case, allylborane would be produced as a nucleophilic agent in situ.

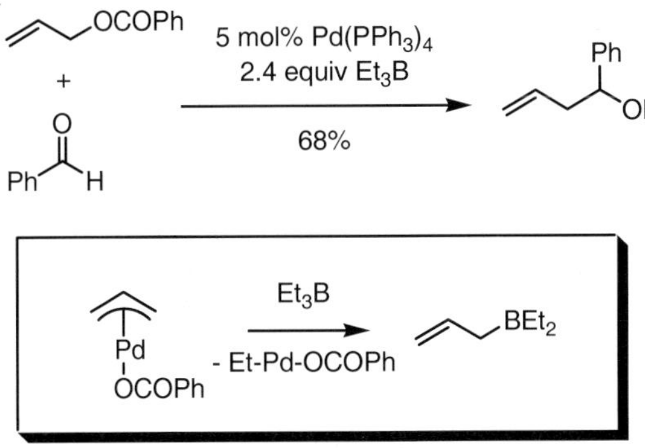

**Scheme 1.33**

Araki and co-workers [39] applied a combination of Pd(PPh$_3$)$_4$ catalyst and a stoichiometric amount of InI for charge reversal of the $\pi$-allylpalladium complexes generated from allyl acetates (Scheme 1.34). Here, allylindium(III) species are generated by reductive transmetalation of a $\pi$-allylpalladium complex with the In$^{I}$ salt. A similar catalyst combination, Pd(OAc)$_2$/(2-furyl)$_3$P and In powder, was applied in the three-component coupling reaction of aryl iodides, allenes, and aldehydes by Kang and co-workers [40] (Scheme 1.35). Here, a $\pi$-allylpalladium intermediate was formed by the Heck reaction of aryl iodides with allenes and the intramolecular allylation of aldehydes afforded the cyclic compounds in good yields.

Oppolzer and co-workers [41] reported cyclization through palladium-catalyzed zinc-ene reactions (Scheme 1.36). These reactions also involved transmetalation between $\pi$-allylpalladium intermediates and Et$_2$Zn to form the nucleophilic allyl zinc species. Cyclization via the zinc-ene reaction, followed by trapping with I$_2$, afforded the cyclized product.

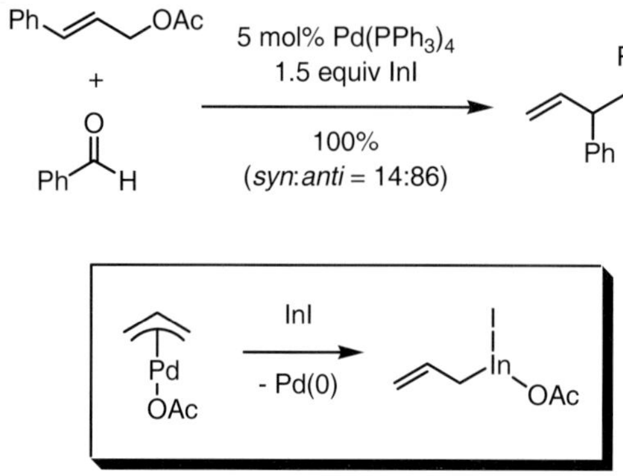

Ph～～OAc

5 mol% Pd(PPh₃)₄
1.5 equiv InI

+

$$Ph \overset{O}{\underset{H}{\bigwedge}}$$

100%
(*syn*:*anti* = 14:86)

$$\text{Ph} \quad \overset{\text{Ph}}{\underset{\text{Ph}}{\bigwedge}} \text{OH}$$

**Scheme 1.34**

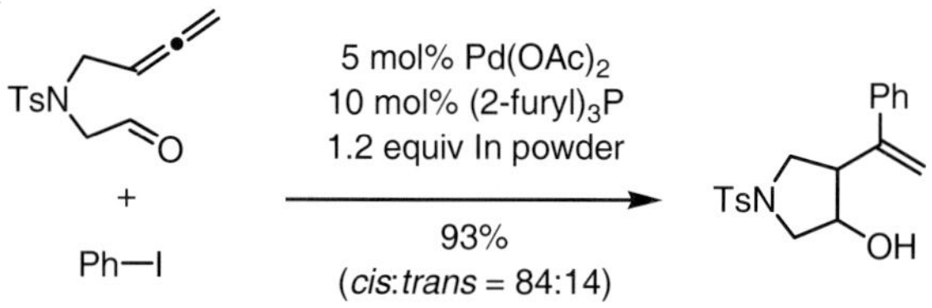

InI

- Pd(0)

～～In-OAc

---

$$TsN \overset{\cdot \cdot}{\underset{O}{\bigwedge}}$$

5 mol% Pd(OAc)₂
10 mol% (2-furyl)₃P
1.2 equiv In powder

+

Ph—I

93%
(*cis*:*trans* = 84:14)

$$TsN \overset{Ph}{\bigwedge} OH$$

**Scheme 1.35**

---

PhO₂S ～ ≡—SiⁱPr₃
PhO₂S ～～OAc

(1) 5 mol% Pd(PPh₃)₄
5 equiv Et₂Zn

(2) I₂
81%

PhO₂S ～ =SiⁱPr₃ I
PhO₂S ～

$$R \overset{R}{\underset{Pd}{\bigwedge}} \text{—SiⁱPr}_3 \longrightarrow R \overset{R}{\bigwedge} \text{—SiⁱPr}_3 \underset{ZnEt}{\longrightarrow} R \overset{R}{\bigwedge} \overset{Si^iPr_3}{\underset{ZnEt}{=}}$$

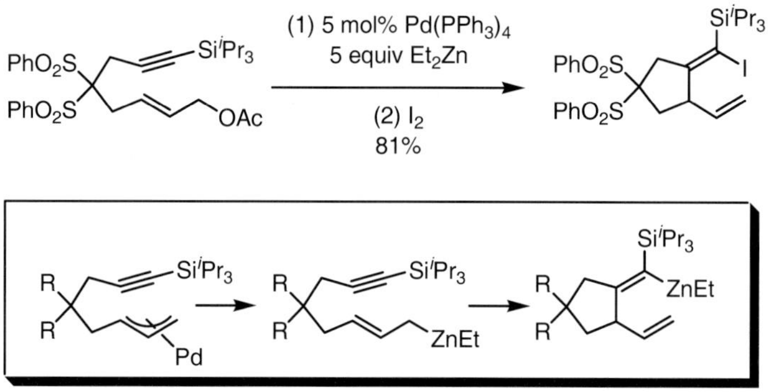

**Scheme 1.36**

We developed a different protocol to generate a nucleophilic π-allylpalladium species and succeeded in the allylation of various aldehydes and imines (Scheme 1.37) [42]. The reactions of allylstannanes with aldehydes or imines in the presence of a catalytic amount of PdCl₂(PPh₃)₂ afforded the corresponding homoallyl alcohols or amines in high yields. The key is to generate a bis-π-allylpalladium complex, which acts as a characteristic nucleophilic agent. Catalytic asymmetric allylation of imines has been achieved using the chiral π-allylpalladium chloride dimer as a source of the palladium catalyst [43]. More recently, it has been reported that in the presence of water the chemical yields and enantioselectivities of catalytic asymmetric

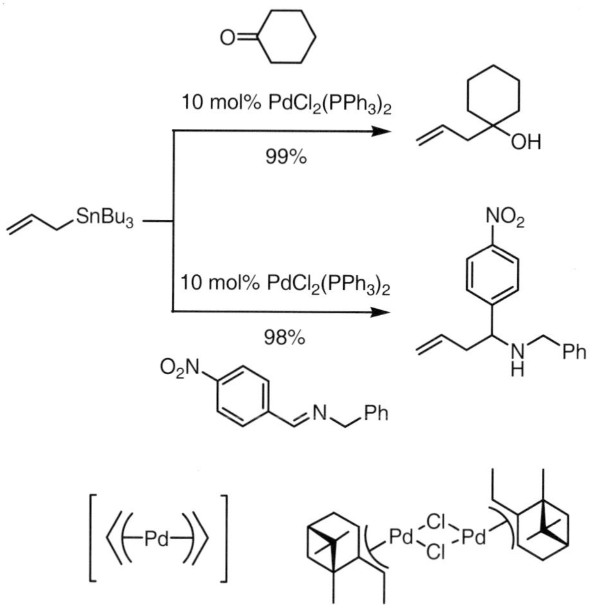

**Scheme 1.37**

allylations of imines were enhanced, the latter up to 91% ee [44]. Not only allylstannanes but also allylsilanes can be used as a nucleophilic allyl source (Scheme 1.38) [45], and in the presence of a chiral palladium catalyst the asymmetric allylation takes place smoothly giving high yields and high ee values.

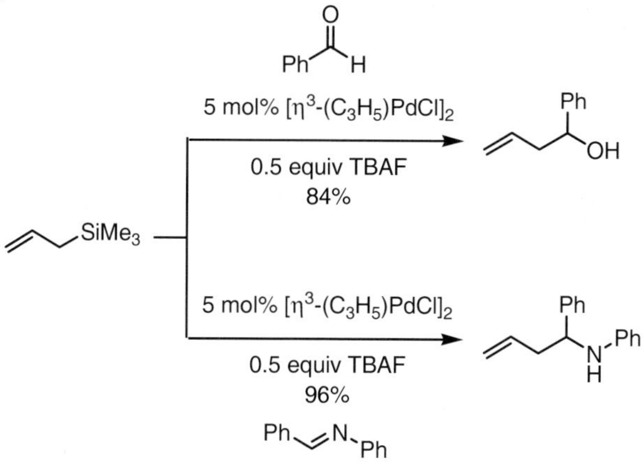

**Scheme 1.38**

1.2.5
### Nickel-Catalyzed Three-Component Coupling Reaction

Mori and co-workers [46] developed a stereoselective cyclization involving a π-allylnickel complex generated from a 1,3-diene and a hydride nickel complex (Scheme 1.39). The combination of a catalytic amount of Ni(cod)$_2$/PPh$_3$ complex and a stoichiometric amount of Et$_3$SiH is the key to generating the active hydride nickel species, insertion of the 1,3-diene moiety into which affords a π-allylnickel intermediate. The cyclization takes place in a stereoselective manner.

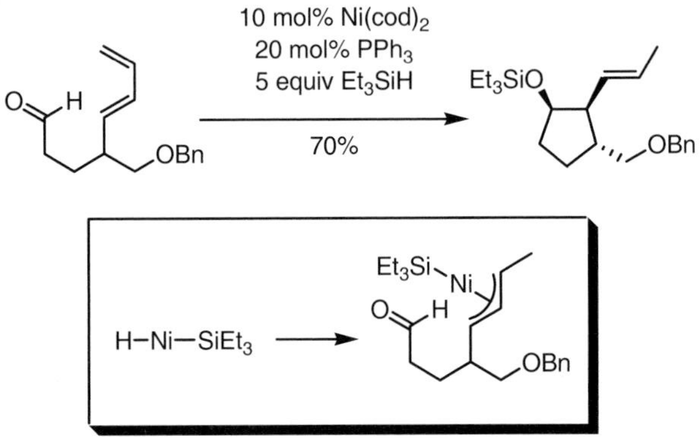

**Scheme 1.39**

Tamaru and co-workers investigated the intermolecular homoallylation of aldehydes with 1,3-dienes. They developed two catalytic systems; one is a combination of a catalytic amount of Ni(acac)$_2$ and a stoichiometric amount of Et$_3$B (Scheme 1.40) [47], and the other is a combination of a Ni(acac)$_2$ catalyst with Et$_2$Zn (Scheme 1.41) [48]. Both systems were found to promote the homoallylation of aldehydes with high regio- and stereoselectivity. The two catalyst systems show complementary reactivities. The Ni–Et$_3$B system is effective for reactive carbonyl compounds such as aromatic and α,β-unsaturated aldehydes, whereas the Ni–Et$_2$Zn combination is effective for saturated aldehydes and ketones. A proposed mechanism involves nucleophilic addition of diene-nickel(0) complexes to carbonyl compounds coordinated by Et$_2$Zn (or Et$_3$B) and ethyl group migration from Zn to Ni to form a π-allylnickel intermediate. Subsequent β-hydrogen elimination from the Et group and reductive elimination furnishes the product and ethylene with regeneration of the Ni$^0$ catalyst.

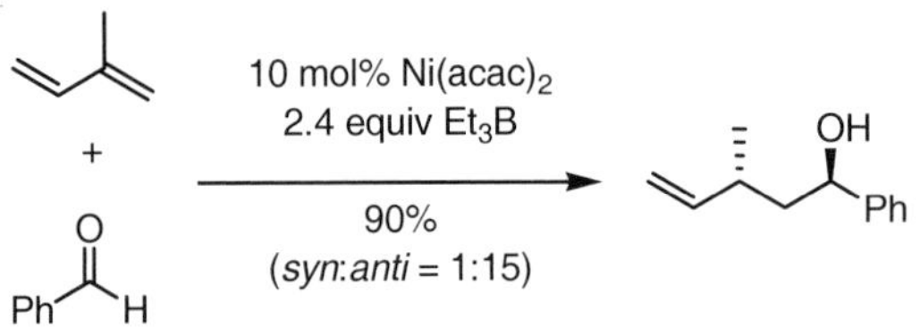

**Scheme 1.40**

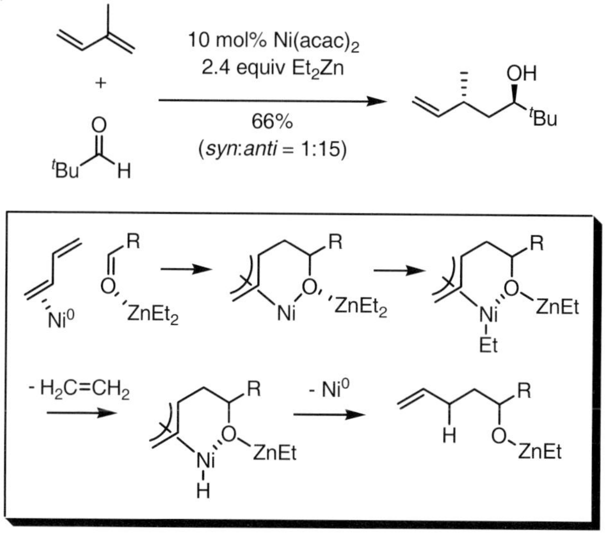

**Scheme 1.41**

Montgomery and co-workers employed a similar combination as a catalyst system, specifically a catalytic amount of Ni(cod)$_2$ and a stoichiometric amount of Me$_2$Zn/ MeZnCl, for the alkylative cyclization of alkynyl enones (Scheme 1.42) [49] and of alkynyl aldehydes (Scheme 1.43) [50]. They also succeeded in cyclizing alkynyl enones with a Ni–AlMe$_3$ system [51], and in cyclizing alkynyl ketones with a Ni–Et$_3$SiH system [52].

**Scheme 1.42**

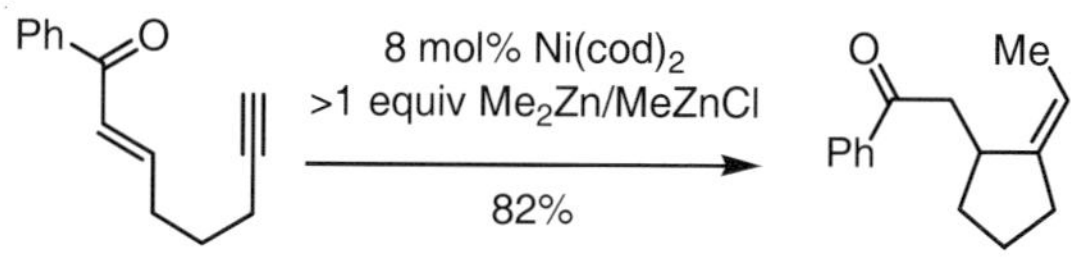

**Scheme 1.43**

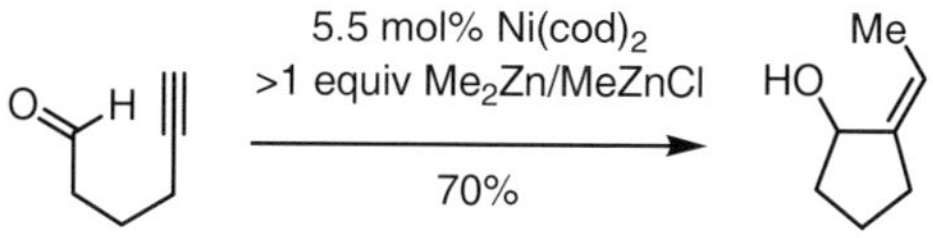

Catalytic asymmetric reductive coupling of alkynes and aldehydes was investigated by Jamison and co-workers [53] utilizing a stoichiometric amount of Et$_3$B and a catalytic amount of Ni(cod)$_2$ combined with a chiral phosphine ligand (Scheme 1.44). High regioselectivity and enantioselectivity were observed. The same authors also employed imines instead of aldehydes to synthesize allylamine derivatives (Scheme 1.45) [54]. The asymmetric reaction using (+)-neomenthyldiphenylphosphine ((+)-NMDPP) resulted in moderate enantioselectivities. Homoallyl alcohols have

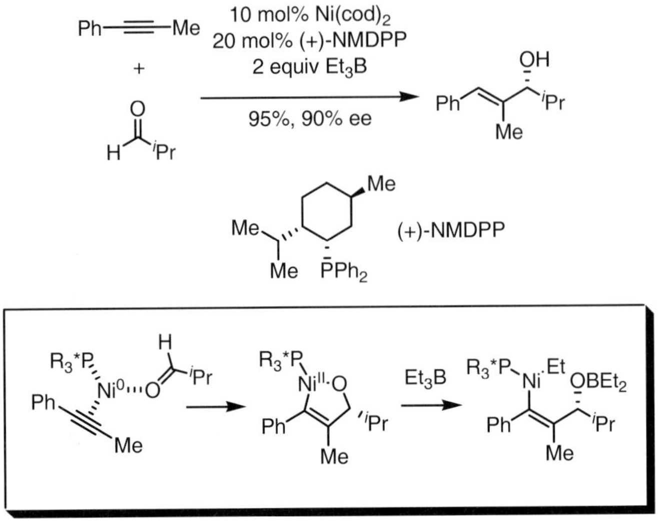

**Scheme 1.44**

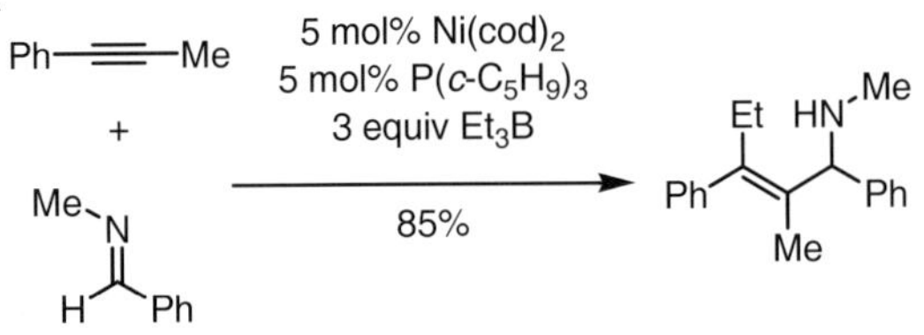

**Scheme 1.45**

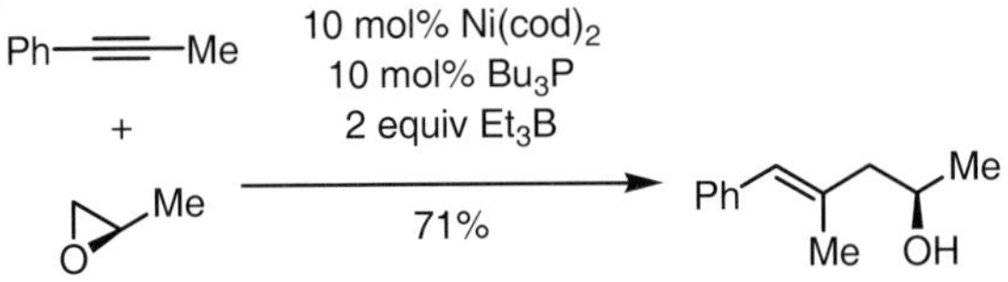

**Scheme 1.46**

been synthesized by carrying out the reactions between alkynes and epoxides in the presence of a catalytic amount of Ni(cod)$_2$/Bu$_3$P and a stoichiometric amount of Et$_3$B (Scheme 1.46) [55].

## 1.2.6
## The Nozaki–Hiyama–Kishi Reaction

Nozaki and co-workers [56] first reported the addition of vinyl halides to aldehydes in the presence of a stoichiometric amount of CrCl$_2$, but the success of the reaction depended strongly on the nature of the Cr source. Later, Nozaki's and Kishi's groups independently discovered that this reaction could be effectively promoted by the

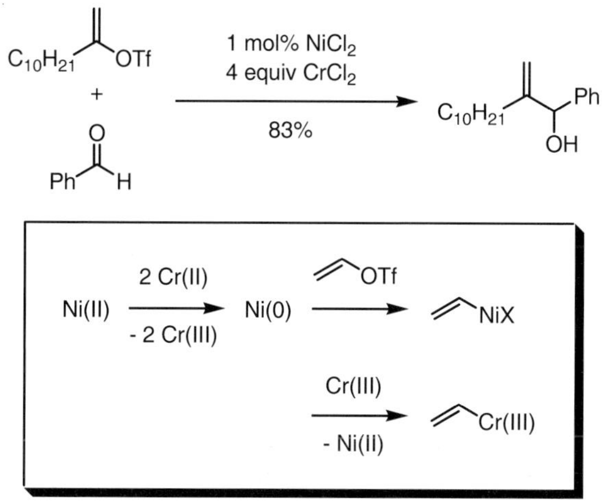

**Scheme 1.47**

addition of a catalytic amount of NiCl$_2$ or Pd(OAc)$_2$ (Scheme 1.47) [57]. The active nucleophilic species is proposed to be a vinylchromium(III) species generated by transmetalation between the vinylnickel intermediate and Cr$^{III}$ salt. The Cr$^{II}$ salt also acts as a reductant for Ni$^{II}$ to generate an Ni$^0$ catalyst.

Fürstner et al. [58] reported that the use of catalytic amounts of NiCl$_2$ and CrCl$_2$ was sufficient for the above transformations when the reactions were carried out in the presence of stoichiometric amounts of Mn powder and Me$_3$SiCl (Scheme 1.48). The Cr$^{III}$ salt produced at the end of the catalytic cycle was efficiently reduced by Mn powder to regenerate the active Cr$^{II}$ salt. However, the use of stoichiometric amounts of Mn and Me$_3$SiCl proved essential for smooth transformations. An extensive investigation of the asymmetric version of this Ni–Cr-mediated reaction has recently been carried out by Kishi and co-workers [59].

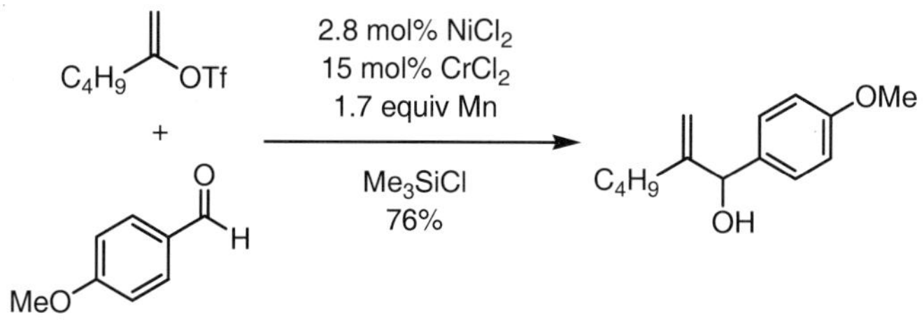

**Scheme 1.48**

## 1.3
### Reactions Promoted by a Combination of Catalytic Amounts of Two Metals

### 1.3.1
### Transition Metal Catalyzed Cross-Coupling Reactions

#### 1.3.1.1 The Stille Reaction

The use of an additional metal salt in the cross-coupling reaction has been a widely adopted approach to enhance the reactivity and sometimes the selectivity of the transformation. The use of a Cu catalyst in combination with a Pd catalyst in the Stille coupling reaction [60] was first investigated by Liebeskind et al. [61] (Scheme 1.49).

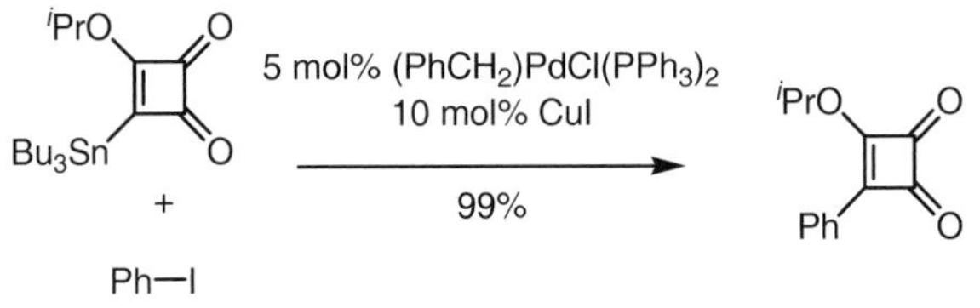

**Scheme 1.49**

They reported therein the cross-coupling of a stannylcyclobutenedione with various aryl iodides in the presence of $(PhCH_2)PdCl(PPh_3)_2$ and CuI catalysts. Using the Pd–Cu bimetallic catalyst system, a rate enhancement was observed and the reaction temperature could be lowered. The authors proposed the following two roles for the Cu additive [62].

First, the Cu catalyst most probably scavenges free phosphine ligand, which retards transmetalation (Eq. 1.1). When triphenylphosphine was used as a ligand for the Pd catalyst, addition of the Cu catalyst led to a rate enhancement as compared to the traditional Stille coupling conditions. On the other hand, little effect was observed when the Cu additive was used in the presence of a soft ligand such as triphenylarsine ($AsPh_3$), which did not inhibit the ligand dissociation from the Pd center.

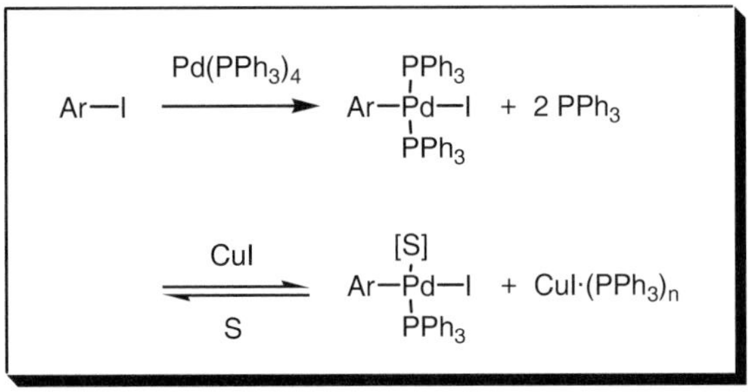

(1.1)

Second, the Cu catalyst probably generates an organocopper species in highly polar solvents such as NMP and DMF in the absence of strongly coordinating ligands (Eq. 1.2). The transmetalation between the Pd intermediate and the derived organocopper species proceeds more easily than the transmetalation with the organostannane itself.

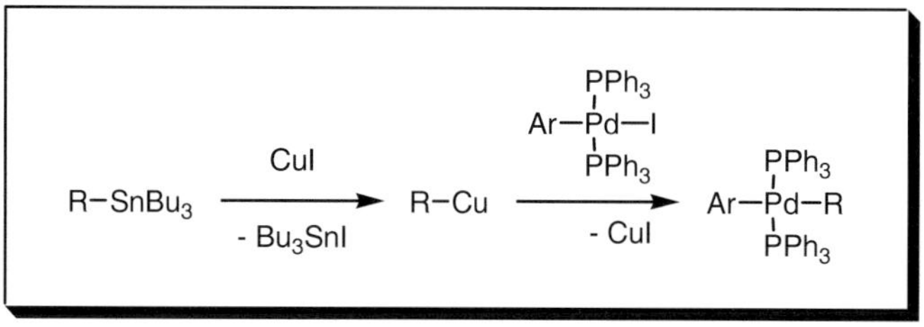

$$R\text{-}SnBu_3 \xrightarrow[- Bu_3SnI]{CuI} R\text{-}Cu \xrightarrow[- CuI]{\overset{PPh_3}{\underset{PPh_3}{Ar\text{-}Pd\text{-}I}}} \overset{PPh_3}{\underset{PPh_3}{Ar\text{-}Pd\text{-}R}} \qquad (1.2)$$

Recently, Cho and co-workers [63] reported on regioselective Stille couplings of 3,5-dibromo-2-pyrone. The reaction in the presence of catalytic amounts of Pd(PPh$_3$)$_4$ and CuI in toluene produced 5-bromo-3-phenyl-2-pyrone (Scheme 1.50a), whereas the reaction in the presence of catalytic Pd(PPh$_3$)$_4$ and a stoichiometric amount of CuI in DMF furnished 3-bromo-5-phenyl-2-pyrone (Scheme 1.50b).

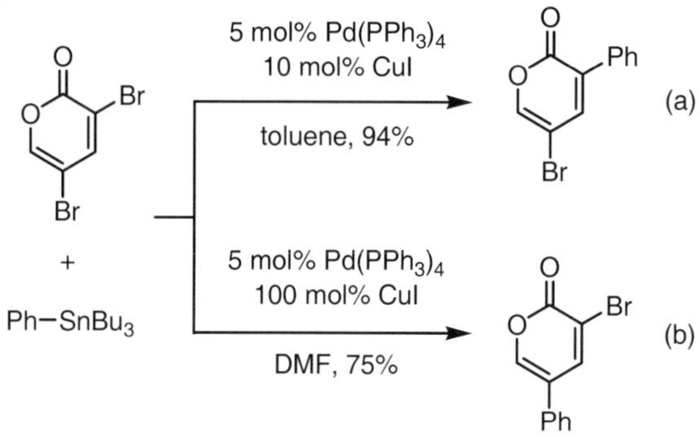

**Scheme 1.50**

Flack and co-workers [64] employed a combination of PdCl$_2$(PPh$_3$)$_2$ and CuCN catalysts for the reaction between acyl chlorides and $\alpha$-heteroatom-substituted stannanes and investigated the stereoselectivity of the coupling reaction (Scheme 1.51). The reaction proceeded with retention of configuration.

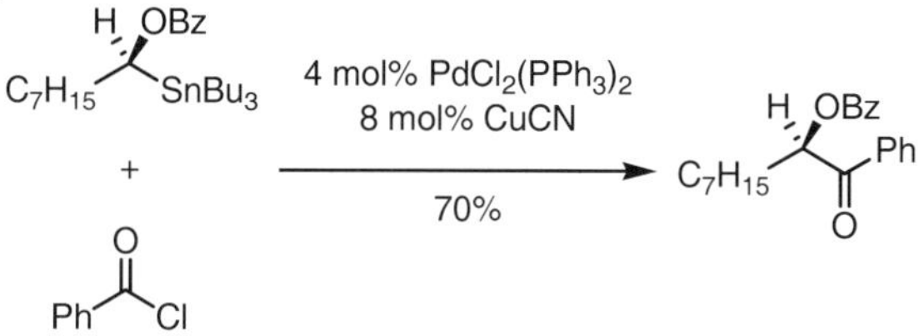

**Scheme 1.51**

### 1.3.1.2 The Hiyama Reaction

Mori and Hiyama et al. [65] developed the cross-coupling reaction of alkynylsilanes with aryl or alkenyl triflates in the presence of a combination of Pd(PPh$_3$)$_4$ and CuCl catalysts (Scheme 1.52). This is a new type of the reaction between sp$^2$- and sp-hybridized carbon centers. The similar coupling reaction between aryl halides and terminal alkynes is known as the Sonogashira reaction; however, alkynylsilanes

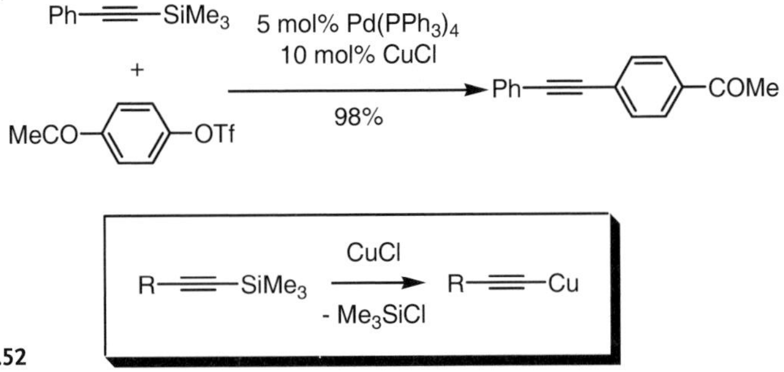

Ph—≡—SiMe₃    5 mol% Pd(PPh₃)₄
            10 mol% CuCl
+
                                    → Ph—≡—⟨aryl⟩—COMe
MeCO—⟨aryl⟩—OTf
            98%

R—≡—SiMe₃  —CuCl→  R—≡—Cu
              - Me₃SiCl

**Scheme 1.52**

are inert under the corresponding conditions. The role of the Cu additive is to generate a reactive alkynylcopper species by transmetalation from silicon to copper in polar solvents such as DMF [66].

### 1.3.1.3 The Sonogashira Reaction

The cross-coupling reaction between terminal alkynes and aryl or vinyl halides was developed by Sonogashira and Hagihara et al. [67] (Scheme 1.53). The reaction was found to proceed smoothly under catalysis by a combination of PdCl₂(PPh₃)₂ and CuCl in the presence of Et₂NH. The role of the Pd catalyst is to form aryl- or vinyl-palladium species by oxidative insertion into the carbon–halogen bond, and the role of the Cu catalyst is to generate a copper-acetylide to facilitate transmetalation with the Pd intermediate. The effect of the amine has also been thoroughly investigated and cyclic amines such as piperidine and pyrrolidine were found to increase the rate of transformation [68].

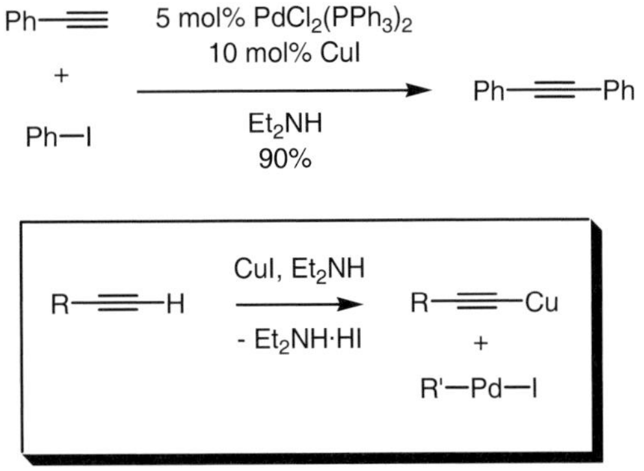

Ph—≡    5 mol% PdCl₂(PPh₃)₂
        10 mol% CuI
+
                            → Ph—≡—Ph
Ph—I
        Et₂NH
        90%

R—≡—H  —CuI, Et₂NH→  R—≡—Cu
          - Et₂NH·HI        +
                           R'—Pd—I

**Scheme 1.53**

The effect of additional additives was also surveyed and the optimal combination of catalysts was found to depend on the starting materials. When an alkynyl epoxide was used as a coupling partner with vinyl triflate, a combination of Pd(PPh₃)₄ and AgI gave better results than a Pd–Cu bimetallic catalyst (Scheme 1.54) [69]. Several ZnX₂ salts were examined as co-catalysts, although a stoichiometric amount of ZnX₂ was usually needed to obtain high yields of adducts [70].

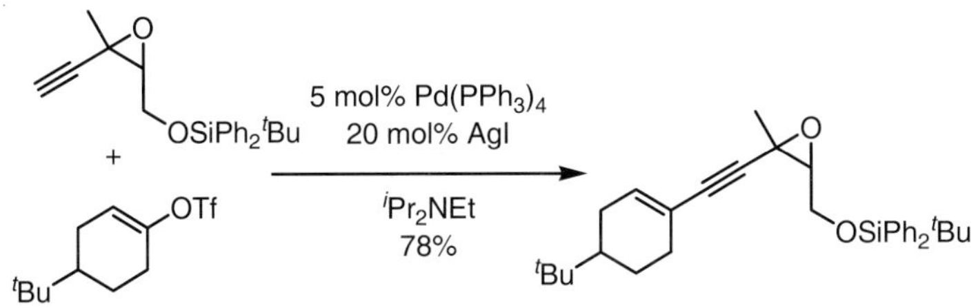

**Scheme 1.54**

The construction of complicated cyclic compounds is often accomplished through sequential reactions involving a Sonogashira coupling (Pd–Cu bimetallic system), where the cross-coupling reaction takes place with concomitant cyclization. We reported the synthesis of benzopyranone derivatives by the reaction of vinyl bromides with terminal alkynes under catalysis by Pd(PPh$_3$)$_4$ and CuI (Scheme 1.55) [71].

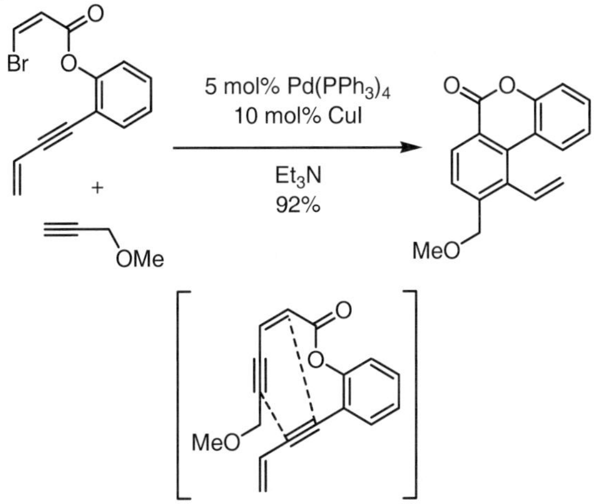

**Scheme 1.55**

The initial Sonogashira coupling reaction afforded an intermediate bearing two conjugated enyne moieties, and subsequent palladium-catalyzed intramolecular benzannulation produced the tricyclic adduct. Cacchi and co-workers [72] reported the synthesis of benzofuran derivatives through the coupling of terminal alkynes with 2-iodophenols in the presence of Pd(OAc)$_2$(PPh$_3$)$_2$ and CuI as a bimetallic catalyst system (Scheme 1.56). The reaction proceeds through the formation of the 2-alkynylphenol intermediate. Yamanaka and co-workers [73] succeeded in synthesizing indoles from terminal alkynes and N-(2-iodophenyl)methanesulfonamide by employing a Pd–Cu bimetallic catalyst (Scheme 1.57). The presence of the methanesulfonyl group on the nitrogen atom is essential for the spontaneous cyclization to afford the indole skeleton. When 2-iodobenzoic acid was used as a coupling partner for terminal alkynes in the presence of a Pd–Cu bimetallic catalyst, consecutive heteroannulation took place to afford phthalides (Scheme 1.58) [74]. The synthesis of γ-alkylidene butenolides has also been achieved by Lu and co-workers [75] employing (Z)-3-bromopropenoic acid as a starting material.

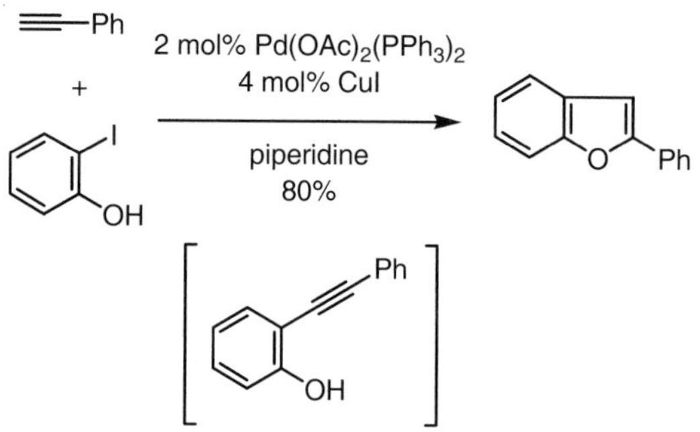

**Scheme 1.56**

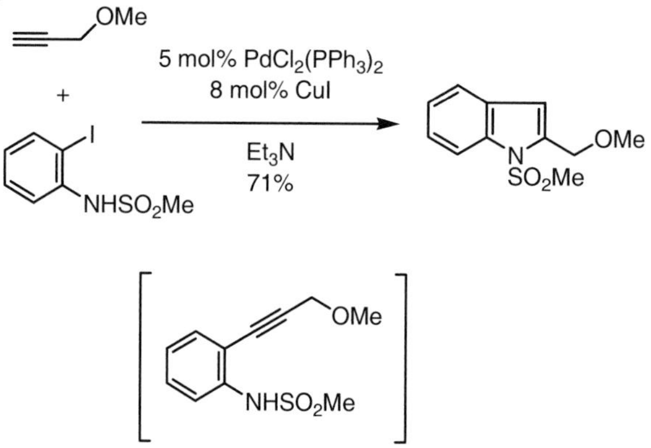

**Scheme 1.57**

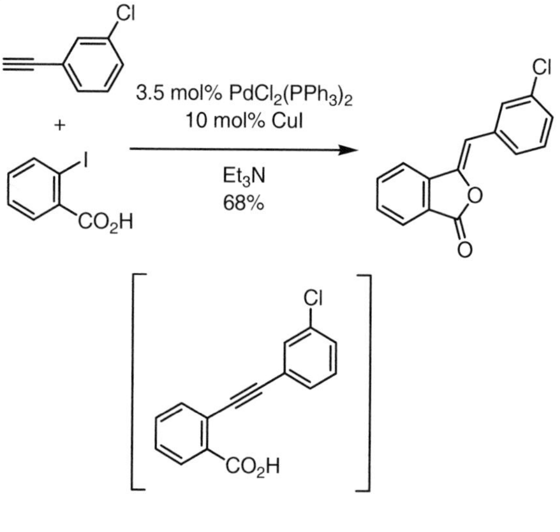

**Scheme 1.58**

1.3.2
**The Wacker Reaction**

Murahashi and Hosokawa et al. [76] reported that the Wacker oxidation could be promoted by a catalytic amount of a Pd–Cu catalyst under $O_2$ atmosphere. They found that the C-C double bonds of allylamides could be directly oxidized with molecular oxygen in the presence of a catalyst system consisting of $PdCl_2(MeCN)_2$ and CuCl together with hexamethylphosphoric triamide (HMPA) as an additive (Scheme 1.59). The oxidation proceeded in a regioselective manner to afford aldehydes from *N*-allylamides, which is in sharp contrast to the usual Wacker oxidation giving methyl ketones. It is proposed that the oxidation is caused by a Pd–OOH species derived from a hydridepalladium intermediate (Pd–H) and molecular oxygen ($O_2$). This hydridepalladium intermediate would be generated in situ by chloropalladation of the alkene followed by β-elimination. The role of the Cu catalyst can be envisaged as one of accelerating the formation of the active Pd–OOH species from the Pd–H species in the presence of HMPA.

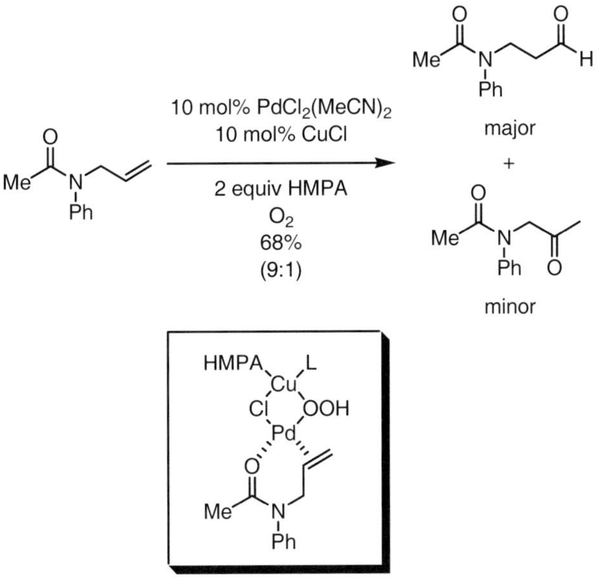

**Scheme 1.59**

The same authors also succeeded in aminating alkenes in the presence of a combination of $PdCl_2$ and CuCl catalysts under $O_2$ atmosphere (Scheme 1.60) [77]. Alkenes having an electron-withdrawing group and styrene could be used as substrates. Stahl and co-workers [78] investigated similar reactions between styrene derivatives and amides and found that the regioselectivity was changed on addition of catalytic amounts of bases such as $Et_3N$ and NaOAc (Scheme 1.61). In this case, Markovnikov-type adducts were obtained exclusively.

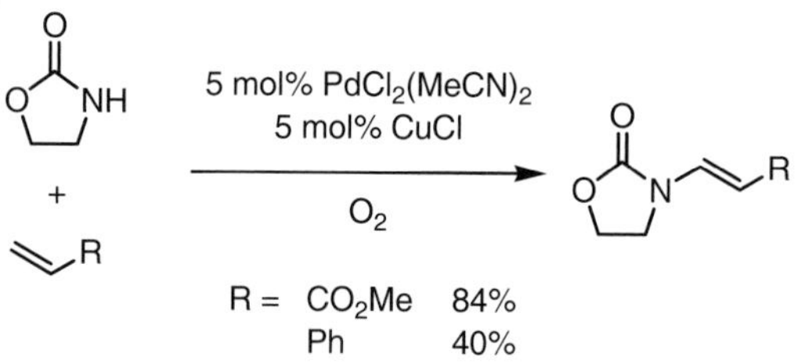
O
NH
O
+
R
5 mol% PdCl2(MeCN)2
5 mol% CuCl
O2
R = CO2Me 84%
Ph 40%
O
O
N
R

**Scheme 1.60**

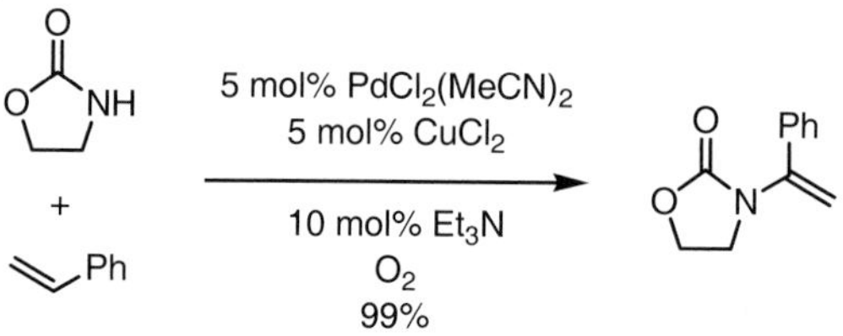
O
NH
O
+
Ph
5 mol% PdCl2(MeCN)2
5 mol% CuCl2
10 mol% Et3N
O2
99%
O
Ph
O
N

**Scheme 1.61**

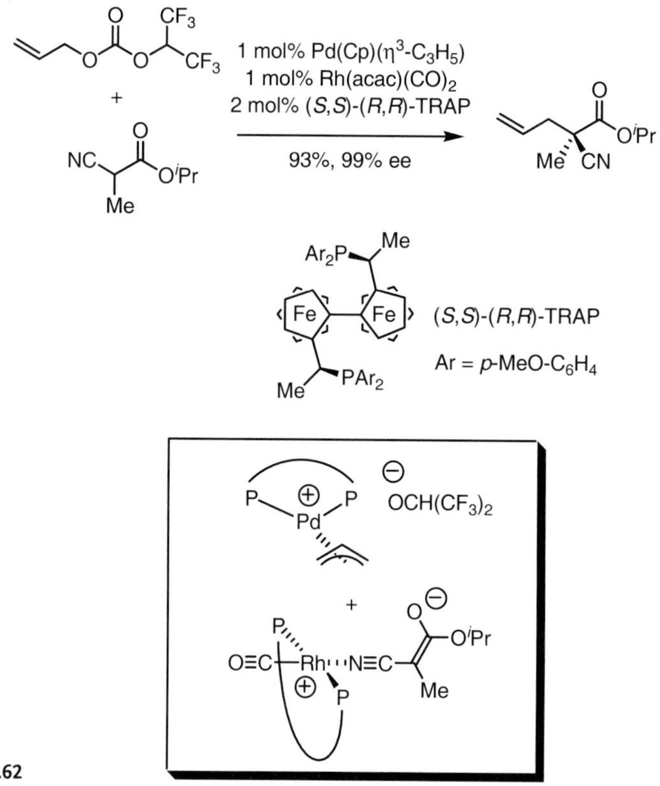
O
CF3
O
O
CF3
+
NC
O
O^iPr
Me
1 mol% Pd(Cp)(η3-C3H5)
1 mol% Rh(acac)(CO)2
2 mol% (S,S)-(R,R)-TRAP
93%, 99% ee
O
O^iPr
Me
CN
Ar2P
Me
Fe
Fe
PAr2
Me
(S,S)-(R,R)-TRAP
Ar = p-MeO-C6H4
P
P
Pd
OCH(CF3)2
+
O
O^iPr
P
O≡C
Rh
N≡C
P
Me

**Scheme 1.62**

## 1.3.3
## Reactions Involving π-Allylpalladium Intermediates

Sawamura and Ito et al. [79] successfully accomplished an enantioselective allylic alkylation reaction by employing a two-component catalyst system consisting of Rh(acac)(CO)$_2$ and Pd(Cp)(η$^3$-C$_3$H$_5$) complexes, combined with (S,S)-(R,R)-2,2″-bis[1-(diarylphosphino)ethyl]-1,1″-biferrocene (TRAP) as a *trans*-chelating chiral phosphine ligand (Scheme 1.62). The use of allyl hexafluoroisopropyl carbonate and active methyne compounds bearing cyano groups as starting materials was essential to obtain the allylated products in high yields and with excellent enantioselectivities. The chiral Rh catalyst coordinates to the cyano group to control the orientation of a prochiral enolate, while the Pd catalyst behaves as a precursor for formation of the π-allylpalladium species.

Tsuji and co-workers [80] reported the allylation of ketones starting from their enol acetates catalyzed by a combination of Pd$_2$(dba)$_3$·CHCl$_3$/dppe and Bu$_3$SnOMe (Scheme 1.63a). When the same reaction was conducted in MeCN, the corresponding α,β-unsaturated ketones were obtained (Scheme 1.63b) [81]. The key to these transformations is the generation of the corresponding stannyl enolates in situ by the reaction of enol acetates with Bu$_3$SnOMe. Transmetalation of stannyl enolates with π-allylpalladium methoxide gives the π-allylpalladium enolates. Reductive elimination affords the allylated ketones, while β-hydrogen elimination produces the α,β-unsaturated ketones as the final products.

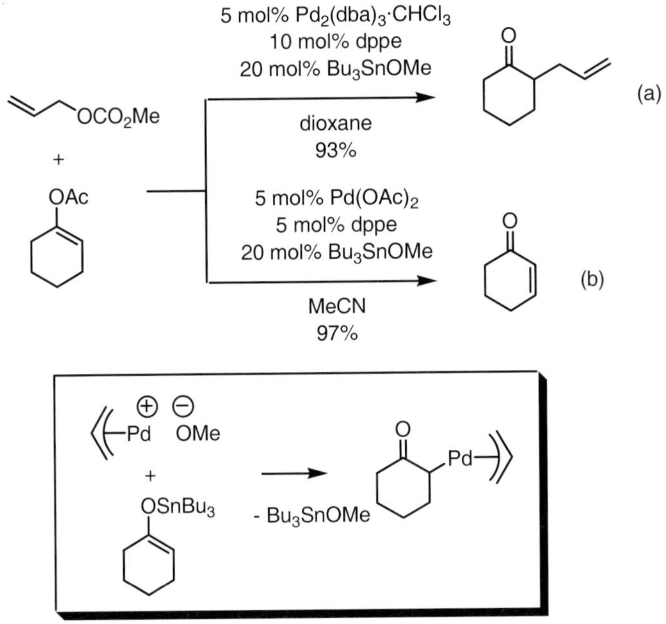

**Scheme 1.63**

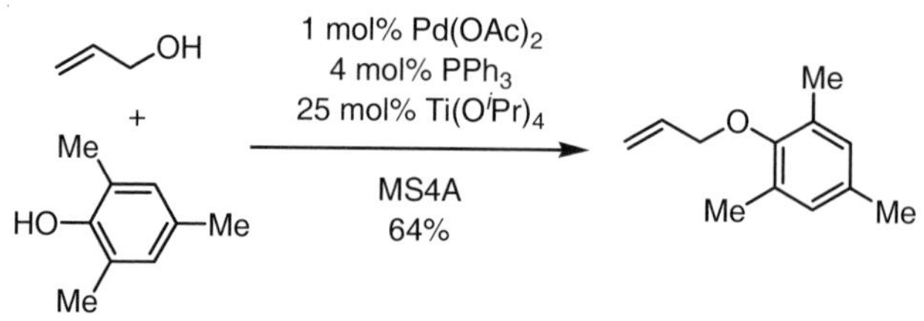

**Scheme 1.64**

Miura and co-workers employed Pd(OAc)$_2$/PPh$_3$ and Ti(O$^i$Pr)$_4$ catalysts in the allylic substitution reaction between allyl alcohol and phenols (Scheme 1.64) [82]. Without the addition of the Ti catalyst, the reaction was sluggish and only a low yield of the desired allyl phenyl ether was obtained. Yang et al. [83] applied the same Pd–Ti catalyst system to the reaction between allyl alcohol and anilines (Scheme 1.65). The role of the Ti additive may be the activation of allyl alcohol by forming allyl titanate, thereby facilitating oxidative addition of the Pd$^0$ catalyst. Alternatively, the Ti catalyst may accelerate the ligand exhange between π-allyl-palladium hydroxide and nucleophiles such as phenols and anilines.

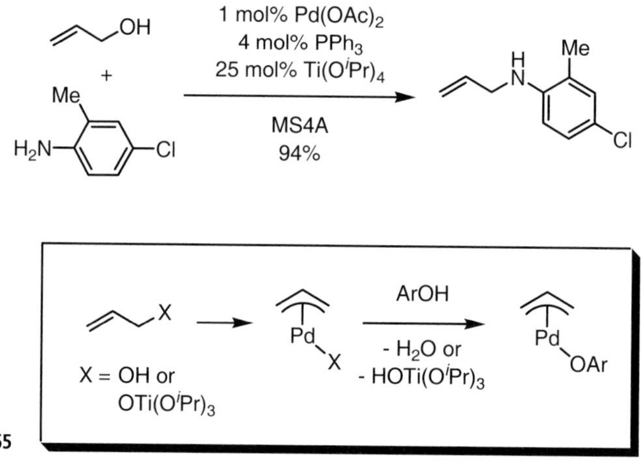

**Scheme 1.65**

### 1.3.4
### Transition Metal Catalyzed Cyclization Reactions

#### 1.3.4.1 [3+2] Cycloaddition Reactions
We developed a synthesis of 1,2,3-triazoles based on Pd–Cu bimetallic-catalyzed three-component coupling (TCC) of non-activated alkynes, allyl methyl carbonate, and trimethylsilyl azide (Scheme 1.66). Regioselective formation of 2-allyl-1,2,3-triazoles was achieved by carrying out the TCC reaction in the presence of a catalyst system of Pd$_2$(dba)$_3 \cdot$ CHCl$_3$/P(OPh)$_3$ and CuCl(PPh$_3$)$_3$ [84]. A regioselective synthesis of 1-allyl-1,2,3-triazoles was achieved by conducting the TCC reaction

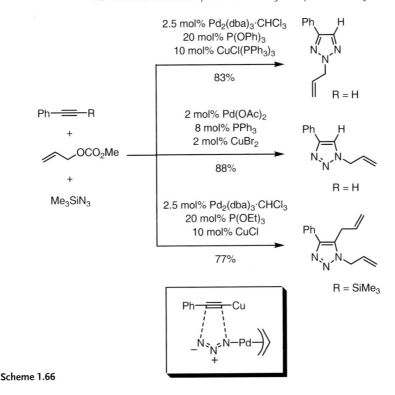

**Scheme 1.66**

under catalysis by a Pd(OAc)$_2$/PPh$_3$ and CuBr$_2$ system [85]. When trimethylsilyl-acetylenes were used as starting materials and the reaction was performed in the presence of Pd$_2$(dba)$_3$· CHCl$_3$/P(OEt)$_3$ and CuCl catalysts, 1,5-diallyltriazoles were formed selectively [86]. The formation of a copper-acetylide, which activates the C-C triple bond of alkynes, and the generation of active azide species such as π-allylpalladium azide complex and/or allyl azide, are proposed as being involved in the catalytic cycles. [3+2] cycloaddition between the copper-acetylide and the active azide species to afford the triazole framework is believed to be a common feature of all of these transformations.

It is well known that the palladium-trimethylenemethane (Pd-TMM) complex generated from 2-(trimethylsilylmethyl)allyl acetate adds to α,β-unsaturated carbonyl compounds in a 1,4-addition fashion to produce methylenecyclopentane derivatives. Trost and co-workers found a dramatic change in the chemoselectivity upon the addition of Bu$_3$SnOAc. A combination of Pd(OAc)$_2$/PPh$_3$ and Bu$_3$SnOAc catalysts promoted 1,2-addition of the Pd-TMM complex to carbonyl compounds and methylenetetrahydrofuran cycloadducts were obtained (Scheme 1.67) [87]. In(acac)$_3$ was also found to be applicable instead of Bu$_3$SnOAc [88]. Two possible roles of the Sn and In additives have been postulated. One is the stabilization of the alkoxide anion generated after addition of the Pd-TMM complex to the carbonyl group, and the other is activation of the carbonyl group by coordination prior to addition of the Pd-TMM complex.

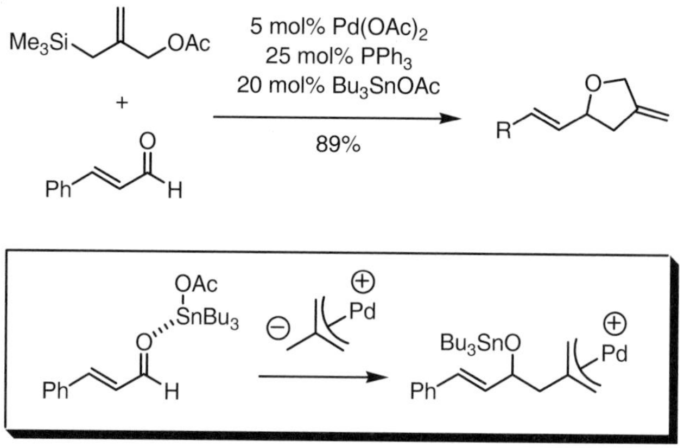

**Scheme 1.67**

Furukawa and co-workers [89] applied a combination of $PdCl_2[(S)$-TolBINAP] and $AgBF_4$ catalysts to the asymmetric 1,3-dipolar cycloaddition of nitrones to alkenes (Scheme 1.68). The desired isoxazoline was obtained in the highest yield and with high ee with $(S)$-TolBINAP as the chiral ligand. The Ag salt was added to generate a reactive cationic Pd complex, thereby accelerating the reaction and giving high enantioselectivity.

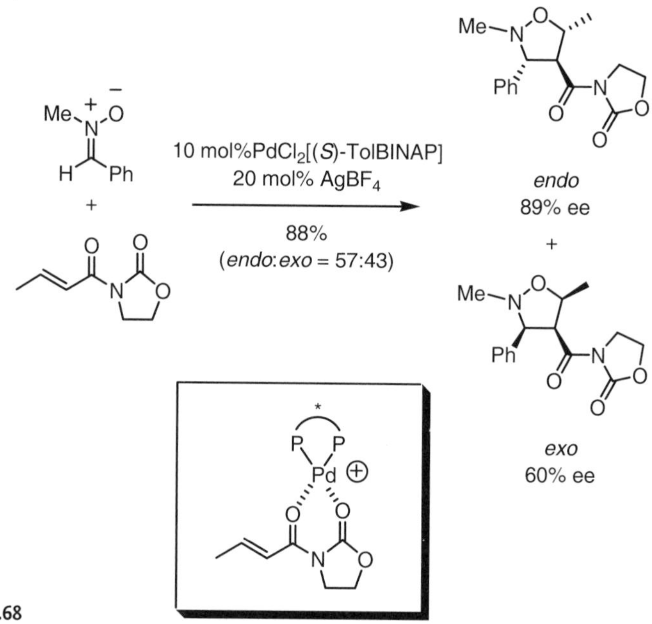

**Scheme 1.68**

### 1.3.4.2 Intramolecular [*n*+2] Cyclization Reactions

Wender and co-workers [90] investigated the reaction of vinylcyclopropanes with a variety of C–C π-systems in the presence of transition metals, and they were the first to achieve intramolecular cycloaddition of alkynyl-vinylcyclopropanes to provide seven-membered rings based on a [5+2] cycloaddition reaction (Scheme 1.69). Essentially, the reaction could be catalyzed by RhCl(PPh$_3$)$_3$ alone, but the addition of AgOTf had a marked accelerating effect. The role of the Ag additive is assumed to be the removal of the chloride ligand from the complex, thereby freeing a coordination site and forming a more reactive cationic Rh center. Similar [5+2] cycloaddition reactions have been accomplished by employing alkenyl-vinylcyclo-propanes [91] and allenyl-vinylcyclopropanes [92] as starting materials. Trost et al. [93] also reported that [5+2] cycloaddition reactions of alkynyl-vinylcyclopropanes proceeded under catalysis by a combination of [CpRu(MeCN)$_3$]PF$_6$ and In(OTf)$_3$ and discussed a detailed analysis of the regioselectivity of the cyclization.

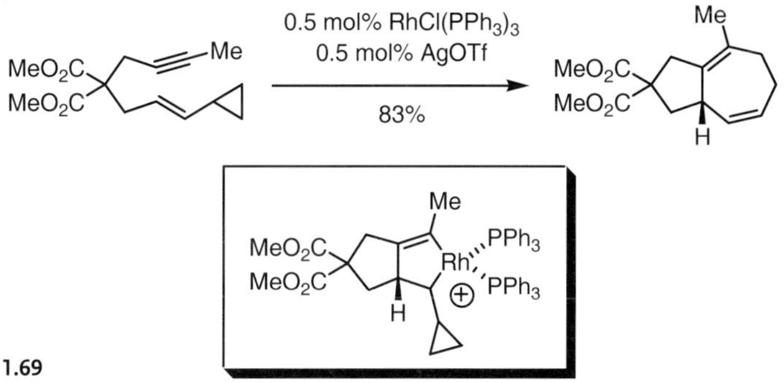

**Scheme 1.69**

Wender and co-workers [94] also successfully accomplished a transition metal catalyzed intramolecular [6+2] cycloaddition through a reaction of 2-vinylcyclo-butanone and alkene moieties (Scheme 1.70). The addition of AgOTf was not a prerequisite for the reaction to proceed when [Rh(CO)$_2$Cl]$_2$ was used as catalyst, but the use of the Rh–Ag bimetallic catalyst did facilitate the cycloaddition. When the reaction was conducted with RhCl(PPh)$_3$ as catalyst, the addition of AgOTf was essential to obtain the desired cycloaddition products in high yields.

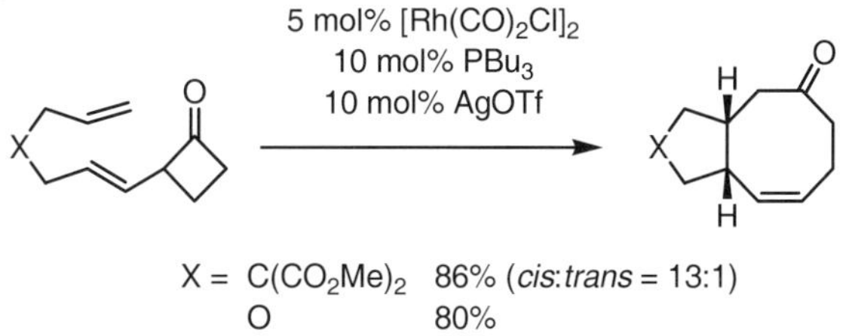

X = C(CO$_2$Me)$_2$  86% (*cis:trans* = 13:1)
O     80%

**Scheme 1.70**

The rhodium-catalyzed intramolecular [4+2] cycloaddition reaction between dienes and alkynes was developed by Zhang and co-workers [95] (Scheme 1.71). They applied a combination of [Rh(dppb)Cl]$_2$ and AgSbF$_6$ as a catalyst system.

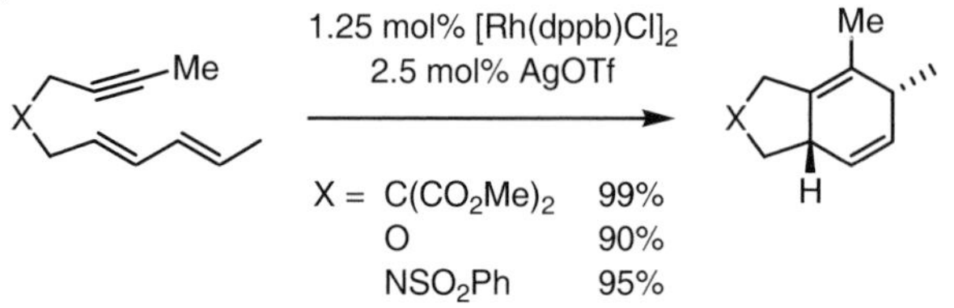

1.25 mol% [Rh(dppb)Cl]₂
2.5 mol% AgOTf

X = C(CO₂Me)₂   99%
        O              90%
        NSO₂Ph     95%

**Scheme 1.71**

Gilbertson et al. [96] achieved the formation of eight-membered rings through a rhodium-catalyzed [4+2+2] cycloaddition between alkynyl-dienes and external alkynes (Scheme 1.72). The employment of a bimetallic catalyst system, [Rh(NBD)Cl]₂/Me-DuPHOS and AgSbF₆, was essential for the [4+2+2] cyclization. Although the products were obtained with high diastereoselectivity, the enantioselectivity was only moderate. With other combinations of catalysts, for example [Rh(NBD)Cl]₂ with AgSbF₆ and [Rh(CO)₂Cl]₂/PPh₃, only the intramolecular [4+2] cycloaddition of alkynyl-dienes took place to afford a six-membered ring.

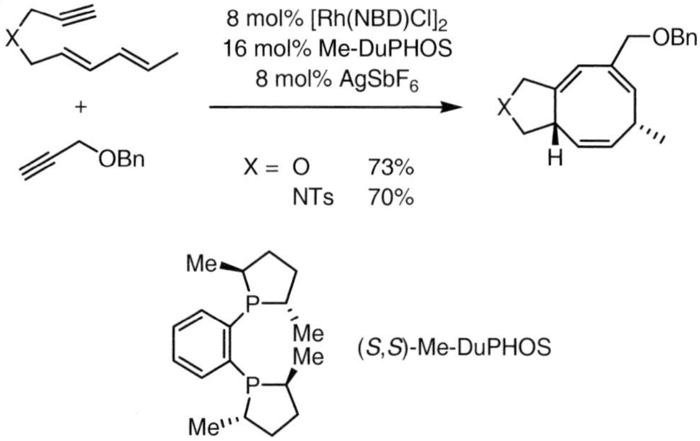

8 mol% [Rh(NBD)Cl]₂
16 mol% Me-DuPHOS
8 mol% AgSbF₆

+

OBn

X = O      73%
      NTs    70%

(S,S)-Me-DuPHOS

**Scheme 1.7**

Evans and co-workers [97] observed a similar type of [4+2+2] cycloaddition reaction between enynes and dienes under catalysis by a RhCl(PPh₃)₃ and AgOTf bimetallic system (Scheme 1.73). The choice of the Ag additive was crucial to obtain the desired product. For example, if AgSbF₆ was used with the Rh catalyst, dimerization of the starting enyne to produce the tricyclic product was the predominant reaction. These authors further developed the reaction to a three-component coupling of propargylamine, allyl methyl carbonate, and a diene, in which the rhodium-catalyzed allylic amination and [4+2+2] cycloaddition occurred successively (Scheme 1.74).

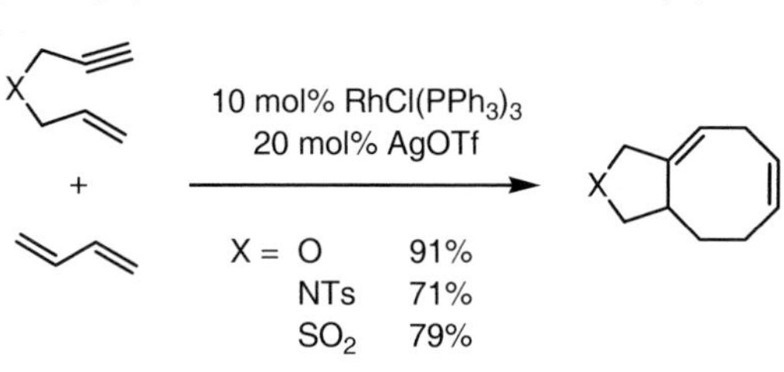

10 mol% RhCl(PPh₃)₃
20 mol% AgOTf

+

X = O      91%
      NTs    71%
      SO₂    79%

**Scheme 1.73**

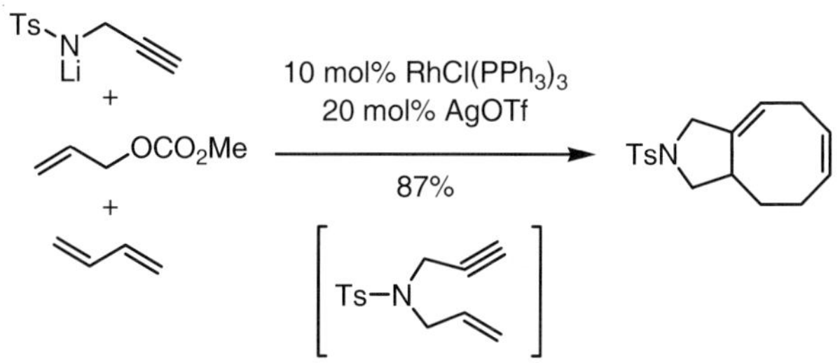

**Scheme 1.74**

### 1.3.4.3 Intermolecular [*n*+2+2] Cyclotrimerization Reactions

Ikeda and co-workers [98] reported the regioselective formation of substituted benzenes through the [2+2+2] cyclotrimerization of α,β-enones with alkynes followed by aerobic oxidation of the intermediate in the presence of DBU (Scheme 1.75). A combination of $Ni(acac)_2/PPh_3$ and $Me_3Al$ catalysts proved effective for this transformation. The Al catalyst is proposed to function as a Lewis acid and activates enones so that they form a nickelacycle intermediate. The Al additive may also serve to generate a reactive $Ni^0$ catalyst in the initial stage of the catalytic cycle.

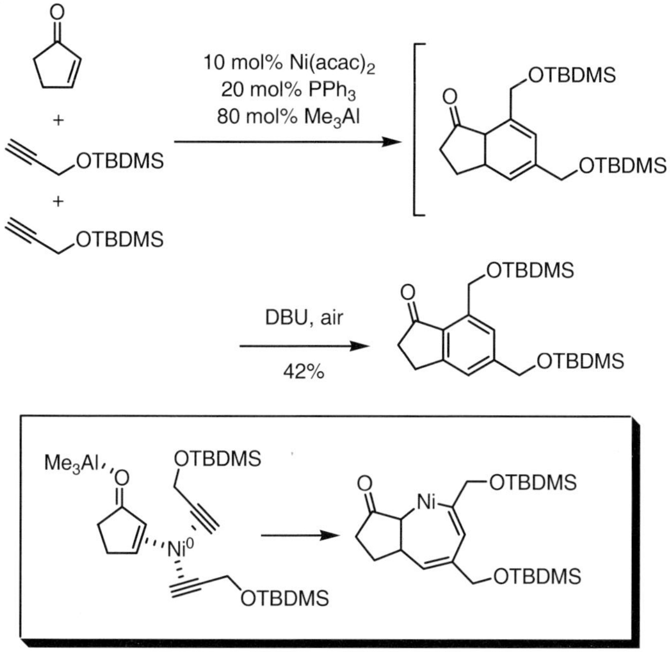

**Scheme 1.75**

Lyons and co-workers [99] originally devised a [2+2+2] cycloaddition reaction between norbornadiene and various acetylenes in the presence of $Co(acac)_3/dppe$ and $Et_2AlCl$ catalysts. Later, Lautens et al. [100] carried out extensive investigations on this reaction (Scheme 1.76). It was suggested that the role of the Al additive is to

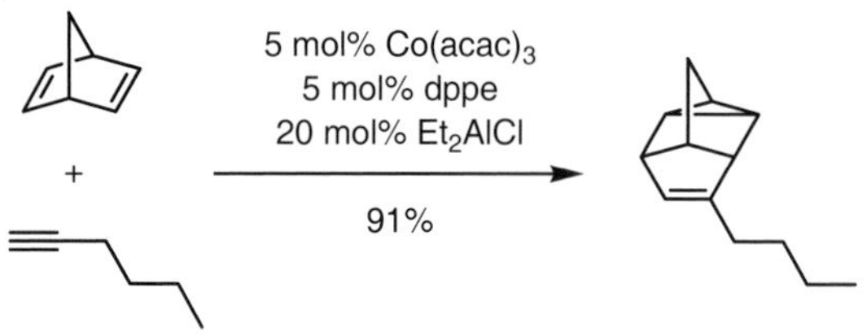

**Scheme 1.76**

reduce the Co complex in the presence of the phosphine ligand. The asymmetric cycloaddition was explored by utilizing chiral bidentate phosphine ligands such as Chiraphos and Prophos [101]. Similar reactions have been reported by other research groups using $CoI_2$/$PPh_3$ and Zn powder as a catalyst system [102]. In this case, the Zn powder serves as a reductant for the Co pre-catalyst. When dienes were employed instead of acetylenes, a [4+2+2] cycloaddition reaction took place to afford a polycyclic compound (Scheme 1.77) [103].

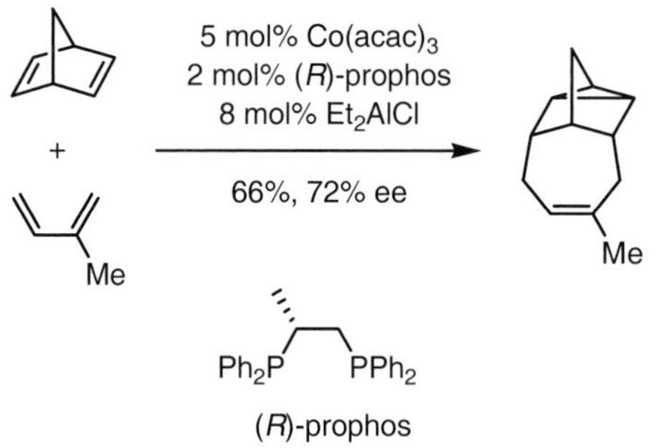

**Scheme 1.77**              (*R*)-prophos

### 1.3.4.4 [2+2+1] Cycloaddition Reactions; The Pauson–Khand Reaction

Jeong and co-workers [104] developed a one-pot preparation of bicyclopentenones from propargyl malonates or propargyl sulfonamides and allylic acetate in the presence of $Pd_2(dba)_3 \cdot CHCl_3$/dppp and [RhCl(CO)(dppp)]$_2$ as a heterobimetallic catalyst system (Scheme 1.78). The overall transformation consists of two con-

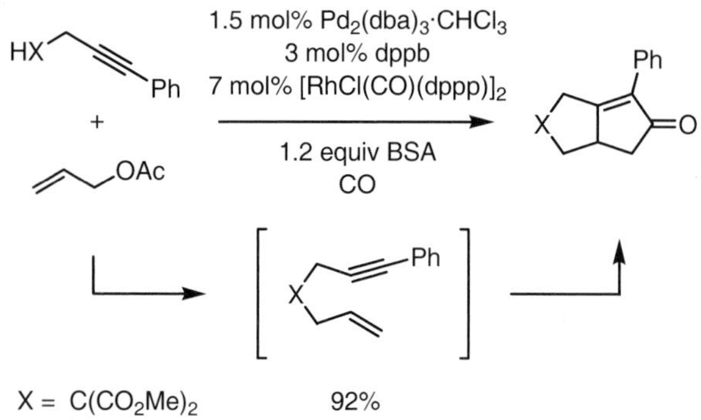

**Scheme 1.78**   X = C(CO$_2$Me)$_2$        92%
                   NTs               90%

secutive reactions. The first reaction is the palladium-catalyzed formation of enynes through allylic substitution, and the second one is the rhodium-catalyzed Pauson–Khand reaction to construct the cyclopentenone framework.

### 1.3.4.5 Cycloisomerization Reactions

Zhang et al. [105] reported the transition metal catalyzed cycloisomerization of 1,6-enynes, which led to the formation of cyclic compounds incorporating a 1,4-diene moiety (Scheme 1.79). The catalyst was prepared in situ from [RhCl(dppb)]$_2$ and AgSbF$_6$. The role of the Ag additive was to generate an active cationic Rh species. Enantioselective rhodium-catalyzed cycloisomerization was also studied by utilizing chiral bidentate phosphine ligands [106]. A similar cycloisomerization using a combination of NiCl$_2$(PPh$_3$)$_2$ and CrCl$_2$ catalysts was reported by Trost et al. [107] (Scheme 1.80).

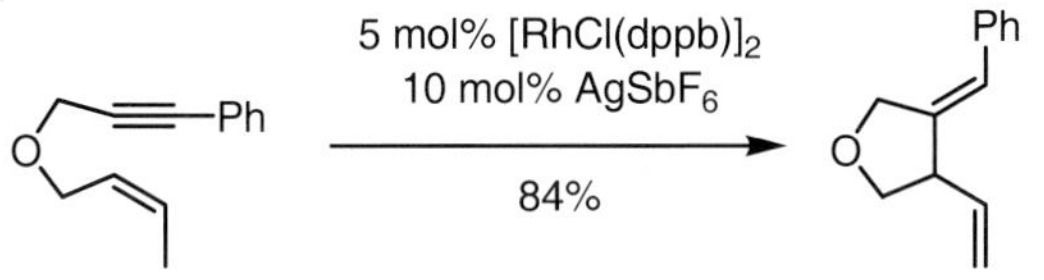

**Scheme 1.79**

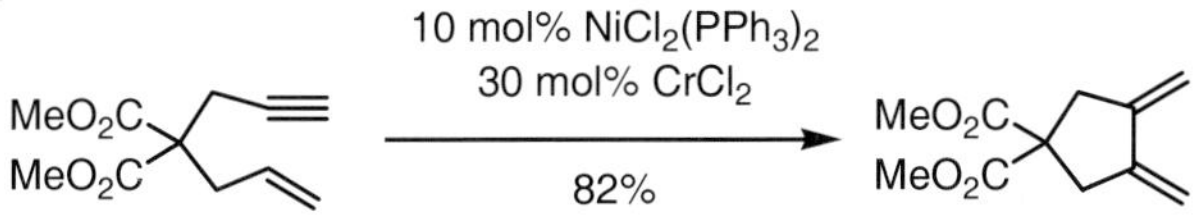

**Scheme 1.80**

### 1.3.4.6 Indole-Forming Reaction

We developed a synthesis of 3-allyl-*N*-(alkoxycarbonyl)indoles from isocyanates and allyl carbonate using a combined catalyst system comprising Pd(PPh$_3$)$_4$ and CuCl (Scheme 1.81) [108]. The Pd catalyst is a precursor of a π-allylpalladium methoxide, generated through the reaction with allyl methyl carbonate. The Cu catalyst is proposed to serve a dual role, activating the isocyanate functional group as a Lewis acid so as to facilitate the addition of the π-allylpalladium methoxide, and simultaneously coordinating to the alkyne moiety as a typical transition metal catalyst to promote the ensuing cyclization to construct the indole core.

Watanabe and co-workers [109] reported an indole synthesis involving reductive *N*-heterocyclization of nitroarenes using PdCl$_2$(PPh$_3$)$_2$ and SnCl$_2$ catalysts under a CO atmosphere (Scheme 1.82). The reaction required both Pd and Sn catalysts, although their roles in the catalytic cycle were obscure. The carbon monoxide (CO) operates as an efficient deoxygenating agent for 2-nitrostyrenes and extrusion of CO$_2$ generates the corresponding palladium-nitrene intermediates. The Sn catalyst might be involved in the reduction process of the nitro group. The electrophilic nitrene probably attacks the olefinic carbon, and subsequent hydrogen transfer via [1,5]-sigmatropic rearrangement gives the indole core.

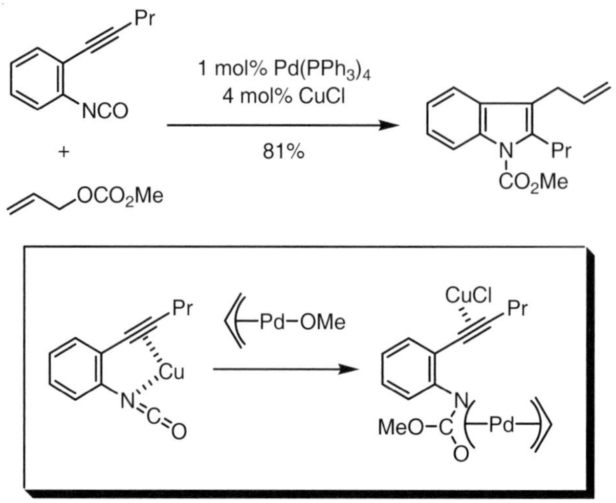

**Scheme 1.81**

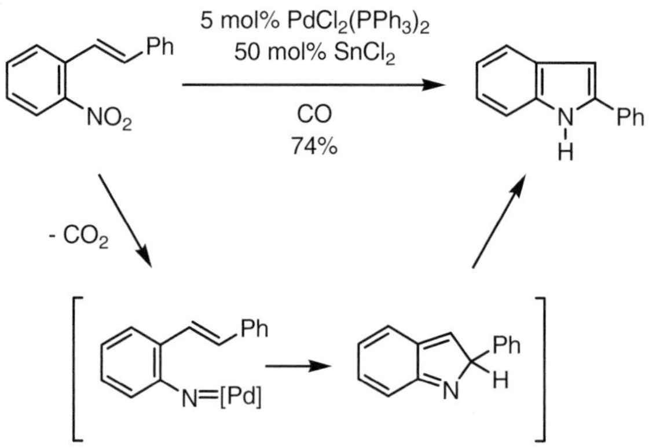

**Scheme 1.82**

### 1.3.4.7 Furan- and Pyrrole-Forming Reactions

Hidai and Uemura et al. [110] disclosed a heterobimetallic-catalyzed synthesis of furans from propargyl alcohols and ketones (Scheme 1.83). This reaction could only be realized with a combination of a thiolate-bridged diruthenium complex, [Cp*RuCl(μ-SMe)$_2$RuCp*Cl], and PtCl$_2$. The overall process consists of two consecutive reactions. Substitution with acetone at the propargyl position most probably takes place in the presence of the Ru complex to afford the corresponding γ-ketoalkyne intermediate. Then, Pt-catalyzed hydration of the alkyne with H$_2$O generated in situ produces the 1,4-diketone, which undergoes intramolecular cyclization catalyzed by the Pt complex to afford the furan derivative. When the above reaction was carried out in the presence of anilines, the corresponding pyrroles were obtained through imine intermediates (Scheme 1.84).

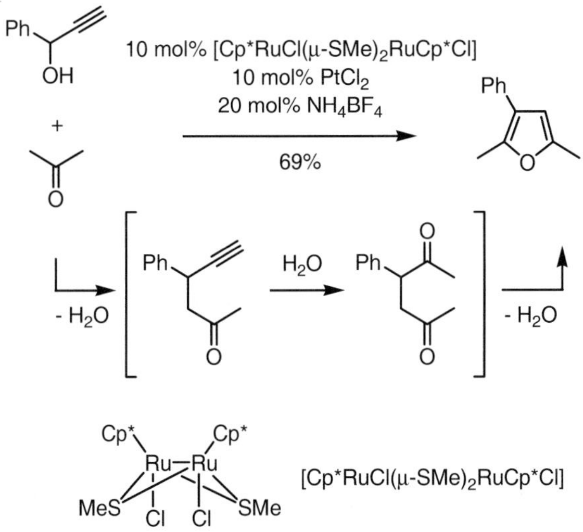

Ph
OH

10 mol% [Cp*RuCl(μ-SMe)₂RuCp*Cl]
10 mol% PtCl₂
20 mol% NH₄BF₄

+

69%

Ph

- H₂O

Ph

H₂O   Ph

- H₂O

Cp*      Cp*
Ru—Ru
MeS   Cl   Cl   SMe

[Cp*RuCl(μ-SMe)₂RuCp*Cl]

**Scheme 1.83**

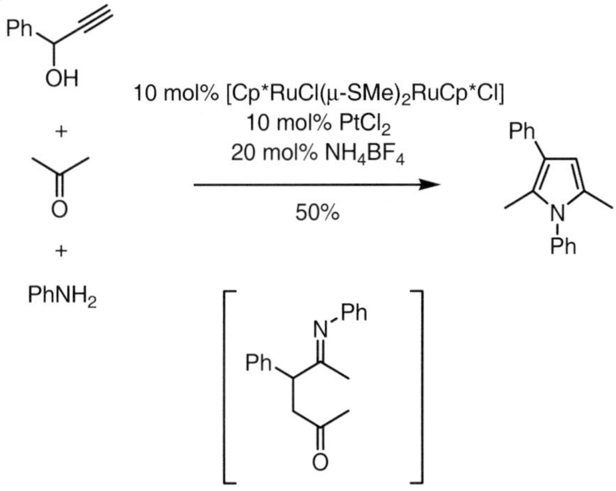

Ph
OH

10 mol% [Cp*RuCl(μ-SMe)₂RuCp*Cl]
10 mol% PtCl₂
20 mol% NH₄BF₄

+

50%

Ph

N
Ph

+

PhNH₂

Ph
N-Ph

O

**Scheme 1.84**

## 1.3.5
## Reactions Involving Nucleophilic Addition of Carbonyl Compounds

### 1.3.5.1   The Aldol Reaction

Shibasaki and Sodeoka et al. [111] investigated a catalytic asymmetric aldol reaction involving a chiral Pd$^{II}$ enolate (Scheme 1.85). The highest enantioselectivity was observed when the reaction was conducted in the presence of a combination of PdCl₂[(R)-BINAP] and AgOTf catalysts in wet DMF. The Ag catalyst abstracts a chloride ion from the Pd center to generate a reactive cationic Pd species. Transmetalation between the Pd catalyst and a silyl enol ether forms the chiral Pd$^{II}$ enolate, which controls the orientation of a prochiral enolate during the course of the reaction with the aldehyde.

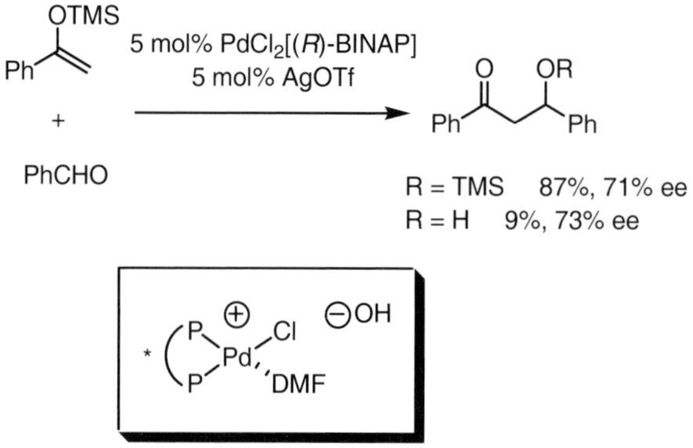

OTMS

Ph⟍═

5 mol% PdCl₂[(*R*)-BINAP]
5 mol% AgOTf

+

PhCHO

→

$$\underset{Ph}{\overset{O}{\parallel}}\wedge\underset{Ph}{\overset{OR}{|}}$$

R = TMS   87%, 71% ee
R = H     9%, 73% ee

**Scheme 1.85**

## 1.3.5.2 Alkynylation Reactions

Li et al. [112] studied the alkynylation of aldehydes through a Grignard-type reaction involving C–H bond activation (Scheme 1.86). The corresponding alkynylated products were obtained from the reaction of terminal alkynes with aldehydes in the presence of RuCl₃ and In(OAc)₃ as a bimetallic catalyst system. The addition of morpholine increased the conversion of the reaction. The key to this transformation is that the In salt functions as a Lewis acid to activate the starting aldehydes even in the presence of water. The Ru catalyst simultaneously inserts into the C–H bond of the alkyne and the derived Ru-acetylide intermediate undergoes Grignard-type addition to the aldehyde activated by coordination of the In salt. When imines were used as reaction partners, a combination of RuCl₃ and CuBr catalysts proved effective for successful reaction (Scheme 1.87) [113]. Without the addition of the Cu salt, no reaction took place. Activation of an imine through coordination of the Cu additive is essential for addition of the Ru-acetylide species.

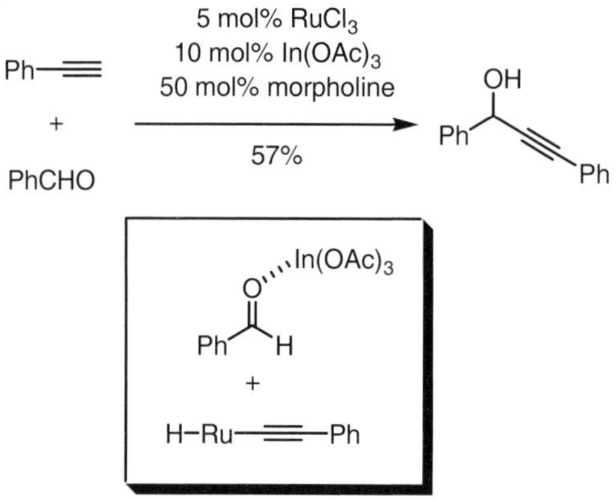

Ph──≡

+

PhCHO

5 mol% RuCl₃
10 mol% In(OAc)₃
50 mol% morpholine

──────→
57%

$$\underset{Ph}{\overset{OH}{|}}\diagdown\diagup Ph$$

**Scheme 1.86**

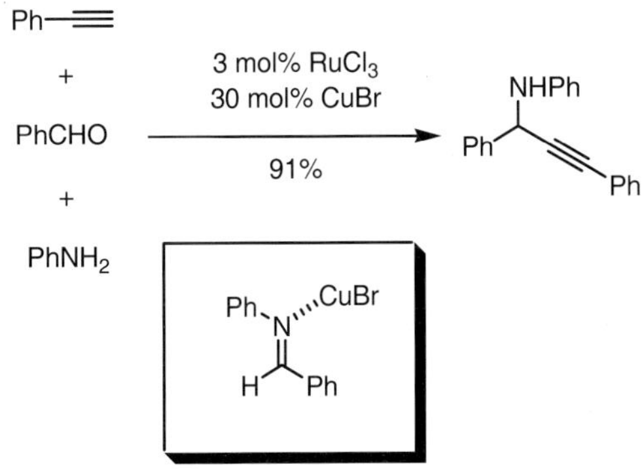

**Scheme 1.87**

### 1.3.5.3 Conjugate Addition Reactions

Trost and co-workers [114] reported a ruthenium-catalyzed three-component addition reaction of terminal alkynes, α,β-unsaturated carbonyl compounds, and nucleophilic reactants. 1,5-Diketones were obtained using $H_2O$ as a nucleophilic agent in the presence of a combination of CpRu(COD)Cl and In(OTf)$_3$ catalysts (Scheme 1.88). The reaction proceeds through attack of $H_2O$ on the alkyne, which is activated by the π-coordination of a cationic ruthenium species. The derived ruthenium-enolate, equivalent to a vinylruthenium intermediate, then adds to the α,β-unsaturated ketone in a 1,4-manner. The role of the In salt remains to be clarified, although it is proposed that it prevents deactivation of the Ru catalyst by facilitating chloride dissociation and that it activates the vinyl ketones by coordination as a Lewis acid catalyst. A similar type of reaction was carried out using Me$_4$NCl as a nucleophilic agent (Scheme 1.89) [115]. In this case, (E)-vinyl chlorides were obtained as the major products. A bimetallic catalyst comprising CpRu(COD)Cl and SnCl$_4$· 5 H$_2$O was found to be the optimal catalyst system for this transformation. When LiBr was used as a nucleophilic agent, (Z)-vinyl bromides were obtained as the major products in the presence of [CpRu(MeCN)$_3$]PF$_6$ and SnBr$_4$ catalysts (Scheme 1.90) [116]. The high Z-selectivity in forming the adducts can be

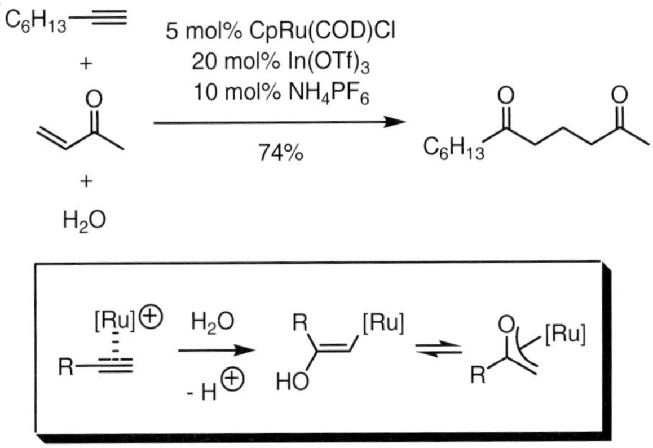

**Scheme 1.88**

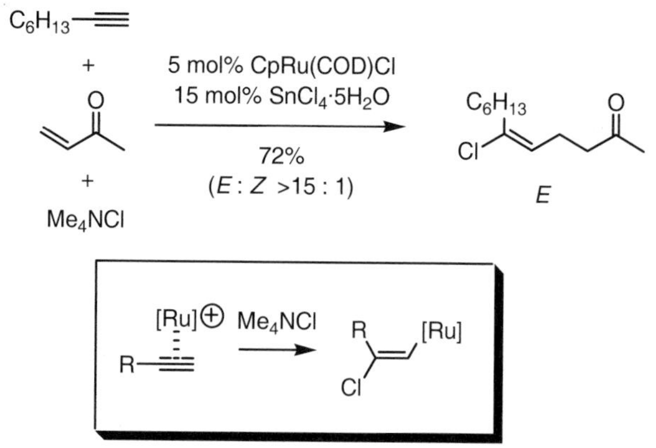

**Scheme 1.89**

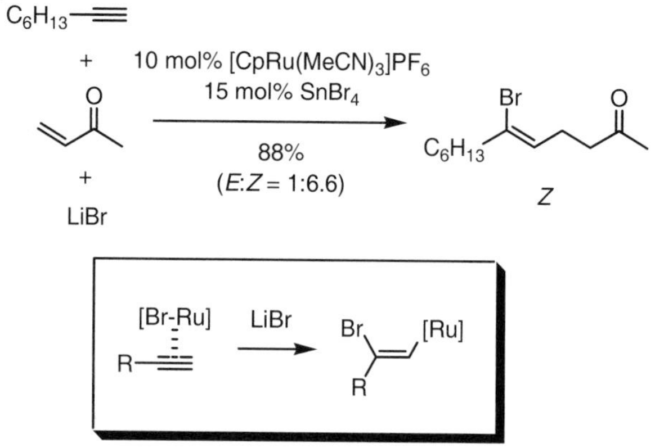

**Scheme 1.90**

rationalized by invoking a mechanism in which the vinylruthenium intermediate is formed by *cis*-addition of a neutral Ru–Br species to the alkyne.

### 1.3.6
### Miscellaneous Reactions

#### 1.3.6.1 Transition Metal Catalyzed Reactions

Shi and co-workers [117] reported that the hydroamination of methylenecyclopropanes (MCPs) with sulfonamides was effectively catalyzed by a combination of Pd(PPh$_3$)$_4$ and Pd(OAc)$_2$ catalysts (Scheme 1.91). The use of a single Pd$^0$ catalyst, such as Pd(PPh$_3$)$_4$ or Pd$_2$(dba)$_3$, or a single Pd$^{II}$ catalyst, such as PdCl$_2$(PPh$_3$)$_2$ or Pd(OAc)$_2$, did not promote the ring-opening reaction of MCPs with toluene-sulfonamide. When the reaction was carried out with the Pd$^0$–Pd$^{II}$ combined catalyst, Pd(PPh$_3$)$_4$ and Pd(OAc)$_2$, the corresponding diallyl tosylamide was obtained in excellent yield. The Pd$^0$ catalyst serves as a precursor of a hydridepalladium amide species. The Pd$^{II}$ catalyst is assumed to act as a weak Lewis acid in coordinating to the double bond of MCP and accelerates the ring-opening of the cyclopropane moiety after the hydropalladation has taken place.

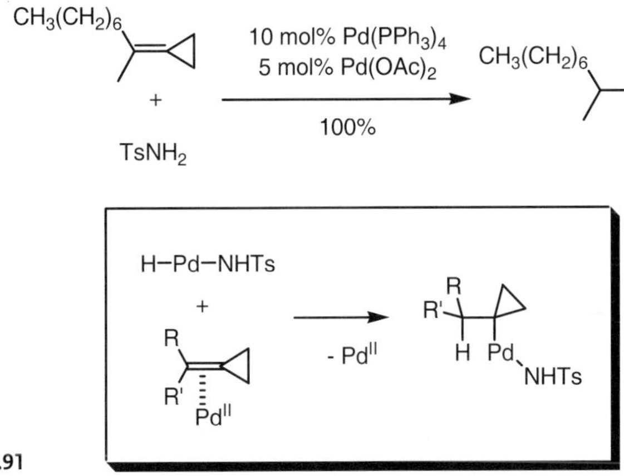

**Scheme 1.91**

Shirakawa and co-workers [118] employed $Pd(OAc)_2/PPh_3$ and $In(OTf)_3$ as a combined catalyst system for the dimerization of vinylarenes (Scheme 1.92). The reaction of styrene gave 1,3-diphenyl-1-butene selectively. The role of the In additive is probably to activate the vinylarenes by making them more susceptible to nucleophilic attack by the $Pd^0$ complex. Then, the palladium intermediate undergoes insertion of another vinylarene to provide the dimerized product.

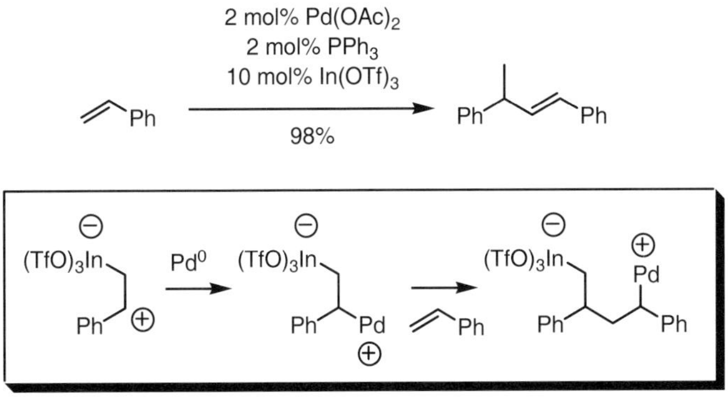

**Scheme 1.92**

Trost et al. [119] investigated redox isomerization of propargyl alcohols to enals and enones, and found that a combination of $(\eta^5\text{-indenyl})RuCl(PPh_3)_2$ and $InCl_3$ catalysts effectively promoted isomerization to afford the corresponding $\alpha,\beta$-unsaturated carbonyl compounds (Scheme 1.93). The reaction proceeded to some extent in the presence of the Ru catalyst alone, albeit only to a low conversion. Two possibilities are proposed for the effect of the In additive. One is that the In catalyst functions as a chloride scavenger, thereby generating a reactive cationic Ru species. The other is the formation of an In-bridged intermediate to release the strain and to accelerate the ensuing hydride migration.

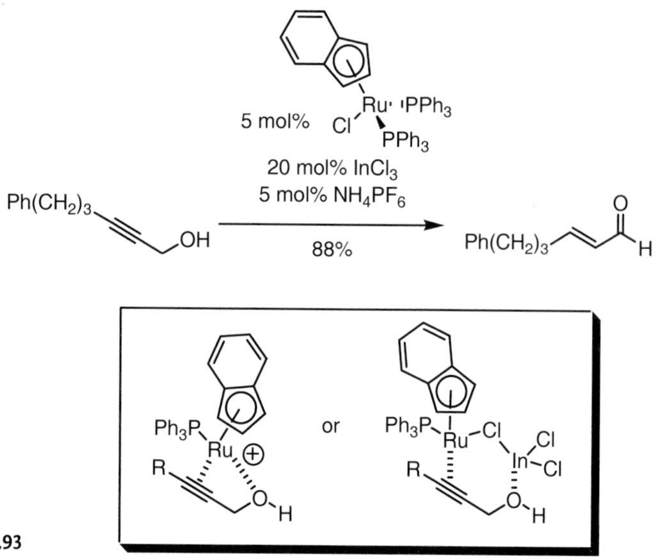

**Scheme 1.93**

Tsukada and Inoue et al. [120] reported a stereoselective *cis*-addition of aromatic C–H bonds to alkynes by using a dinuclear palladium catalyst, $Pd_2(p\text{-}CH_3\text{-}C_6H_4)_2(\mu\text{-}OH)(\mu\text{-}dpfam)$ (dpfam = *N,N'*-bis[2-(diphenylphosphino)phenyl]formamidinate) (Scheme 1.94). A mechanistic rationale was not provided, but this reaction did not take place in the presence of typical palladium complexes such as $Pd(OAc)_2$ or $Pd_2(dba)_3 \cdot CHCl_3$, nor with mononuclear PdMe(dpfam).

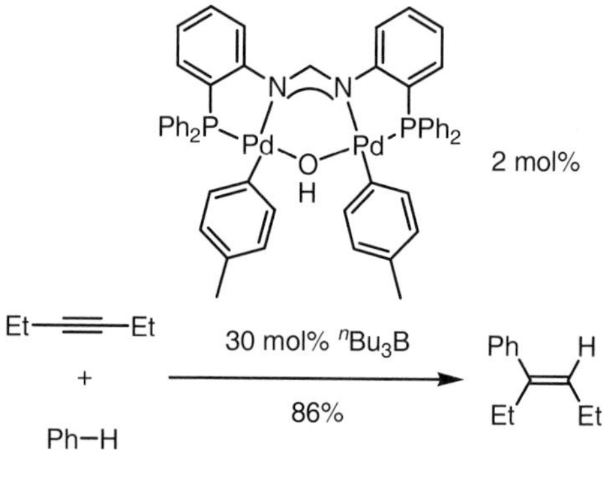

**Scheme 1.94**

### 1.3.6.2 Lewis Acid Catalyzed Reactions

Maruoka and co-workers designed a homobimetallic titanium system as a bidentate Lewis acid catalyst. The simultaneous coordination of a carbonyl group in a σ,σ-complexation mode with a bidentate Lewis acid enhances the reactivity and selectivity of the carbonyl substrate compared to coordination by a corresponding monodentate Lewis acid. The allylation of benzaldehyde with tetraallyltin is a typical example where such reactivity enhancement is observed (Scheme 1.95) [121].

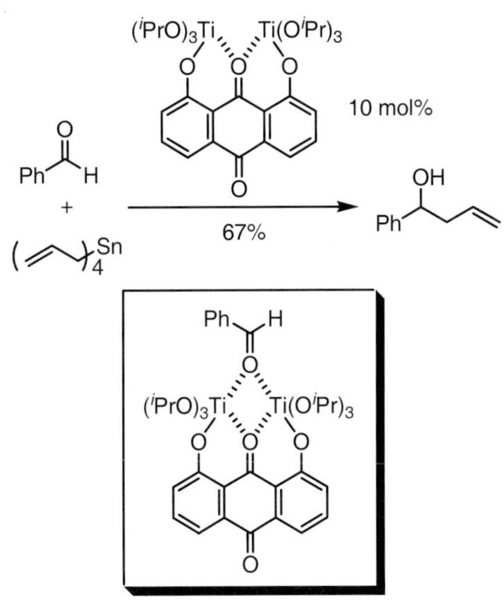

**Scheme 1.95**

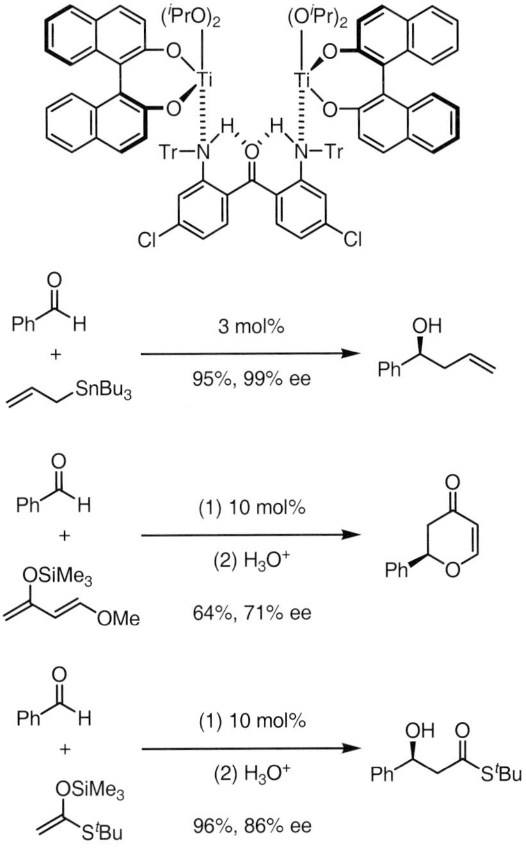

**Scheme 1.96**

Use of the bis-Ti complex led to the corresponding adduct in good yield, whereas the use of the mono-Ti complex gave only a trace amount of the product. Further developments have involved the use of bis-Ti complexes bearing binaphthol ligands, and these systems have enabled asymmetric allylations [122], enantioselective hetero-Diels–Alder reactions, and aldol reactions of aldehydes (Scheme 1.96) [123].

### 1.3.6.3 Sequential Reactions

Doye and co-workers [124] developed a one-pot procedure for the synthesis of indoles through two sequential reactions, namely the hydroamination of alkynes and an aromatic amination reaction, without isolating the intermediates (Scheme 1.97). The Ti complex Cp$_2$TiMe$_2$ was found to promote the hydroamination of alkynes to produce imine intermediates, and the Pd complex Pd$_2$(dba)$_3$, in combination with an imidazolium ligand, catalyzed the intramolecular aromatic amination reaction of the corresponding enamine under basic conditions to form the indole skeleton.

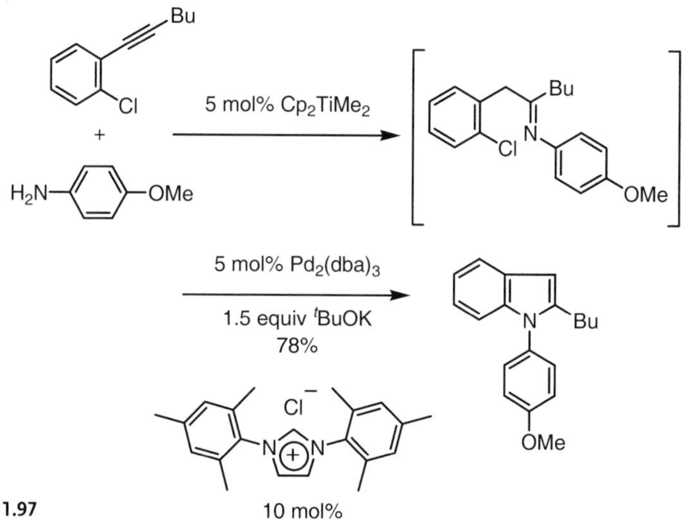

**Scheme 1.97**

Trost et al. [125] reported a one-pot enantio- and diastereoselective synthesis of heterocycles using ruthenium and palladium catalysts (Scheme 1.98). The Ru complex [CpRu(MeCN)$_3$]PF$_6$ was found to catalyze the coupling between the starting alkyne and alkene to form the diene intermediate bearing suitable functional groups in the appropriate positions for the second step. After the formation of the diene intermediate, the Pd complex Pd$_2$(dba)$_3$ · CHCl$_3$, together with the chiral bidentate phosphine ligand, was added to the reaction mixture to promote the asymmetric allylic alkylation. An enantioselective cyclization took place to afford pyrrolidine derivatives. These authors succeeded in obtaining not only N-containing but also O-containing cyclic compounds in good yields and with high enantioselectivities.

Hayashi and co-workers [126] successfully accomplished an asymmetric synthesis of 1-aryl-1,2-ethanediols from arylacetylenes through stepwise hydrosilylation of

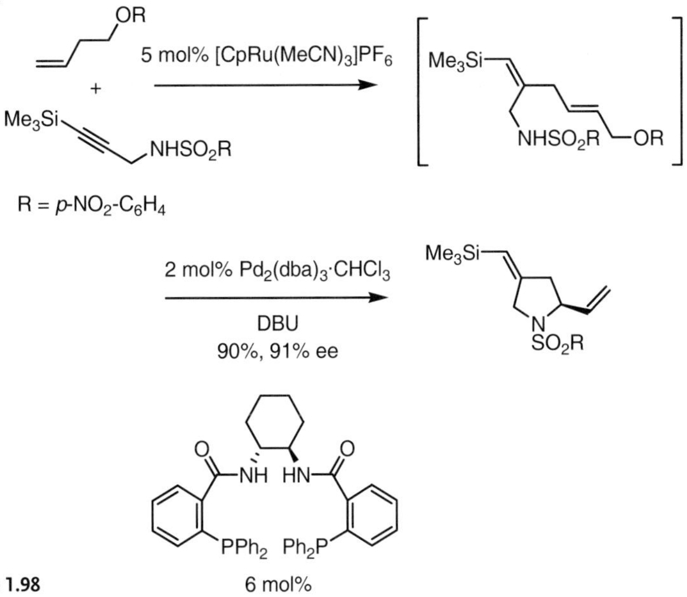

Scheme 1.98

the alkynes using platinum and palladium catalysts (Scheme 1.99). The Pt catalyst, $[PtCl_2(C_2H_4)]_2$, promoted hydrosilylation of the alkyne to afford the vinylsilane intermediate in a regioselective manner, while the Pd catalyst, $[(\eta^3\text{-}C_3H_5)PdCl]_2$, together with the chiral phosphine ligand, promoted asymmetric hydrosilylation of the derived vinylsilane to afford the 1,2-bis(silyl)ethane derivative. Subsequent oxidation using hydrogen peroxide produced the corresponding diol as the final product with high enantioselectivity.

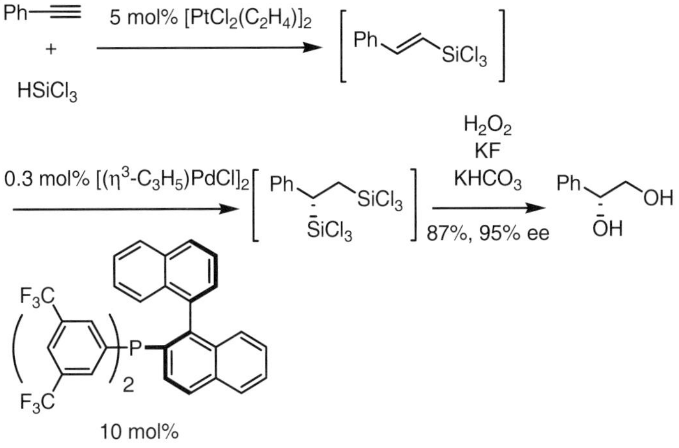

Scheme 1.99

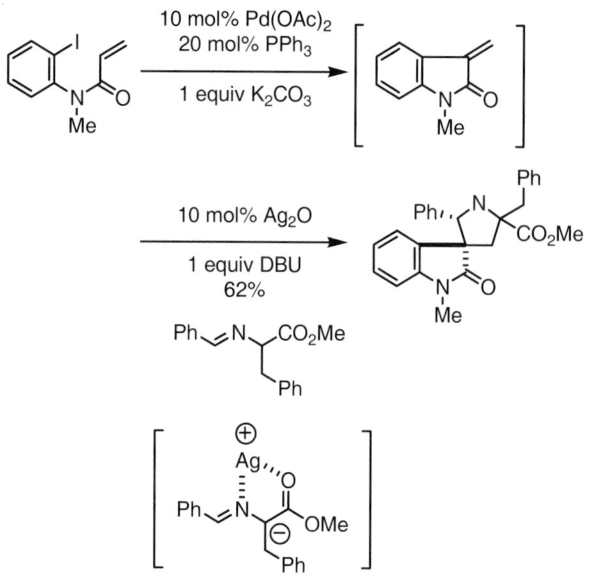

Scheme 1.100

Grigg and co-workers [127] investigated the synthesis of *spiro*-oxindoles by a sequence of intramolecular Heck reaction followed by 1,3-dipolar cycloaddition using palladium and silver catalysts (Scheme 1.100). The Pd complex, Pd(OAc)$_2$/PPh$_3$, promotes the intramolecular Heck reaction to afford a relatively unstable 3-methyleneoxindole intermediate. The introduction of the Ag salt, Ag$_2$O, together with the imine and DBU, provides the corresponding azomethine ylide, which undergoes [3+2] cycloaddition with the methyleneoxindole intermediate to furnish the *spiro*-oxindole regiospecifically.

# References

1   N. KRAUSE (Ed.), *Modern Organocopper Chemistry*, Wiley-VCH, Weinheim, 2002; (b) R. J. K. TAYLOR (Ed.), *Organocopper Reagents*, Oxford University Press, Oxford, 1994.

2   (a) F. DIEDERICH, P. J. STANG (Eds.), *Metal-Catalyzed Cross-Coupling Reactions*, Wiley-VCH, Weinheim, 1998; (b) J. TSUJI, *Transition Metal Reagents and Catalysts*, Wiley, Chichester, 2000.

3   (a) S. GRONOWITZ, A. MESSMER, G. TIMÁRI, *J. Heterocycl. Chem.* 1992, *29*, 1049; (b) S. GRONOWITZ, P. BJÖRK, J. MALM, A.-B. HÖRNFELDT, *J. Organomet. Chem.* 1993, *460*, 127.

4   M. KOSUGI, Y. NEGISHI, M. KAMEYAMA, T. MIGITA, *Bull. Chem. Soc. Jpn.* 1985, *58*, 3383.

5   X. HAN, B. M. STOLTZ, E. J. COREY, *J. Am. Chem. Soc.* 1999, *121*, 7600.

6   R. WITTENBERG, J. SROGL, M. EGI, L. S. LIEBESKIND, *Org. Lett.* 2003, *5*, 3033.

7   M. EGI, L. S. LIEBESKIND, *Org. Lett.* 2003, *5*, 801.

8   F.-A. ALPHONSE, F. SUZENET, A. KEROMNES, B. LEBRET, G. GUILLAUMET, *Org. Lett.* 2003, *5*, 803.

9   (a) M. KOSUGI, I. HAGIWARA, T. SUMIYAMA, T. MIGITA, *J. Chem. Soc., Chem. Commun.* 1983, 344;

(b) M. Kosugi, I. Hagiwara, T. Migita, *Chem. Lett.* **1983**, 839; (c) M. Kosugi, I. Hagiwara, T. Sugiyama, T. Migita, *Bull. Chem. Soc. Jpn.* **1984**, *57*, 242.

10  I. Kuwajima, H. Urabe, *J. Am. Chem. Soc.* **1982**, *104*, 6831.

11  (a) K. Hirabayashi, J. Kawashima, Y. Nishihara, A. Mori, T. Hiyama, *Org. Lett.* **1999**, *1*, 299; (b) K. Hirabayashi, A. Mori, J. Kawashima, M. Suguro, Y. Nishihara, T. Hiyama, *J. Org. Chem.* **2000**, *65*, 5342.

12  S. Chang, S. H. Yang, P. H. Lee, *Tetrahedron Lett.* **2001**, *42*, 4833.

13  (a) L. S. Hegedus, *Comprehensive Organic Synthesis*, Vol. 4 (Ed.: B. M. Trost, I. Fleming), Pergamon, Oxford, 1991, p. 551; (b) J. Tsuji, *Synthesis* **1984**, 369.

14  T. Hosokawa, H. Hirata, S. Murahashi, A. Sonoda, *Tetrahedron Lett.* **1976**, *21*, 1821.

15  A. I. Roshchin, S. M. Kel'chevski, N. A. Bumagin, *J. Organomet. Chem.* **1998**, *560*, 163.

16  (a) M. F. Semmelhack, C. Bodurow, *J. Am. Chem. Soc.* **1984**, *106*, 1496; (b) M. F. Semmelhack, C. Kim, N. Zhang, C. Bodurow, M. Sanner, W. Doubler, M. Meier, *Pure Appl. Chem.* **1990**, *62*, 2035.

17  Y. Tamaru, T. Kobayashi, S. Kawamura, H. Ochiai, M. Hojo, Z. Yoshida, *Tetrahedron Lett.* **1985**, *26*, 3207.

18  M. F. Semmelhack, W. R. Epa, *Tetrahedron Lett.* **1993**, *34*, 7205.

19  (a) D. Lathbury, P. Vernon, T. Gallagher, *Tetrahedron Lett.* **1986**, *27*, 6009; (b) D. N. A. Fox, T. Gallagher, *Tetrahedron* **1990**, *46*, 4697; (c) T. Gallagher, I. W. Davies, S. W. Jones, D. Lathbury, M. F. Mahon, K. C. Molloy, R. W. Shaw, P. Vernon, *J. Chem. Soc., Perkin Trans. 1* **1992**, 433.

20  (a) M. Kimura, N. Saeki, S. Uchida, H. Harayama, S. Tanaka, K. Fugami, Y. Tamaru, *Tetrahedron Lett.* **1993**, *34*, 7611; (b) Y. Tamaru, M. Kimura, *Synlett* **1997**, 749.

21  Y. Tamaru, M. Hojo, Z. Yoshida, *J. Org. Chem.* **1988**, *53*, 5731.

22  (a) T. Pei, R. A. Widenhoefer, *Chem. Commun.* **2002**, 650; (b) T. Pei, X. Wang, R. A. Widenhoefer, *J. Am. Chem. Soc.* **2003**, *125*, 648.

23  D. Yang, J.-H. Li, Q. Gao, Y.-L. Yan, *Org. Lett.* **2003**, *5*, 2869.

24  (a) J. T. Link, *Org. React.* **2002**, *60*, 157; (b) R. H. Heck, *Org. React.* **1982**, *27*, 345.

25  (a) K. Karabelas, A. Hallberg, *Tetrahedron Lett.* **1985**, *26*, 3131; (b) K. Karabelas, C. Westerlund, A. Hallberg, *J. Org. Chem.* **1985**, *50*, 3896; (c) K. Karabelas, A. Hallberg, *J. Org. Chem.* **1986**, *51*, 5286; (d) K. Karabelas, A. Hallberg, *J. Org. Chem.* **1988**, *53*, 4909.

26  M. M. Abelman, T. Oh, L. E. Overman, *J. Org. Chem.* **1987**, *52*, 4130.

27  (a) T. Jeffery, *Tetrahedron Lett.* **1991**, *32*, 2121; (b) T. Jeffery, *J. Chem. Soc., Chem. Commun.* **1991**, 324.

28  (a) Y. Sato, M. Sodeoka, M. Shibasaki, *J. Org. Chem.* **1989**, *54*, 4738; (b) Y. Sato, M. Sodeoka, M. Shibasaki, *Chem. Lett.* **1990**, 1953; (c) M. Shibasaki, C. D. J. Boden, A. Kojima, *Tetrahedron* **1997**, *53*, 7371; (d) A. B. Dounay, L. A. Overman, *Chem. Rev.* **2003**, *103*, 2945.

29  (a) R. Grigg, V. Loganathan, V. Santhakumar, V. Sridharan, A. Teasdale, *Tetrahedron Lett.* **1991**, *32*, 687; (b) W. Cabri, I. Candiani, A. Bedeschi, S. Penco, *J. Org. Chem.* **1992**, *57*, 1481; (c) W. Carbi, I. Candiani, *Acc. Chem. Res.* **1995**, *28*, 2.

30  D. Flubacher, G. Helmchen, *Tetrahedron Lett.* **1999**, *40*, 3867.

31  (a) R. C. Larock, J. M. Zenner, *J. Org. Chem.* **1995**, *60*, 482; (b) J. M. Zenner, R. C. Larock, *J. Org. Chem.* **1999**, *64*, 7312.

32  G. Poli, G. Giambastiani, A. Mordini, *J. Org. Chem.* **1999**, *64*, 2962.

33  K. Fuji, N. Kinoshita, K. Tanaka, *Chem. Commun.* **1999**, 1895.

34  (a) T. Tabuchi, J. Inanaga, M. Yamaguchi, *Tetrahedron Lett.* **1986**, *27*, 1195; (b) T. Tabuchi, J. Inanaga, M. Yamaguchi, *Tetrahedron Lett.* **1987**, *28*, 215.

35  Y. Masuyama, N. Kinugawa, Y. Kurusu, *J. Org. Chem.* **1987**, *52*, 3702.

36  (a) J. P. Takahara, Y. Masuyama, Y. Kurusu, *J. Am. Chem. Soc.* **1992**, *114*, 2577; (b) Y. Masuyama, R. Hayashi, K. Otake, Y. Kurusu, *J. Chem. Soc., Chem. Commun.* **1988**, 44;

(c) Y. Masuyama, J. P. Takahara, Y. Kurusu, *J. Am. Chem. Soc.* **1988**, *110*, 4473.

**37** (a) K. Yasui, Y. Goto, T. Yajima, Y. Taniseki, K. Fugami, A. Tanaka, Y. Tamaru, *Tetrahedron Lett.* **1993**, *34*, 7619; (b) Y. Tamaru, A. Tanaka, K. Yasui, S. Goto, S. Tanaka, *Angew. Chem. Int. Ed. Engl.* **1995**, *34*, 787.

**38** (a) M. Kimura, I. Kiyama, T. Tomizawa, Y. Horino, S. Tanaka, Y. Tamaru, *Tetrahedron Lett.* **1999**, *40*, 6795; (b) Y. Tamaru, *J. Organomet. Chem.* **1999**, *576*, 215.

**39** (a) S. Araki, T. Kamei, T. Hirashita, H. Yamamura, M. Kawai, *Org. Lett.* **2000**, *2*, 847; (b) S. Araki, K. Kameda, J. Tanaka, T. Hirashita, H. Yamamura, M. Kawai, *J. Org. Chem.* **2001**, *66*, 7919.

**40** S.-K. Kang, S.-W. Lee, J. Jung, Y. Lim, *J. Org. Chem.* **2002**, *67*, 4376.

**41** (a) W. Oppolzer, F. Schröder, *Tetrahedron Lett.* **1994**, *35*, 7939; (b) W. Oppolzer, J. Ruiz-Montes, *Helv. Chim. Acta* **1993**, *76*, 1266.

**42** (a) H. Nakamura, N. Asao, Y. Yamamoto, *J. Chem. Soc., Chem. Commun.* **1995**, 1273; (b) H. Nakamura, H. Iwama, Y. Yamamoto, *Chem. Commun.* **1996**, 1459; (c) H. Nakamura, H. Iwama, Y. Yamamoto, *J. Am. Chem. Soc.* **1996**, *118*, 6641.

**43** H. Nakamura, K. Nakamura, Y. Yamamoto, *J. Am. Chem. Soc.* **1998**, *120*, 4242.

**44** R. A. Fernandes, A. Stimac, Y. Yamamoto, *J. Am. Chem. Soc.* **2003**, *125*, 14133.

**45** K. Nakamura, H. Nakamura, Y. Yamamoto, *J. Org. Chem.* **1999**, *64*, 2614.

**46** (a) Y. Sato, M. Takimoto, K. Hayashi, T. Katsuhara, K. Takagi, M. Mori, *J. Am. Chem. Soc.* **1994**, *116*, 9771; (b) Y. Sato, M. Takimoto, M. Mori, *Tetrahedron Lett.* **1996**, *37*, 887.

**47** (a) M. Kitamura, A. Ezoe, K. Shibata, Y. Tamaru, *J. Am. Chem. Soc.* **1998**, *120*, 4033; (b) M. Kimura, A. Ezoe, S. Tanaka, Y. Tamaru, *Angew. Chem. Int. Ed.* **2001**, *40*, 3600.

**48** (a) M. Kimura, H. Fujimatsu, A. Ezoe, K. Shibata, M. Shimizu, S. Matsumoto, Y. Tamaru, *Angew. Chem. Int. Ed.* **1999**,

*38*, 397; (b) K. Shibata, M. Kimura, M. Shimizu, Y. Tamaru, *Org. Lett.* **2001**, *3*, 2181.

**49** (a) J. Montgomery, A. V. Savchenko, *J. Am. Chem. Soc.* **1996**, *118*, 2099; (b) J. Montgomery, E. Oblinger, A. Savchenko, *J. Am. Chem. Soc.* **1997**, *119*, 4911.

**50** E. Oblinger, J. Montgomery, *J. Am. Chem. Soc.* **1997**, *119*, 9065.

**51** M. V. Chevliakov, J. Montgomery, *Angew. Chem. Int. Ed.* **1998**, *37*, 3144.

**52** X.-Q. Tang, J. Montgomery, *J. Am. Chem. Soc.* **1999**, *121*, 6098.

**53** (a) K. M. Miller, W.-S. Huang, T. F. Jamison, *J. Am. Chem. Soc.* **2003**, *125*, 3442; (b) E. A. Colby, T. F. Jamison, *J. Org. Chem.* **2003**, *68*, 156.

**54** S. J. Patel, T. F. Jamison, *Angew. Chem. Int. Ed.* **2003**, *42*, 1364.

**55** C. Molinaro, T. F. Jamison, *J. Am. Chem. Soc.* **2003**, *125*, 8076.

**56** K. Takai, K. Kimura, T. Kuroda, T. Hiyama, H. Nozaki, *Tetrahedron Lett.* **1983**, *24*, 5281.

**57** (a) K. Takai, M. Tagashira, T. Kuroda, K. Oshima, K. Utimoto, H. Nozaki, *J. Am. Chem. Soc.* **1986**, *108*, 6048; (b) H. Jin, J. Uenishi, W. J. Christ, Y. Kishi, *J. Am. Chem. Soc.* **1986**, *108*, 5644; (c) Y. Kishi, *Pure Appl. Chem.* **1992**, *64*, 343.

**58** (a) A. Fürstner, N. Shi, *J. Am. Chem. Soc.* **1996**, *118*, 12349; (b) A. Fürstner, *Chem. Rev.* **1999**, *99*, 991.

**59** (a) H.-w. Choi, K. Nakajima, D. Demeke, F.-A. Kang, H.-S. Jun, Z.-K. Wan, Y. Kishi, *Org. Lett.* **2002**, *4*, 4435; (b) Z.-K. Wan, H.-w. Choi, F.-A. Kang, K. Nakajima, D. Demeke, Y. Kishi, *Org. Lett.* **2002**, *4*, 4431.

**60** (a) V. Farina, V. Krishnamurthy, W. J. Scott, *Org. React.* **1997**, *50*, 3; (b) V. Farina, *Pure Appl. Chem.* **1996**, *68*, 73.

**61** (a) L. S. Liebeskind, R. W. Fengl, *J. Org. Chem.* **1990**, *55*, 5359; (b) L. S. Liebeskind, J. Wand, *Tetrahedron Lett.* **1990**, *31*, 4293; (c) L. S. Liebeskind, M. S. Yu, R. H. Yu, J. Wang, K. S. Hagen, *J. Am. Chem. Soc.* **1993**, *115*, 9048.

**62** V. Farina, S. Kapadia, B. Krishnan, C. Wang, L. S. Liebeskind, *J. Org. Chem.* **1994**, *59*, 5905.

63 W.-S. Kim, H.-J. Kim, C.-G. Cho,
*J. Am. Chem. Soc.* **2003**, *125*, 14288.

64 (a) J. Ye, R. K. Bhatt, J. R. Falck,
*J. Am. Chem. Soc.* **1994**, *116*, 1;
(b) J. R. Falck, R. K. Bhatt, J. Ye,
*J. Am. Chem. Soc.* **1995**, *117*, 5973.

65 (a) Y. Nishihara, K. Ikegashira,
A. Mori, T. Hiyama, *Chem. Lett.* **1997**,
1233; (b) Y. Nishihara, K. Ikegashira,
K. Hirabayashi, J. Ando, A. Mori,
T. Hiyama, *J. Org. Chem.* **2000**, *65*, 1780;
(c) Y. Nishihara, J. Ando, T. Kato,
A. Mori, T. Hiyama, *Macromolecules*
**2000**, *33*, 2779.

66 Y. Nishihara, M. Takemura, A. Mori,
K. Osakada, *J. Organomet. Chem.* **2001**,
*620*, 282.

67 (a) K. Sonogashira, Y. Tohda,
N. Hagihara, *Tetrahedron Lett.* **1975**, *50*,
4467; (b) K. Sonogashira, T. Yatake,
Y. Tohda, S. Takahashi, N. Hagihara,
*J. Chem. Soc., Chem. Commun.* **1977**, 291;
(c) K. Sonogashira, *Comprehensive
Organic Synthesis*, Vol. 3 (Eds.: B. M.
Trost, I. Fleming), Pergamon, Oxford,
1991, p. 521; (d) K. Sonogashira,
*J. Organomet. Chem.* **2002**, *653*, 46.

68 (a) M. Alami, G. Linstrumelle, *Tetra-
hedron Lett.* **1991**, *32*, 6109; (b) M. Alami,
F. Ferri, G. Linstrumelle, *Tetrahedron
Lett.* **1993**, *34*, 6403.

69 P. Bertus, P. Pale, *Tetrahedron Lett.* **1996**,
*37*, 2019.

70 (a) G. T. Crisp, P. D. Turner,
K. A. Stephens, *J. Organomet. Chem.*
**1998**, *570*, 219; (b) Y. Azuma, A. Sato,
M. Morone, *Heterocycles* **1996**, *42*, 789;
(c) L. Anastasia, E. Negishi, *Org. Lett.*
**2001**, *3*, 3111.

71 T. Kawasaki, Y. Yamamoto, *J. Org. Chem.*
**2002**, *67*, 5138.

72 A. Arcadi, F. Marinelli, *Synthesis* **1986**,
749.

73 T. Sakamoto, Y. Kondo, S. Iwashita,
T. Nagano, H. Yamanaka, *Chem. Pharm.
Bull.* **1988**, *36*, 1305.

74 N. G. Kundu, M. Pal, *J. Chem. Soc.,
Chem. Commun.* **1993**, 86.

75 (a) X. Liu, X. Huang, S. Ma, *Tetrahedron
Lett.* **1993**, *34*, 5963; (b) X. Lu, G. Chen,
L. Xia, G. Guo, *Tetrahedron: Asymmetry*
**1997**, *8*, 3067.

76 T. Hosokawa, S. Aoki, M. Takano,
T. Nakahira, Y. Yoshida, S. Murahashi,
*J. Chem. Soc., Chem. Commun.* **1991**,
1559.

77 T. Hosokawa, M. Takano, Y. Kuroki,
S. Murahashi, *Teterahedron Lett.* **1992**,
*33*, 6643.

78 V. I. Timokhin, N. R. Anastasi,
S. S. Stahl, *J. Am. Chem. Soc.* **2003**, *125*,
12996.

79 M. Sawamura, M. Sudoh, Y. Ito,
*J. Am. Chem. Soc.* **1996**, *118*, 3309.

80 J. Tsuji, I. Minami, I. Shimizu, *Tetrahedron Lett.* **1983**, *24*, 4713.

81 (a) J. Tsuji, I. Minami, I. Shimizu,
*Tetrahedron Lett.* **1983**, *24*, 5639;
(b) I. Minami, K. Takahashi,
I. Shimizu, T. Kimura, J. Tsuji,
*Tetrahedron* **1986**, *42*, 2971.

82 T. Satoh, M. Ikeda, M. Miura,
M. Nomura, *J. Org. Chem.* **1997**, *62*,
4877.

83 S.-C. Yang, C.-W. Hung, *J. Org. Chem.*
**1999**, *64*, 5000.

84 S. Kamijo, T. Jin, Z. Huo, Y. Yamamoto,
*J. Am. Chem. Soc.* **2003**, *125*, 7786.

85 S. Kamijo, T. Jin, Z. Huo, Y. Yamamoto,
*J. Org. Chem.* **2004**, *69*, 2386.

86 S. Kamijo, T. Jin, Y. Yamamoto,
*Tetrahedron Lett.* **2004**, *45*, 689.

87 (a) B. M. Trost, S. A. King, *Tetrahedron
Lett.* **1986**, *27*, 5971; (b) B. M. Trost,
S. A. King, T. Schmidt, *J. Am. Chem.
Soc.* **1989**, *111*, 5902; (c) B. M. Trost,
S. A. King, *J. Am. Chem. Soc.* **1990**, *112*,
408.

88 (a) B. M. Trost, S. Sharma, T. Schmidt,
*J. Am. Chem. Soc.* **1992**, *114*, 7903;
(b) B. M. Trost, S. Sharma, T. Schmidt,
*Tetrahedron Lett.* **1993**, *34*, 7183.

89 (a) K. Hori, H. Kodama, T. Ohta,
I. Furukawa, *Tetrahedron Lett.* **1996**, *37*,
5947; (b) K. Hori, H. Kodama, T. Ohta,
I. Furukawa, *J. Org. Chem.* **1999**, *64*,
5017.

90 (a) P. A. Wender, H. Takahashi,
B. Witulski, *J. Am. Chem. Soc.* **1995**,
*117*, 4720; (b) P. A. Wender,
A. J. Dyckman, C. O. Husfeld,
D. Kadereit, J. A. Love, H. Rieck,
*J. Am. Chem. Soc.* **1999**, *121*, 10442.

91 (a) P. A. Wender, C. O. Husfeld,
E. Langkopf, J. A. Love, *J. Am. Chem.
Soc.* **1998**, *120*, 1940; (b) P. A. Wender,
C. O. Husfeld, E. Langkopf, J. A. Love,
N. Pleuss, *Tetrahedron* **1998**, *54*, 7203.

**92** P. A. WENDER, F. GLORIUS, C. G. HUS-
FELD, E. LANGKOPF, J. A. LOVE, *J. Am.
Chem. Soc.* **1999**, *121*, 5348.

**93** B. M. TROST, H. C. SHEN, *Org. Lett.* **2000**,
*2*, 2523.

**94** P. A. WENDER, A. G. CORREA, Y. Sato,
R. Sun, *J. Am. Chem. Soc.* **2000**, *122*,
7815.

**95** B. WANG, P. CAO, X. ZHANG, *Tetrahedron
Lett.* **2000**, *41*, 8041.

**96** S. R. GILBERTSON, B. DEBOEF, *J. Am.
Chem. Soc.* **2002**, *124*, 8784.

**97** A. P. EVANS, J. E. ROBINSON, E. W. BAUM,
A. N. FAZAL, *J. Am. Chem. Soc.* **2002**, *124*,
8782.

**98** (a) S. IKEDA, N. MORI, Y. SATO, *J. Am.
Chem. Soc.* **1997**, *119*, 4779; (b) S. IKEDA,
H. WATANABE, Y. SATO, *J. Org. Chem.*
**1998**, *63*, 7026.

**99** J. E. LYONS, H. K. MYERS, A. SCHNEIDER,
*J. Chem. Soc., Chem. Commun.* **1978**, 636.

**100** (a) M. LAUTENS, C. M. CRUDDEN,
*Organometallics* **1989**, *8*, 2733;
(b) M. LAUTENS, W. TAM, L. G. EDWARDS,
*J. Org. Chem.* **1992**, *57*, 8.

**101** M. LAUTENS, J. C. LAUTENS, A. C. SMITH,
*J. Am. Chem. Soc.* **1990**, *112*, 5627.

**102** (a) I.-F. DUAN, C.-H. CHENG, J.-S. SHAW,
S.-S. CHENG, K. F. LIOU, *J. Chem. Soc.,
Chem. Commun.* **1991**, 1346;
(b) O. PARDIGON, A. TENAGLIA,
G. BUONO, *J. Org. Chem.* **1995**, *60*, 1868.

**103** (a) M. LAUTENS, W. TAM., J. C. LAUTENS,
L. G. EDWARDS, C. M. CRUDDEN,
A. C. SMITH, *J. Am. Chem. Soc.* **1995**, *117*,
6863; (b) Y. CHEN, R. KIATTANSAKUL,
B. MA, J. K. SNYDER, *J. Org. Chem.* **2001**,
*66*, 6932.

**104** N. JEONG, S. D. SEO, J. Y. SHIN, *J. Am.
Chem. Soc.* **2000**, *122*, 10220.

**105** P. CAO, B. WANG, X. ZHANG, *J. Am.
Chem. Soc.* **2000**, *122*, 6490.

**106** P. CAO, X. ZHANG, *Angew. Chem. Int. Ed.*
**2000**, *39*, 4104.

**107** B. M. TROST, J. M. TOUR, *J. Am. Chem.
Soc.* **1987**, *109*, 5268.

**108** (a) S. KAMIJO, Y. YAMAMOTO, *Angew.
Chem. Int. Ed.* **2002**, *41*, 3230;
(b) S. KAMIJO, Y. YAMAMOTO, *J. Org.
Chem.* **2003**, *68*, 4764.

**109** M. AKAZOME, T. KONDO, Y. WATANABE,
*J. Org. Chem.* **1994**, *59*, 3375.

**110** Y. NISHIBAYASHI, M. YOSHIKAWA,
Y. INADA, M. D. MILTON, M. HIDAI,
S. UEMURA, *Angew. Chem. Int. Ed.* **2003**,
*42*, 2681.

**111** M. SODEOKA, K. OHRAI, M. SHIBASAKI,
*J. Org. Chem.* **1995**, *60*, 2648.

**112** C. WEI, C.-J. LI, *Green Chem.* **2002**, *4*, 39.

**113** C.-J. LI, C. WEI, *Chem. Commun.* **2002**, 268.

**114** B. M. TROST, M. PORTNOY, H. KURIHARA,
*J. Am. Chem. Soc.* **1997**, *119*, 836.

**115** B. M. TROST, A. B. PINKERTON, *J. Am.
Chem. Soc.* **1999**, *121*, 1988.

**116** B. M. TROST, A. B. PINKERTON, *Angew.
Chem. Int. Ed.* **2000**, *39*, 360.

**117** M. SHI, Y. CHEN, B. XU, *Org. Lett.* **2003**,
*5*, 1225.

**118** T. TSUCHIMOTO, S. KAMIYAMA,
R. NEGORO, E. SHIRAKAWA, Y. KAWAKAMI,
*Chem. Commun.* **2003**, 852.

**119** B. M. TROST, R. C. LIVINGSTON, *J. Am.
Chem. Soc.* **1995**, *117*, 9586.

**120** N. TSUKADA, T. MITSUBOSHI, H. SETO-
GUCHI, Y. INOUE, *J. Am. Chem. Soc.* **2003**,
*125*, 12102.

**121** N. ASAO, S. KII, H. HANAWA, K. MARUOKA,
*Tetrahedron Lett.* **1998**, *39*, 3729.

**122** S. KII, K. MARUOKA, *Tetrahedron Lett.*
**2001**, *42*, 1935.

**123** S. KII, T. HASHIMOTO, K. MARUOKA,
*Synlett* **2002**, 931.

**124** H. SIEBENEICHER, I. BYTSCHKOV, S. DOYE,
*Angew. Chem. Int. Ed.* **2003**, *42*, 3042.

**125** B. M. TROST, M. R. MACHACEK, *Angew.
Chem. Int. Ed.* **2002**, *41*, 4693.

**126** T. SHIMADA, K. MUKAIDE,
A. SHINOHARA, J. W. HAN, T. HAYASHI,
*J. Am. Chem. Soc.* **2002**, *124*, 1584.

**127** R. GRIGG, E. L. MILLINGTON,
M. THORNTON-PETT, *Tetrahedron Lett.*
**2002**, *43*, 2605.

# 2
# Zinc Polymetallic Asymmetric Catalysis

*Naoya Kumagai and Masakatsu Shibasaki*

## 2.1
## Introduction

Zinc is a naturally abundant and environmentally benign metallic element, located on the far right of the first series of transition metals. Since the discovery and characterization of organozinc species, various organozinc reagents have been developed for use in organic synthesis, mainly as a result of the mild and unique reactivity of the metal. In contrast to the widespread usage of organozinc reagents, the development of Zn-catalyzed asymmetric reactions has remained a relatively unexplored area of research. Surprisingly, it is only quite recently that truly catalytic, both in Zn metal and chiral ligands, asymmetric reactions have been realized, especially in the case of C–C bond-forming reactions. In catalytic reactions, Zn polymetallic catalysts, which contain two or more Zn atoms, enable simultaneous activation of both reaction partners and high catalytic efficiencies. In this chapter, recent advances in catalytic asymmetric reactions promoted by Zn polymetallic asymmetric catalysts are reviewed.

## 2.2
## Asymmetric Alternating Copolymerization with Dimeric Zn Complexes

Asymmetric synthesis polymerization has attracted much attention as an efficient and economical method for accessing optically active polymers. Since the first report on alternating copolymerization of epoxides with $CO_2$ in the presence of achiral Zn catalysts [1], considerable effort has been devoted to the development of an asymmetric version of this useful reaction [2]. Because the ring-opening of a *meso*-epoxide proceeds with inversion at one of the two chiral centers, successful asymmetric ring-opening by a chiral catalyst can afford an optically active aliphatic polycarbonate with an (R,R)- or (S,S)-*trans*-1,2-diol unit. Although this concept of enantioselective desymmetrization of *meso* materials represents an attractive and powerful method in asymmetric synthesis, there have been few reports on its

*Multimetallic Catalysts in Organic Synthesis.* Edited by M. Shibasaki and Y. Yamamoto
Copyright © 2004 WILEY-VCH Verlag GmbH & Co. KGaA, Weinheim
ISBN: 3-527-30828-8

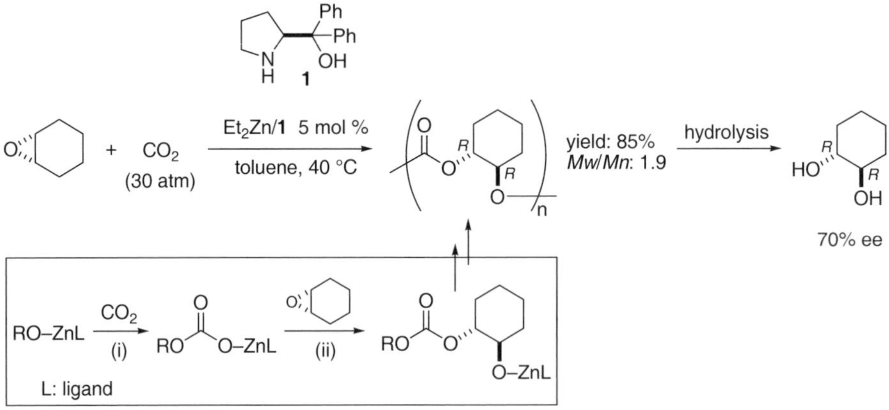

**Scheme 2.1** Asymmetric alternating copolymerization catalyzed by $Et_2Zn/1$.

application to asymmetric polymerization [3]. The first successful example of asymmetric alternating copolymerization of cyclohexene oxide with $CO_2$ was reported in 1999 by Nozaki et al. [4a, b], using a dimeric Zn catalyst prepared from an equimolar mixture of $Et_2Zn$ and (S)-diphenyl(pyrrolidin-2-yl) (1) [5]. The reaction is initiated by (i) $CO_2$ insertion into a Zn–alkoxide bond, followed by (ii) ring-opening of the epoxide by nucleophilic attack of the resulting carbonate (Scheme 2.1). With 5 mol % of the catalyst in toluene, cyclohexene oxide underwent copolymerization with $CO_2$ (30 atm) at 40 °C to give poly[cyclohexane-oxide-*alt*-$CO_2$] in 85% yield with an $M_w/M_n$ ratio of 1.9 (Scheme 2.1).

The fully alternating nature of the copolymer was established by $^1H$ NMR analysis. Hydrolysis of the copolymer gave *trans*-1,2-cyclohexane diol with 70% ee. X-ray crystallographic analysis of the catalyst revealed that the complex forms a dimeric structure, in which two zinc centers are coordinated in a distorted tetrahedral geometry, as illustrated in Figure 2.1. The two ethyl groups on the zinc centers in the complex are oriented in a *syn* fashion, as is commonly found, as for example in the dimer of $R_2Zn$-(–)-DAIB. The isolated complex initiated and catalyzed the copolymerization of cyclohexene oxide with $CO_2$ (30 atm) at 40 °C to give the copolymer in 57% yield with a high $M_w/M_n$ ratio of 15.7, and the enantiomeric excess of the *trans*-1,2-cyclohexane diol unit was determined to be 49% after hydrolysis. Thus, the dimeric Zn complex **2** can promote asymmetric alternating copolymerization, although its catalytic activity and enantioselectivity are lower than

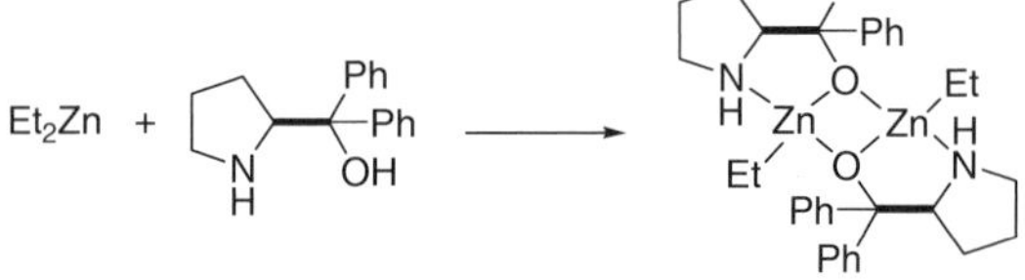

**Figure 2.1** The structure of the dimeric Zn complex derived from an $Et_2Zn/1$ mixture.

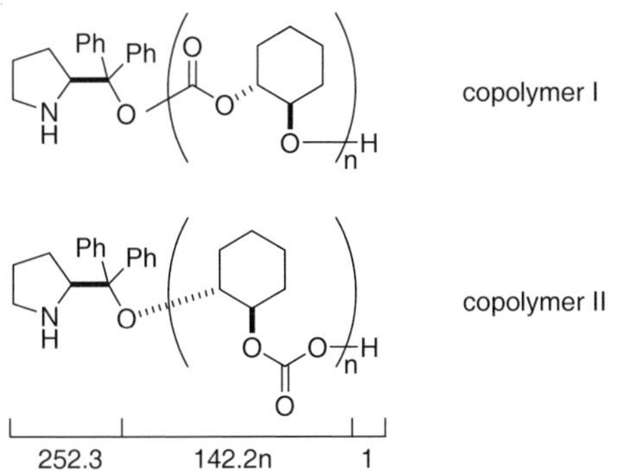

Figure 2.2 Structures of copolymers I and II.

those achieved with a mixture of $Et_2Zn$ and ligand **1**. To clarify the reason for the different catalytic activities, several mechanistic studies were carried out on the reaction with the isolated complex **2**. A MALDI-TOF mass spectrum of the copolymer showed only one series of signals with regular intervals of 142.2 mass units (repeating unit) and the mass number of [252.3 (ligand **1**) + 142.2n (repeating unit) + 1.0 (H) + 23 ($Na^+$)], suggesting that copolymerization was initiated by ligand **1** on a Zn center and terminated by protolysis to give either copolymer I or II (Figure 2.2). Judging by the instability of the carbonic monoester in copolymer II, copolymer I is the more reasonable structure, and the initiation reaction should be $CO_2$ insertion into the Zn–alkoxide bond of catalyst **2**. On the basis of these results, a possible reaction mechanism is illustrated in Scheme 2.2.

During the copolymerization, one chiral ligand dissociates from the catalytically active Zn center and the remaining chiral ligand controls the enantioselectivity of the epoxide ring opening. To prevent ligand dissociation, the addition of alcohols

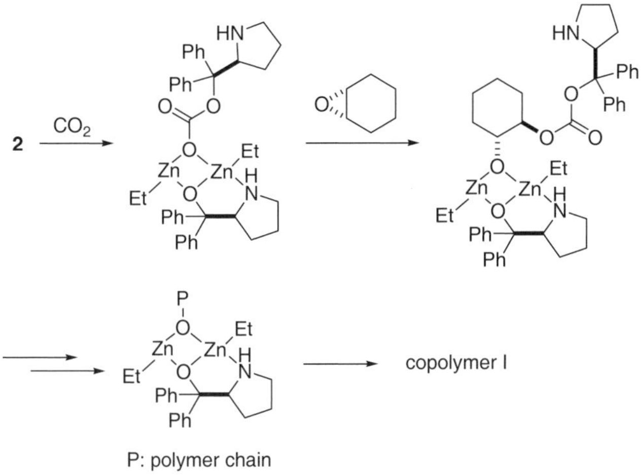

P: polymer chain

Scheme 2.2 A possible reaction mechanism initiated by complex **2**.

**Table 2.1** Copolymerization of cyclohexane oxide and $CO_2$ with a mixture of complex **2** and ethanol.[a]

| Entry | EtOH (equiv. to 2) | Yield (%) of polymer | $M_n$ ($M_w/M_n$) | % ee (R,R) |
|---|---|---|---|---|
| 1 | 0.0 | 57 | 11800 (15.70) | 49 |
| 2 | 0.2 | 94 | 12300 (7.18) | 64 |
| 3 | 0.4 | 95 | 12000 (1.30) | 74 |
| 4 | 0.6 | > 99 | 9900 (1.39) | 72 |
| 5 | 0.8 | 91 | 6300 (1.43) | 75 |
| 6 | 1.0 | 71 | 4500 (1.82) | 76 |
| 7 | 2.0 | trace | | |

[a] Cyclohexane oxide (10 mmol) was treated with $CO_2$ (30 atm) in the presence of a mixture of Zn complex **2** (0.25 mmol) and a designated amount of EtOH in toluene (17 mL) at 40 °C for 19 h.

was investigated. The alcohol additive should replace the ethyl group on the zinc center, so that copolymerization would be initiated not by the chiral ligand **1** but by the resulting alkoxide ligand. Addition of 0.2–0.8 equiv. of ethanol to catalyst **2** was found to improve the chemical yield to quantitive, to increase the enantioselectivity to almost 75% ee, and achieved control of the molecular weight (Table 2.1). The introduction of an ethoxy end group derived from the ethoxy ligand on the Zn center was confirmed by MALDI-TOF mass spectra of the obtained copolymers (Figure 2.3). With 0.8 equiv. of ethanol, mass peaks with the mass number of [45.1 (EtO) + 142.2$n$ (repeating unit) + 1.0 (H) + 23 ($Na^+$)] were observed at regular intervals of 142.2 mass units, which can be assigned to copolymer III bearing an ethoxy group as the initiating group.

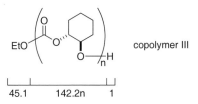

copolymer III

**Figure 2.3** Structure of copolymer III.

The most likely explanation for the effect of ethanol is a reaction involving a dimeric Zn complex **3**, as shown in Scheme 2.3. Because of the lack of chelate stabilization, $CO_2$ insertion into the Zn–OEt bond of **3** occurs selectively, and copolymerization proceeds without chiral ligand dissociation. The narrower molecular weight distribution also reflects the smooth initiation reaction by a Zn–OEt complex, which leads to higher catalytic activity, lower $M_w/M_n$, and higher stereocontrol. The improved results obtained with the complex **3** are comparable to those achieved with a mixture of $Et_2Zn$ and ligand **1** (Scheme 2.1). Thus, the mixture of $Et_2Zn$ and ligand **1** might contain both **2** and **3**, the latter being derived from an impurity in the $Et_2Zn$, most probably EtZn-OEt. In fact, the MALDI-TOF mass spectrum of the copolymer obtained with the $Et_2Zn/1$ mixture exhibited two series of signals attributable to copolymers I and III.

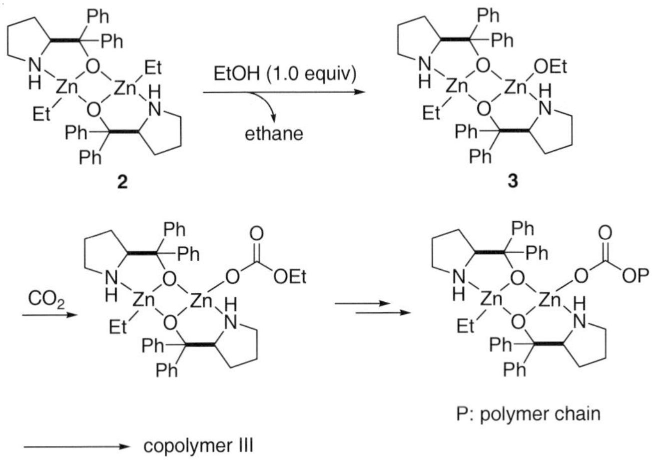

Scheme 2.3 A plausible reaction mechanism with complex **3**.

## 2.3
## Direct Catalytic Asymmetric Aldol Reaction with Zn Polymetallic Catalysts

### 2.3.1
### Introduction

The aldol reaction has established its position in organic synthesis as a remarkably useful synthetic tool, providing access to β-hydroxy carbonyl units. Intensive efforts have raised this classical process to a highly enantioselective transformation employing only tiny amounts of chiral catalysts. While various effective applications have been reported, almost all of these reactions require a pre-conversion of a ketone or ester moiety into a more reactive species, such as an enol silyl ether or a ketene silyl acetal, using no less than stoichiometric amounts of base and silylating reagents (Scheme 2.4 a). Considering an increasing demand for environmentally benign and atom-economic processes [6], such stoichiometric amounts of reagents, which inevitably afford wastes such as salts, should be excluded from the procedure. Thus, the development of a *direct* catalytic asymmetric aldol reaction (Scheme 2.4 b), which employs unmodified ketone as a donor, is desired.

An indication of how the direct enolization of an unmodified ketone with a catalytic amount of reagent could be successfully achieved was provided by an enzyme, namely class II aldolase. The operation of this enzyme is thought to involve co-catalysis by a $Zn^{2+}$ cation and a basic functional group in its active site [7]. The $Zn^{2+}$ cation functions as a Lewis acid in activating a carbonyl group, while the basic part abstracts an α-proton to form a Zn-enolate. This bifunctional mechanism of the aldolases was highly attractive to synthetic organic chemists as a promising strategy to achieve direct catalytic asymmetric aldol reactions with an artificial small molecular catalyst. This section focuses on notable recent advances that have been achieved by employing Zn polymetallic catalysts, which mimic the class II aldolases [8, 9].

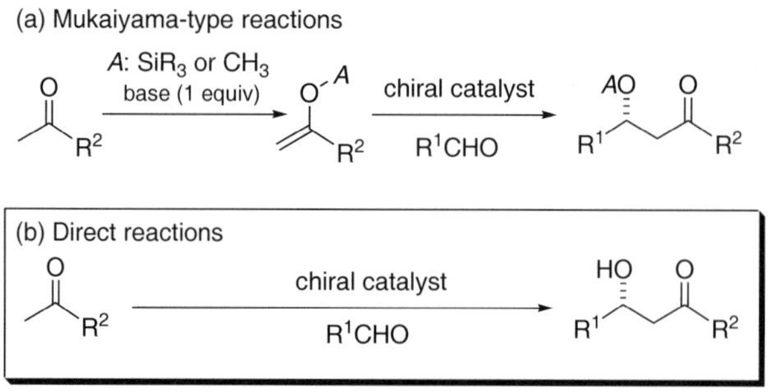

(a) Mukaiyama-type reactions

(b) Direct reactions

**Scheme 2.4** (a) Mukaiyama-type reactions, (b) direct reactions.

2.3.2
**Direct Catalytic Asymmetric Aldol Reaction with Methyl Ketones**

After the first successful direct catalytic asymmetric aldol reaction using the heterobimetallic catalyst LLB · KOH (LaLi$_3$tris(binaphthoxide)·KOH) by Shibasaki [10] in 1997, proline catalysts devised by List [8c] and dinuclear Zn catalysts devised by Trost [11] were also introduced for this purpose. The structure of Trost's dinuclear Zn complex, which was prepared from Et$_2$Zn and chiral ligand **4a**, was proposed to be Zn$_2$-**4a** on the basis of ethane gas emission measurements and ESI-MS analysis (Scheme 2.5). As shown in Table 2.2, the Zn complex proved to be effective in promoting the direct catalytic asymmetric aldol reaction with various methyl aryl ketones (**6a–6e**: 10 equiv.). Excellent enantioselectivity was achieved (up to 99% ee) using 5 mol % of **4a**, although excess ketone was used and yields were moderate in some instances. It is noteworthy that high ee was achieved at a relatively high reaction temperature (5 °C). Bifunctional Zn catalysis is proposed, where two Zn centers work in concert: one Zn acts as a Lewis acid and the other Zn-alkoxide functions as a Brønsted base to generate a Zn-enolate (Figure 2.4).

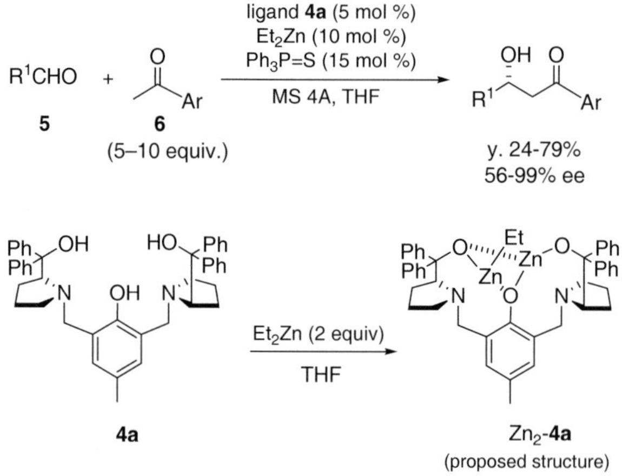

**Scheme 2.5** Direct catalytic asymmetric aldol reactions of methyl ketone **6** promoted by dinuclear Zn$_2$-**4a** complex.

**Table 2.2** Direct catalytic asymmetric aldol reactions of methyl ketone **6** promoted by dinuclear Zn$_2$-4a complex.[a]

| Entry | Aldehyde R$^1$ = | | Ketone (equiv.) Ar = | | Temp (°C) | Time (h) | Yield (%) | ee (%) |
|---|---|---|---|---|---|---|---|---|
| 1 | CH$_3$(CH$_2$)$_2$ | 5a | Ph | 6a (10) | −5 | 24 | 33 | 56 |
| 2 | | 5a | Ph | 6a (10) | −15 | 24 | 24 | 74 |
| 3 | (CH$_3$)$_2$CHCH$_2$ | 5b | Ph | 6a (10) | −5 | 24 | 49 | 68 |
| 4 | (CH$_3$)$_2$CH | 5c | Ph | 6a (10) | 5 | 24 | 62 | 98 |
| 5 | c-hex | 5d | Ph | 6a (10) | 5 | 24 | 60 | 98 |
| 6 | Ph$_2$CH | 5e | Ph | 6a (10) | 5 | 24 | 79 | 99 |
| 7 | rac-Ph(CH$_3$)CH | 5f | Ph | 6a (10) | 5 | 24 | 67 (dr: 2/1) | 94 |
| 8 | TBSOCH$_2$(CH$_3$)$_2$C | 5g | Ph | 6a (10) | 5 | 96 | 61 | 93 |
| 9 | (CH$_3$)CH | 5c | 2-furyl | 6b (10) | 5 | 24 | 66 | 97 |
| 10 | | 5c | 2-MeO-C$_6$H$_4$ | 6c (10) | 5 | 24 | 48 | 97 |
| 11 | | 5c | 4-MeO-C$_6$H$_4$ | 6d (5) | 5 | 24 | 36 | 98 |
| 12 | | 5c | 2-naphthyl | 6e (5) | 5 | 24 | 40 | 96 |

[a] Reaction conditions: ligand **4a** (5 mol %), Et$_2$Zn (10 mol %), Ph$_3$P=S (15 mol %), THF.

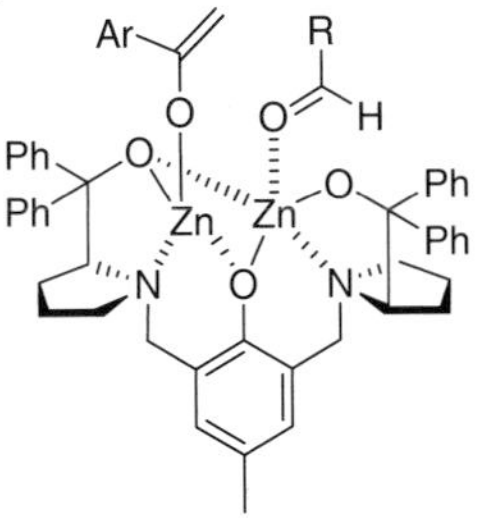

**Figure 2.4** Proposed transition state for the direct aldol reaction of methyl ketone **6**.

Trost also reported that manipulation of ligand **4a** led to better results in some cases. By using ligand **4b** instead of **4a**, the direct aldol reaction of acetone (10–15 equiv.) proceeded smoothly with good enantioselectivity (Scheme 2.6) [12]. The results are summarized in Table 2.3. Aldol adducts were obtained in good yields (59–89%) and with high ee (78–94%). In general, self-condensations of aldehydes need to be suppressed in order to achieve good yields, e.g. by using α-unsubstituted aldehydes. Thus, it is noteworthy that good yields and high ee values were achieved with α-unsubstituted aldehydes (entries 5–7, yields 59–76%, 82–89% ee).

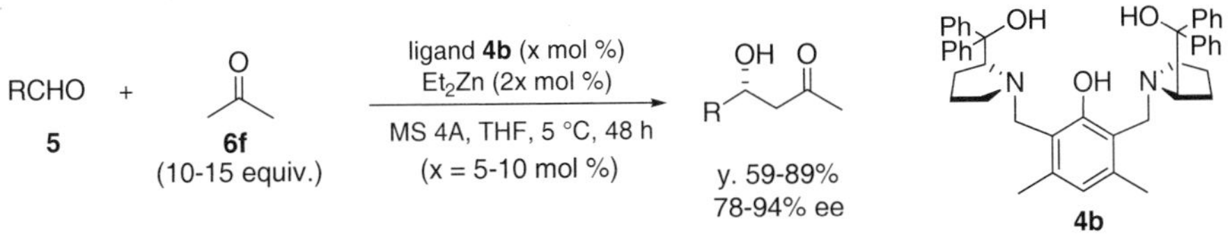

**Scheme 2.6** Direct catalytic asymmetric aldol reactions of acetone (**6f**) promoted by dinuclear Zn$_2$-4b complex.

**Table 2.3** Direct catalytic asymmetric aldol reactions of acetone (**6f**) promoted by dinuclear Zn$_2$-**4b** complex.[a]

| Entry | Aldehyde R$^1$ = | | Catalyst (mol %) | Yield (%) | ee (%) |
|---|---|---|---|---|---|
| 1 | c-hex | **5d** | 10 | 89 | 92 |
| 2 | (CH$_3$)$_2$CH | **5c** | 10 | 89 | 91 |
| 3 | (CH$_3$)$_3$C | **5h** | 10 | 72 | 94 |
| 4 | Ph$_2$CH | **5e** | 10 | 84 | 91 |
| 5 | (CH$_3$)$_2$CHCH$_2$ | **5b** | 10 | 59 | 84 |
| 6 | PhCH$_2$CH$_2$ | **5i** | 10 | 76 | 82 |
| 7 | CH$_3$(CH$_2$)$_2$ | **5a** | 10 | 69 | 89 |
| 8 | Ph | **5j** | 5 | 78 | 83 |
| 9 | 4-NO$_2$-C$_6$H$_4$ | **5k** | 5 | 62 | 78 |

[a] Reaction conditions: ligand **4b** (x mol %), Et$_2$Zn (2x mol %), THF, MS 4A, 5 °C, 48 h.

### 2.3.3
### Direct Catalytic Asymmetric Aldol Reaction with α-Hydroxy Ketones

α-Hydroxy ketones also serve as aldol donors in direct catalytic asymmetric aldol reactions. In contrast to standard Lewis acid catalyzed aldol reactions, protection of the OH group is no longer necessary. The versatility of the resulting chiral 1,2-diols as building blocks makes this process attractive. The first successful result was reported by List [13] by means of proline catalysis. L-Proline catalyzed a highly chemo-, diastereo-, and enantioselective aldol reaction between hydroxyacetone and aldehydes to provide chiral *anti*-1,2-diols. Trost [14] and Shibasaki [15] have also reported important contributions in the field of Zn polymetallic catalysis.

Trost reported direct aldol reactions with 2-hydroxyacetophenone (**7a**) and 2-hydroxyacetylfuran (**7b**) using the dinuclear Zn$_2$-**4a** catalyst mentioned in Section 2.3.2 [14] (Scheme 2.7). Aldol reactions between various aldehydes and 1.5 equiv. of ketone proceeded smoothly at −35 °C with 2.5–5 mol % of catalyst in the presence of 4 Å molecular sieves to afford the products *syn*-selectively (*syn*/*anti* = 3 : 1–100 : 0) in 62–98% yield and with 81–98% ee (Table 2.4). Strikingly, the absolute configuration of the stereocenter derived from the aldehyde is opposite to that obtained using acetophenone (Scheme 2.5) or acetone (Scheme 2.6) as the donor, possibly due to the bidentate coordination of the α-hydroxy ketone to the catalyst as depicted in Figure 2.5. In some cases, the reaction was performed with a ketone/aldehyde ratio of 1.1 : 1.0, albeit at the expense of conversion, which gave

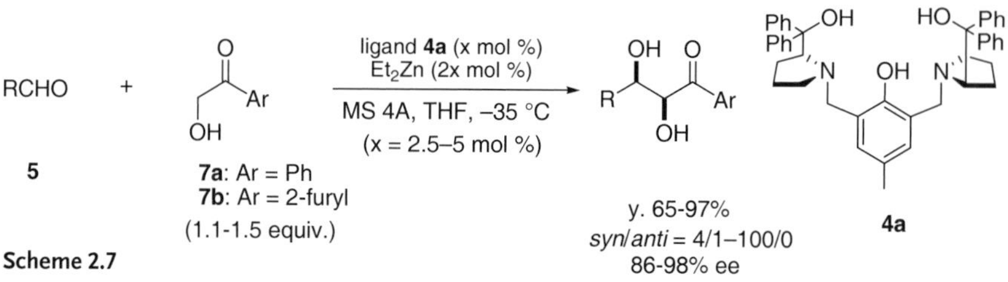

Scheme 2.7

**Table 2.4** Direct catalytic asymmetric aldol reactions of hydroxy ketone **7** promoted by dinuclear Zn$_2$-**4a** complex.[a]

| Entry | Aldehyde R$^1$ = | | Ketone (equiv.) | Catalyst (x mol %) | Yield (%) | dr (syn/anti) | ee (%) (syn) |
|---|---|---|---|---|---|---|---|
| 1 | c-hex | **5d** | **7a** (1.5) | 2.5 | 83 | 30/1 | 92 |
| 2 | | **5d** | **7a** (1.5) | 5 | 97 | 5/1 | 90 |
| 3 | | **5d** | **7b** (1.3) | 5 | 90 | 6/1 | 96 |
| 4 | | **5d** | **7b** (1.1) | 5 | 77 | 6/1 | 98 |
| 5 | (CH$_3$)$_2$CH | **5c** | **7a** (1.5) | 2.5 | 89 | 13/1 | 93 |
| 6 | | **5c** | **7a** (1.1) | 5 | 72 | 6/1 | 93 |
| 7 | Ph$_2$CH | **5e** | **7a** (1.5) | 2.5 | 74 | 100/0 | 96 |
| 8 | (CH$_3$)$_2$CHCH$_2$ | **5b** | **7a** (1.5) | 2.5 | 65 | 35/1 | 94 |
| 9 | | **5b** | **7a** (1.1) | 5 | 79 | 4/1 | 93 |
| 10 | PhCH$_2$CH$_2$ | **5i** | **7a** (1.5) | 2.5 | 78 | 9/1 | 91 |
| 11 | CH$_3$(CH$_2$)$_6$ | **5j** | **7a** (1.5) | 5 | 89 | 5/1 | 86 |
| 12 | CH$_2$=CH(CH$_2$)$_8$ | **5k** | **7a** (1.5) | 5 | 91 | 5/1 | 87 |

[a] Reaction conditions: ligand **4a** (x mol %), Et$_2$Zn (2x mol %), THF, MS 4A, −35 °C, 24 h.

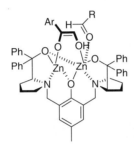

**Figure 2.5** Proposed transition state for the direct aldol reaction of hydroxy ketone **7**.

the best results in terms of achieving an ideal atom-economical process. The reaction with **7b** gave higher ee, and, moreover, the furan moiety is suitable for further conversion of the resulting chiral 1,2-diol. For example, oxidative cleavage of the furan ring was successfully utilized in the asymmetric synthesis of (+)-boronolide [16].

Shibasaki [15] developed direct catalytic asymmetric aldol reactions of 2-hydroxyacetophenones **7** to provide either *anti* or *syn* chiral 1,2-diols by using two types of multifunctional catalysts, (S)-LLB · KOH for *anti*-diols and an Et$_2$Zn/(S,S)-linked-BINOL **8a** complex for *syn*-diols. The Et$_2$Zn/(S,S)-linked-BINOL **8a** complex, which was prepared by simply mixing Et$_2$Zn and (S,S)-linked-BINOL **8a**, was found to promote aldol reactions of 2-hydroxyacetophenones **7** to afford *syn*-1,2-diols in good yields (Scheme 2.8) [15a–c, 17]. The reactivity and stereoselectivity were dependent on the substituent on the aromatic ring of the 2-hydroxyacetophenone **7**. 2-Hydroxy-2′-methoxyacetophenone (**7c**) gave the best results, affording the product in 94% yield and with high stereoselectivity (*syn/anti* = 89 : 11, *syn* = 92% ee, *anti* = 89% ee) with as little as 1 mol % of catalyst [15b]. The catalyst proved to be applicable to various aldehydes, including α-unsubstituted aldehydes. As summarized in Table 2.5, the reactions reached completion within 24 hours with 1 mol % of catalyst

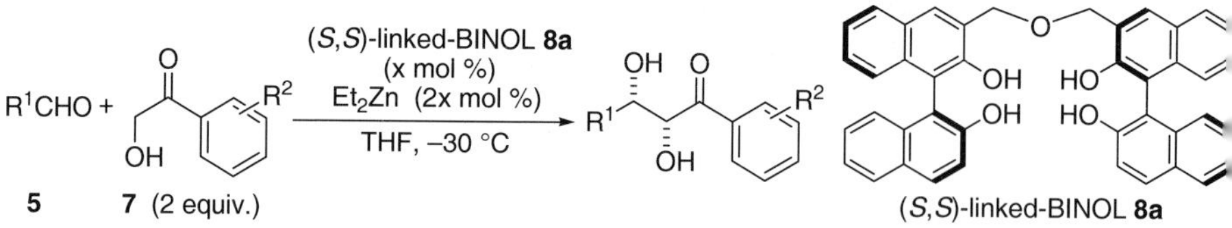

R¹ = PhCH₂CH₂, R² = H (**7a**), x = 10: 48 h, y. 81%, *syn/anti* = 67/33, 78% ee (*syn*)
R¹ = PhCH₂CH₂, R² = 2-MeO (**7c**), x = 3: 4 h, y. 94%, *syn/anti* = 90/10, 90% ee (*syn*)
R¹ = PhCH₂CH₂, R² = 2-MeO (**7c**), x = 1: 16 h, y. 94%, *syn/anti* = 87/13, 93% ee (*syn*)

**Scheme 2.8** Direct catalytic asymmetric aldol reactions of hydroxy ketone **7** promoted by Et₂Zn/(*S*,*S*)-linked-BINOL **8a** complex.

**Table 2.5** Direct catalytic asymmetric aldol reactions of hydroxy ketone **7c** promoted by Et₂Zn/(*S*,*S*)-linked-BINOL **8a** complex.[a]

| Entry | Aldehyde R¹ = | | Time (h) | Yield (%) | dr (anti/syn) | ee (%) (anti/syn) |
|---|---|---|---|---|---|---|
| 1 | PhCH₂CH₂ | 5i | 20 | 94 | 89/11 | 92/89 |
| 2 | CH₃(CH₂)₄ | 5l | 18 | 88 | 88/12 | 95/91 |
| 3 | (CH₃)₂CHCH₂ | 5b | 18 | 84 | 84/16 | 93/87 |
| 4 | CH₃C(O)CH₂CH₂ | 5m | 12 | 91 | 93/7 | 95/– |
| 5 | *trans*-4-decenal | 5n | 24 | 94 | 86/14 | 87/92 |
| 6 | BnOCH₂CH₂ | 5o | 18 | 81 | 86/14 | 95/90 |
| 7 | BnOCH₂ | 5p | 16 | 84 | 72/28 | 96/93 |
| 8 | BOMOCH₂CH₂ | 5q | 14 | 93 | 84/16 | 90/84 |
| 9 | (CH₃)₂CH | 5c | 24 | 83 | 97/3 | 98/– |
| 10 | (CH₃CH₂)₂CH | 5r | 16 | 92 | 96/4 | 99/– |
| 11 | *c*-hex | 5d | 18 | 95 | 97/3 | 98/– |

[a] Reaction conditions: (*S*,*S*)-linked-BINOL **8a** (1 mol %), Et₂Zn (2 mol %), –30 °C, THF.

to give *syn*-1,2-diols in excellent yields and with high stereoselectivities (yields 81–95%, *syn/anti* = 72 : 28–97 : 3, *syn* = 87–99% ee).

Mechanistic investigations by means of kinetic studies, X-ray crystallography, ¹H NMR, and cold-spray ionization mass spectrometry (CSI-MS) shed light on the reaction mechanism and on the structure of the actual active species [15c]. X-ray analysis of a crystal obtained from a 2 : 1 mixture of Et₂Zn and (*S*,*S*)-linked-BINOL **8a** in THF solution revealed that the complex consisted of Zn and ligand **8a** in a 3 : 2 ratio [trinuclear Zn₃(linked-BINOL)₂thf₃] with $C_2$ symmetry (**9**, Figure 2.6). CSI-MS analysis and kinetic studies revealed complex **9** to be a pre-catalyst and indicated that an oligomeric Zn-**8a**-**7c** complex would serve as the actual active species [15c]. The proposed catalytic cycle is shown in Figure 2.7. The product dissociation step is rate-determining.

The identical absolute configuration (*R*) was expressed at the α-positions of both the *syn* and *anti* aldol products (Figure 2.8), suggesting that the present catalyst differentiates the enantioface of the enolate well and that aldehydes approach from the *Re*-face of the zinc enolate (Figure 2.8 A). The *syn*-selectivity can be rationalized in terms of the transition state shown in Figure 2.8 (B). The positive effects of an

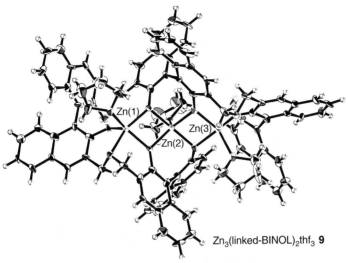

Figure 2.6 X-ray structure of preformed complex Zn$_3$(linked-BINOL)$_2$thf$_3$ **9**.

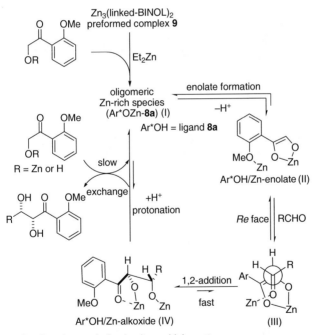

**Figure 2.7** Postulated catalytic cycle for the direct aldol reaction with Et$_2$Zn/(*S,S*)-linked-BINOL **8a** complex.

*ortho* MeO group were interpreted in terms of coordination of this group to one of the Zn centers in the oligomeric Zn complex, thereby affecting the stereoselection step. The electron-donating MeO group also has a beneficial effect on further conversion of the products into esters and amides by regioselective rearrangement [15c].

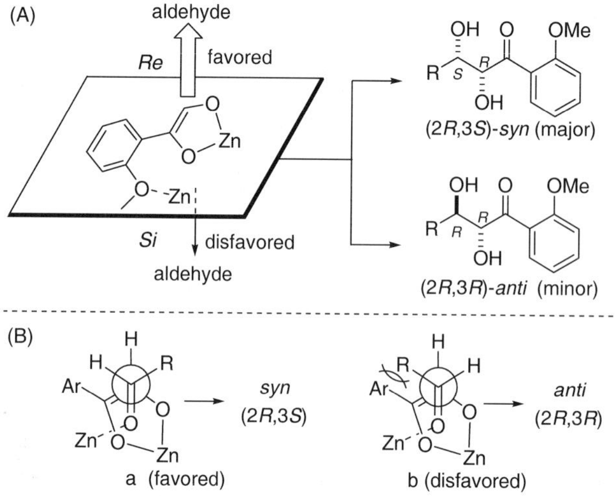

**Figure 2.8** Stereochemical course of direct aldol reaction of hydroxy ketone **7c**.

The mechanistic studies suggested that further addition of Et$_2$Zn and 3 Å molecular sieves would accelerate the reaction. In the presence of 3 Å molecular sieves, the second-generation Zn catalyst, prepared from Et$_2$Zn/linked-BINOL **8a** in a 4 : 1 ratio, smoothly promoted the direct aldol reaction of hydrocinnamaldehyde (**5i**) and 1.1 equiv. of ketone **7c** with a reduced catalyst loading (0.25–0.1 mol %; Scheme 2.9) [15c]. The practical utility of the present reaction was demonstrated by a large-scale reaction performed on a 200 mmol scale using 0.25 mol % of **8a** (0.5 mmol, 307 mg), which afforded 53.7 g of product (yield 96%) with high *dr* (*syn/anti* = 98 : 2) and ee (94%) after 12 h (Scheme 2.9). Considering that the standard catalyst loading for the direct catalytic asymmetric aldol reaction is 2.5 to 20 mol %, the exceptionally low catalyst loading in the present asymmetric zinc catalysis is remarkable.

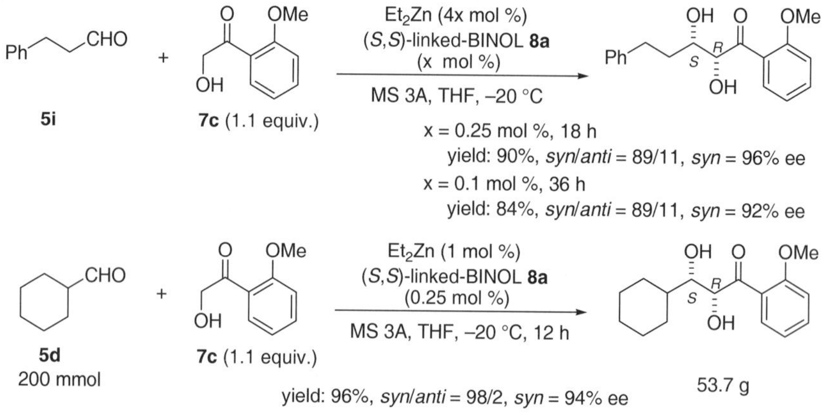

**Scheme 2.9** Direct catalytic asymmetric aldol reactions of hydroxy ketone **7c** promoted by Et$_2$Zn/(*S*,*S*)-linked-BINOL **8a** = 4/1 complex with MS 3A.

The catalyst system of 4 : 1 Et$_2$Zn/linked-BINOL **8a** with 3 Å molecular sieves enabled direct aldol reaction of 2-hydroxy-2′-methoxypropiophenone (**10**), which led to the construction of a chiral tetrasubstituted carbon stereocenter (Scheme 2.10) [15c]. Although a higher catalyst loading and 5 equiv. of ketone **10** were required, various α-unsubstituted aldehydes afforded the corresponding products *syn*-selectively (*syn/anti* = 59 : 41–71 : 29) in moderate to good yields and with high ee (yields 72–97%, *syn* = 68–87% ee, *anti* = 86–97% ee) with (*S,S*)-linked-BINOL **8a** (Table 2.6). Interestingly, the reaction using (*S,S*)-sulfur-linked-BINOL **8b**, a linked-BINOL analogue with sulfur incorporated into the linker instead of oxygen, showed opposite diastereoselectivity (Scheme 2.10). The reactivity with **8b** was somewhat lower than that with **8a**, and aldol adducts were obtained in moderate to good yields (56–82%) by using 10 equiv. of ketone **10**. The major *anti* isomers were obtained with high ee (81–93%), although the ee values of the minor *syn* isomers were rather low (45–60%) (Table 2.6).

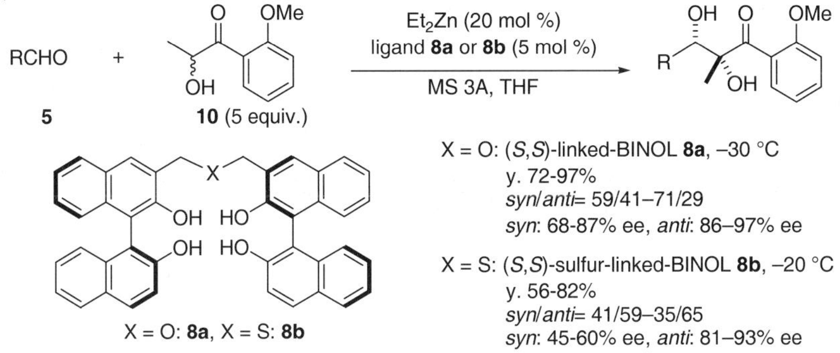

**Scheme 2.10** Direct catalytic asymmetric aldol reactions of 2-hydroxy-2′-methoxypropiophenone (10) promoted by Et$_2$Zn/(*S,S*)-linked-BINOL complex.

**Table 2.6** Direct catalytic asymmetric aldol reactions with 2-hydroxy-2′-methoxypropiophenone (10) promoted by Et$_2$Zn/(*S,S*)-linked-BINOL complex.[a]

| Entry | Aldehyde R$^1$ = | | Ketone 10 (equiv.) | Catalyst (x mol %) | Temp. (°C) | Yield (%) | dr (anti/syn) | ee (%) (anti/syn) |
|---|---|---|---|---|---|---|---|---|
| 1 | PhCH$_2$CH$_2$ | 5i | 5 | 8a (5) | −30 | 97 | 62/38 | 87/96 |
| 2 | | 5i | 10 | 8b (10) | −20 | 82 | 35/65 | 92/60 |
| 3 | PhCH$_2$(CH$_2$)$_2$ | 5s | 5 | 8a (5) | −30 | 72 | 64/36 | 78/90 |
| 4 | | 5s | 10 | 8b (10) | −20 | 63 | 41/59 | 86/45 |
| 5 | CH$_3$CH$_2$ | 5t | 5 | 8a (5) | −30 | 88 | 71/29 | 68/86 |
| 6 | | 5t | 10 | 8b (10) | −20 | 56 | 41/59 | 87/48 |
| 7 | PMBOCH$_2$CH$_2$ | 5u | 5 | 8a (5) | −30 | 89 | 59/41 | 86/95 |
| 8 | | 5u | 10 | 8b (10) | −20 | 73 | 41/59 | 93/58 |
| 9 | BOMOCH$_2$CH$_2$ | 5q | 5 | 8a (5) | −30 | 92 | 69/31 | 87/97 |
| 10 | | 5q | 10 | 8b (10) | −20 | 72 | 39/61 | 81/52 |
| 11 | (CH$_3$)$_2$CH | 5b | 5 | 8a (5) | −30 | 80 | 68/32 | 72/87 |
| 12 | BnOCH$_2$ | 5p | 5 | 8a (5) | −30 | 80 | 65/35 | 85/92 |

[a] Reaction conditions: Et$_2$Zn (4x mol %), ligand (x mol %), MS 3A, THF.

## 2.4
### Direct Catalytic Asymmetric Mannich-Type Reactions

Catalytic asymmetric Mannich-type reactions have been regarded as useful transformations for obtaining chiral β-amino carbonyl compounds from a limited amount of a chiral source. In most cases, it is mandatory to preform a latent enolate (enol silyl ether or ketene silyl acetal) as the reaction partner for imines. Following the same logic as for the *direct* aldol reaction, *direct* catalytic asymmetric Mannich-type reaction, in which an unmodified ketone is used as a reaction partner for an imine, is desirable in terms of atom economy [6]. Polymetallic Zn catalysts, which are described in Section 2.3, have been successfully applied to direct catalytic asymmetric Mannich reactions of α-hydroxy ketones, affording β-amino alcohols in a highly diastereo- and enantioselective manner [18]. Two adjacent stereocenters were thereby constructed at the carbon-carbon bond-forming stage without protection of the hydroxy ketone.

Trost reported *syn*-selective direct catalytic asymmetric Mannich-type reactions with the dinuclear Zn catalyst Zn₂-4a, which was developed for the direct aldol reaction (Scheme 2.11) [19]. Direct Mannich-type reaction of glyoxalate imine **11a** and hydroxyacetophenone (**7a**) proceeded smoothly at –5 °C with 10 mol % of catalyst to afford the β-amino alcohol *syn*-selectively (*syn*/*anti* = 6.5 : 1) in 76% yield and with 95% ee (Table 2.7, entry 1). With the bulkier imine **11b** (R = Me), an increase in both the diastereo- and enantioselectivity was observed with a lower catalyst amount (2.5 mol %) (entry 2). Employment of the biphenyl ligand **4c** further increased the diastereomeric ratio and the yield (entry 3). It was found that the use of 1.1 equiv. of ketone could complete the reaction in the presence of the "zinc-aphilic" additive Ph₃P=S without any deleterious effect on conversion or stereo-control (entry 4). With the *o*-MeO-substituted ketone **7c**, complete stereoselectivity was observed (entry 5).

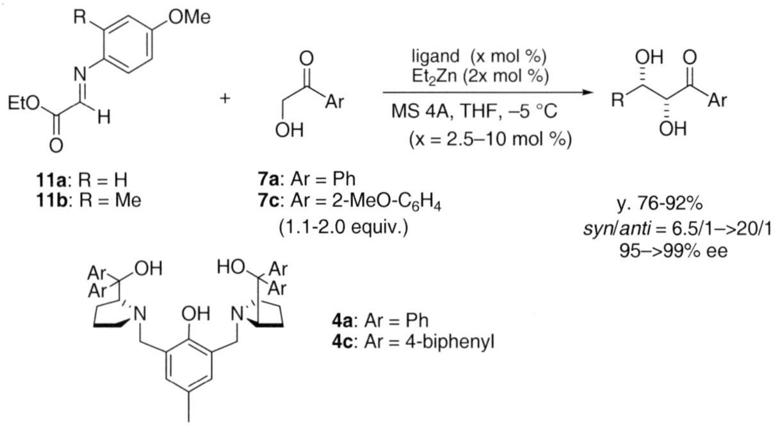

**Scheme 2.11** Direct catalytic asymmetric Mannich-type reaction with dinuclear Zn complex.

**Table 2.7** Direct catalytic asymmetric Mannich-type reaction of hydroxy ketones promoted by dinuclear Zn complex.[a]

| Entry | Imine 11 R = | | Ketone (equiv.) Ar = | | | Catalyst ligand (x mol %) | Yield (%) | dr (syn/anti) | ee (%) (syn) |
|---|---|---|---|---|---|---|---|---|---|
| 1 | H | 11a | $C_6H_5$ | 7a | 2 | 4a | 10 | 76 | 6.5/1 | 95 |
| 2 | Me | 11b | $C_6H_5$ | 7a | 2 | 4a | 2.5 | 79 | 8/1 | > 98 |
| 3 | Me | 11b | $C_6H_5$ | 7a | 2 | 4c | 2.5 | 92 | 12/1 | > 99 |
| 4[b] | Me | 11b | $C_6H_5$ | 7a | 1.1 | 4c | 2.5 | 97 | 8.6/1 | 98 |
| 5 | Me | 11b | 2-MeO-$C_6H_4$ | 7c | 2 | 4c | 5 | 81 | > 20/1 | > 99 |

[a] Reaction conditions: −5 °C, THF, MS 4A.
[b] In the presence of 3.7 mol % of $Ph_3P=S$.

This catalytic system also proved applicable to the aromatic imines **12** (Scheme 2.12). With 5 mol % of $Zn_2$-**4a**, N-p-methoxyphenylimine **12a** was converted to the corresponding β-amino alcohol with 99% ee, albeit with low diastereoselectivity (1.7 : 1) (Table 2.8, entry 1). Modification of the ligand structure to **4d** led to higher diastereoselectivity (entry 2). With N-o-methoxyphenylimine **12b**, a dramatic increase in diastereoselectivity (> 15 : 1) was observed, possibly due to two-point binding through the imine nitrogen and methoxy group (entry 3). A near atom-economical protocol (1.1 equiv. of ketone) is realized with $Ph_3As=O$ as the "zincaphilic" additive (entry 4).

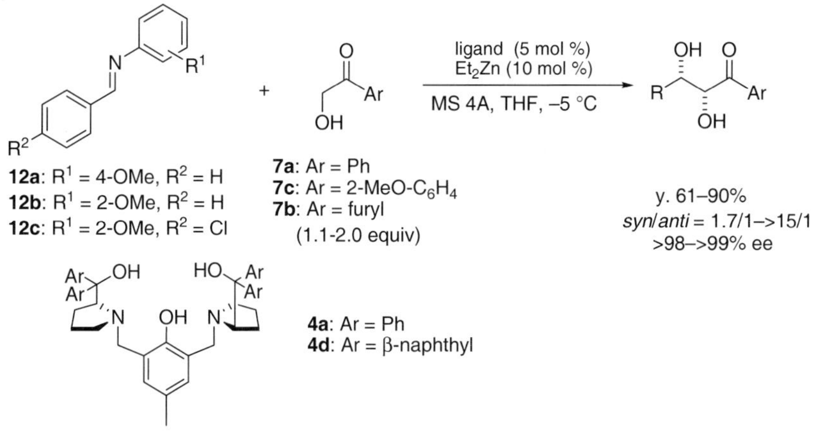

**Scheme 2.12** Direct catalytic asymmetric Mannich-type reaction promoted by dinuclear Zn complex.

As an antipode of Trost's case, Shibasaki [20] reported anti-selective direct catalytic asymmetric Mannich-type reaction by the 4 : 1 $Et_2Zn$/linked-BINOL **8a** complex with 3 Å molecular sieves. Direct Mannich-type reactions of various N-diphenyl-phosphinoyl (Dpp) imines **13** and 2-hydroxy-2'-methoxyacetophenone (**7c**) proceeded smoothly in the presence of 0.25–1 mol % of the chiral source to afford the anti-β-amino alcohols in excellent yields (95–99%) with good diastereo- and enantioselectivity (anti/syn = 76 : 24 − > 98 : 1, 98 − > 99.5% ee).

**Table 2.8** Direct catalytic asymmetric Mannich-type reaction of hydroxy ketones promoted by dinuclear Zn complex.[a]

| Entry | Imine 12 R¹ = | R² = | | Ketone (equiv.) Ar = | | | Catalyst ligand (mol %) | | Yield (%) | dr (syn/anti) | ee (%) (syn) |
|---|---|---|---|---|---|---|---|---|---|---|---|
| 1 | 4-MeO | H | 12a | C₆H₅ | 7a | 2.0 | 4a | 5 | 61 | 1.7/1 | 99 |
| 2 | 4-MeO | H | 12a | C₆H₅ | 7a | 2.0 | 4c | 5 | 64 | 4.3/1 | 99 |
| 3 | 2-MeO | H | 12b | C₆H₅ | 7a | 2.0 | 4c | 5 | 70 | > 15/1 | 99 |
| 4[b] | 2-MeO | Cl | 12c | C₆H₅ | 7a | 1.1 | 4a | 5 | 90 | > 15/1 | > 99 |
| 5 | 2-MeO | Cl | 12c | 2-MeO-C₆H₄ | 7c | 2.0 | 4a | 5 | 68 | > 15/1 | > 98 |
| 6 | 2-MeO | Cl | 12c | 2-furyl | 7b | 2.0 | 4a | 5 | 74 | > 15/1 | 99 |

[a] Reaction conditions: –5 °C, MS 4A, THF.
[b] In the presence of 7.5 mol % of Ph₃As=O.

As shown in Table 2.9, imines derived from aromatic aldehydes bearing various substituents **13a–j** afforded products with high *anti*-selectivity (94 : 6 – > 98 : 2, entries 1–10, 14). Although imines derived from α,β-unsaturated aldehydes showed less *anti*-selectivity, diastereoselectivity was improved by conducting the reaction at a lower temperature (entries 11–13). The reaction can be performed on a gram scale with as little as 0.25 mol % of **8a** (6.2 mg) to afford the product in 99% yield and with complete stereocontrol (*dr* > 98/2, 99% ee) (entry 14). Commercial availability of both Et₂Zn solution and linked-BINOL **8a** also makes the protocol advantageous from a practical viewpoint [17b]. It has been speculated that the high *anti*-selectivity stems from the bulky Dpp group on the imine nitrogen. To avoid steric repulsion, the Mannich-type reaction would proceed via the transition state

**Table 2.9** Direct catalytic asymmetric Mannich-type reaction of Dpp-imines **13** promoted by Et₂Zn/(S,S)-linked-BINOL complex.[a]

| Entry | Imine R = | | Ligand 8a (x mol %) | Temp. (°C) | Time (h) | Yield (%) | dr (anti/syn) | ee (%) (anti) |
|---|---|---|---|---|---|---|---|---|
| 1 | 4-MeC₆H₄ | 13a | 1 | –20 | 9 | 98 | 96/4 | 98 |
| 2 | 2-MeC₆H₄ | 13b | 1 | –20 | 6 | 99 | > 98/2 | 99 |
| 3 | C₆H₅ | 13c | 1 | –20 | 6 | 98 | 96/4 | 99 |
| 4 | 4-MeOC₆H₄ | 13d | 1 | –20 | 6 | 97 | 95/5 | 99 |
| 5 | 4-NO₂C₆H₄ | 13e | 1 | –20 | 9 | 96 | 97/3 | 98 |
| 6 | 4-ClC₆H₄ | 13f | 1 | –20 | 4 | 97 | 97/3 | 98 |
| 7 | 4-BrC₆H₄ | 13g | 1 | –20 | 4 | 97 | 95/5 | 98 |
| 8 | 1-naphthyl | 13h | 1 | –20 | 6 | 97 | 98/2 | > 99.5 |
| 9 | 2-naphthyl | 13i | 1 | –20 | 7 | 95 | 94/6 | 99 |
| 10 | 2-furyl | 13j | 1 | –20 | 7 | 98 | 96/4 | > 99.5 |
| 11 | (E)-cinnam | 13k | 1 | –20 | 4 | 98 | 76/24 | > 99.5 |
| 12 | | 13k | 1 | –30 | 7 | 97 | 81/19 | > 99.5 |
| 13 | cyclo-propyl | 13l | 1 | –30 | 5 | 98 | 80/20 | 99 |
| 14[b] | 2-MeC₆H₄ | 13b | 0.25 | –20 | 6 | 99 | > 98/2 | 99 |

[a] Reaction conditions: Et₂Zn/8a = 4/1, MS 3A, hydroxy ketone/imine = 2/1.
[b] 1.28 g of imine was used.

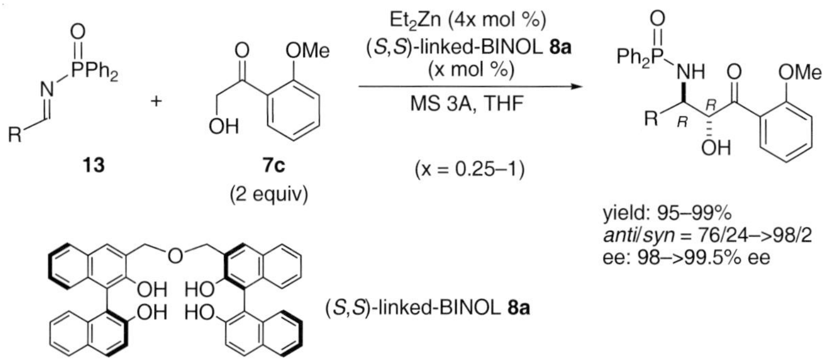

**Scheme 2.13** Direct catalytic asymmetric Mannich-type reaction promoted by Et₂Zn/(S,S)-linked-BINOL complex.

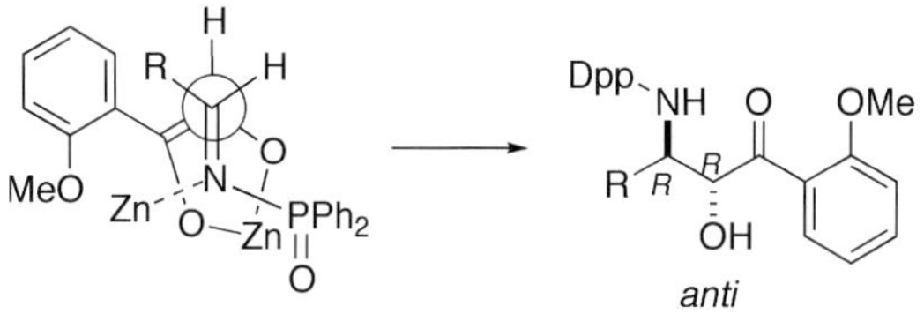

**Figure 2.9** Working transition state model to afford *anti*-product.

shown in Figure 2.9, preferentially affording the *anti* product. Facile transformation of the product to a β-amino-α-hydroxy carboxylic acid derivative (easy removal of the N-Dpp group under acidic conditions and Baeyer–Villiger oxidation of the ketone moiety to an ester) makes this process synthetically useful [20].

## 2.5
## Direct Catalytic Asymmetric Michael Reaction

The Michael reaction is one of the most widely used carbon-carbon bond-forming reactions in synthetic organic chemistry, and a number of catalytic asymmetric variants have been developed over the past decade [21]. Almost all of them utilize either active methylene compounds or reactive nucleophiles prepared by using no less than stoichiometric amounts of reagents. *Direct* catalytic asymmetric Michael reactions, in which unmodified ketones are used directly as donors, are very rare, despite the increasing demand for the development of atom-economical synthetic protocols [22]. In 2001, Shibasaki demonstrated that polymetallic Zn catalysis can be successfully applied in direct catalytic asymmetric Michael reactions by simultaneous activation of both reaction partners [23]. Direct Michael reaction of 2-hydroxy-2'-methoxyacetophenone (**7c**) with vinyl ketones **14**, which usually tend to polymerize under harsh reaction conditions, proceeded efficiently at 4 °C by using 2 : 1 Et₂Zn/(S,S)-linked-BINOL complex (Scheme 2.14). As shown in Table 2.10, aryl vinyl ketones (entries 1–4) and alkyl vinyl ketones (entries 5, 6) were successfully converted to the Michael adducts in good yields (83–90%) and

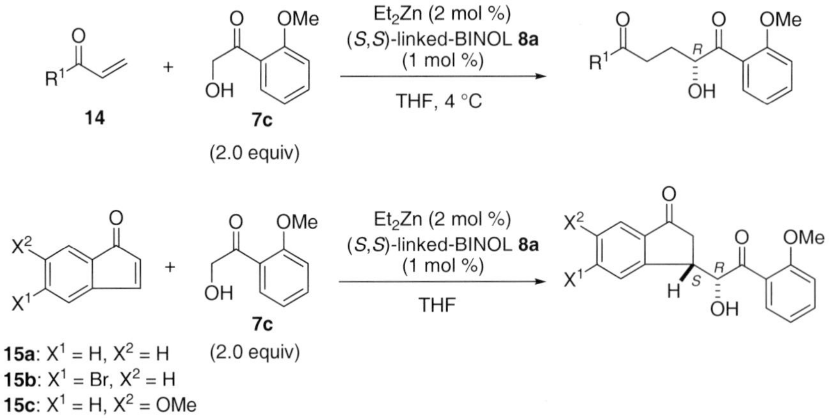

**15a**: $X^1$ = H, $X^2$ = H    (2.0 equiv)
**15b**: $X^1$ = Br, $X^2$ = H
**15c**: $X^1$ = H, $X^2$ = OMe

**Scheme 2.14** Direct catalytic asymmetric Michael reaction of vinyl ketones **14** and indenones **15** promoted by $Et_2Zn/(S,S)$-linked-BINOL = 2/1 complex.

**Table 2.10** Direct catalytic asymmetric Michael reaction of vinyl ketones **14** promoted by $Et_2Zn/(S,S)$-linked-BINOL = 2/1 complex.[a]

| Entry | Vinyl ketone $R^1$ = | | Time (h) | Yield (%) | ee (%) |
|-------|-----------------|-----|----------|-----------|--------|
| 1 | $p$-MeOC$_6$H$_4$ | **14a** | 8 | 83 | 95 |
| 2 | C$_6$H$_5$ | **14b** | 4 | 86[b] | 93 |
| 3 | $o$-MeOC$_6$H$_4$ | **14c** | 12 | 90 | 94 |
| 4 | $p$-ClC$_6$H$_4$ | **14d** | 12 | 84[b] | 92 |
| 5 | CH$_3$ | **14e** | 4 | 86 | 93 |
| 6 | CH$_3$CH$_2$ | **14f** | 4 | 82 | 91 |

[a] Reaction conditions: $Et_2Zn$ (2 mol %), linked-BINOL **8a** (1 mol %), vinyl ketone/hydroxy ketone = 1/2.
[b] Determined by $^1$H NMR analysis.

with high enantioselectivities (91–95% ee) with the use of 1 mol % of catalyst. With indenones **15**, lower reaction temperatures led to higher diastereoselectivity (Table 2.11). By conducting the Michael reaction at −20 °C, excellent diastereo- (97 : 3–98 : 2) and enantioselectivities (97–99%) were observed (entries 2, 4, 5), although the chemical yields were modest due to polymerization.

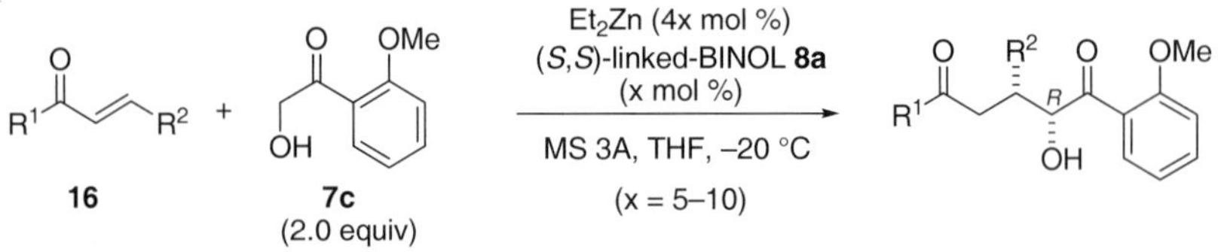

**Scheme 2.15** Direct catalytic asymmetric Michael reaction of acyclic enones **16** promoted by $Et_2Zn/(S,S)$-linked-BINOL = 4/1 complex with MS 3A.

**Table 2.11** Direct catalytic asymmetric Michael reaction of indenones **15** promoted by Et$_2$Zn/(S,S)-linked-BINOL = 2/1 complex.[a]

| Entry | Indenone X$^1$ = | X$^2$ = | | Temp. (°C) | Time (h) | Yield (%) | dr | ee (%) |
|---|---|---|---|---|---|---|---|---|
| 1 | H | H | 15a | 4 | 1.5 | 68 | 95/5 | 97 |
| 2 | H | H | 15a | −20 | 4 | 74 | 98/2 | 99 |
| 3 | Br | H | 15b | 4 | 2 | 76 | 86/14 | 99 |
| 4 | Br | H | 15b | −20 | 4 | 74 | 98/2 | 99 |
| 5 | H | MeO | 15c | −20 | 4 | 65 | 97/3 | 97 |

[a] Reaction conditions: Et$_2$Zn (2 mol %), (S,S)-linked-BINOL **8a** (1 mol %), indenone/hydroxy ketone = 1/2.

**Table 2.12** Direct catalytic asymmetric Michael reaction of acyclic enones **16** promoted by Et$_2$Zn/(S,S)-linked-BINOL = 4/1 complex with MS 3A.[a]

| Entry | Enone R$^1$ = | R$^2$ = | | Cat. (x mol %) | Time (h) | Yield (%) | dr (syn/anti) | ee (%) (syn/anti) |
|---|---|---|---|---|---|---|---|---|
| 1 | Ph | Ph | 16a | 5 | 3 | 93 | 78/22 | 95/93 |
| 2 | 4-ClC$_6$H$_4$ | Ph | 16b | 5 | 3 | 95 | 79/21 | 97/83 |
| 3 | 4-FC$_6$H$_4$ | Ph | 16c | 5 | 3 | 96 | 81/19 | 97/80 |
| 4 | 4-MeO-C$_6$H$_4$ | Ph | 16d | 5 | 6 | 99 | 85/15 | 97/52 |
| 5 | Ph | 4-ClC$_6$H$_4$ | 16e | 5 | 5 | 96 | 76/24 | 95/71 |
| 6 | Me | Ph | 16f | 10 | 16 | 82 | 86/14 | 99/7 |
| 7 | i-Pr | Ph | 16g | 10 | 17 | 73 | 86/14 | 87/– |
| 8 | t-Bu | Ph | 16h | 10 | 24 | 58 | 93/7 | 74/– |
| 9 | Ph | Et | 16i | 10 | 7 | 85 | 81/19 | 97/79 |
| 10 | Ph | CH$_2$CH(CH$_3$)$_2$ | 16j | 5 | 7 | 95 | 69/31 | 97/65 |
| 11 | Ph | CH$_2$OTBDPS | 16k | 10 | 24 | 93 | 61/39 | 81/52 |
| 12 | Ph | (CH$_2$)$_2$OBOM | 16l | 10 | 24 | 72 | 77/23 | 80/– |
| 13 | Me | (CH$_2$)$_4$CH$_3$ | 16m | 10 | 24 | 39 | 68/32 | 93/86 |

[a] Reaction conditions: Et$_2$Zn (4x mol %), (S,S)-linked-BINOL **8a** (x mol %), enone/hydroxy ketone = 1/2.

As mentioned in Section 2.3.3, the second-generation catalytic system, 4 : 1 Et$_2$Zn/(S,S)-linked-BINOL complex with 3 Å molecular sieves, showed much higher reactivity in direct aldol reactions. This improved catalytic system was also applicable to direct Michael reaction and enabled the reaction with less reactive β-substituted enones (Scheme 2.15) [24]. As shown in Table 2.12, direct Michael reactions of various acyclic enones **16** with ketone **7c** proceeded *syn*-selectively at −20 °C. Enones with electron-donating or -withdrawing substituents on the aromatic ring afforded the products in good yields (93–99%) and with high enantioselectivities (95–97% ee) (Table 2.12, entries 1–5). α-Enolizable enones such as **16f** and **16g** were also suitable substrates (entries 6 and 7). Sterically demanding enone **16h** exhibited enhanced *syn*-selectivity (entry 8). The results with cyclic enones are shown in Scheme 2.16. Cyclic enones **17** afforded the corresponding Michael adducts as single diastereomers in moderate to good yields (45–81%) with high ee (85–99% ee), although

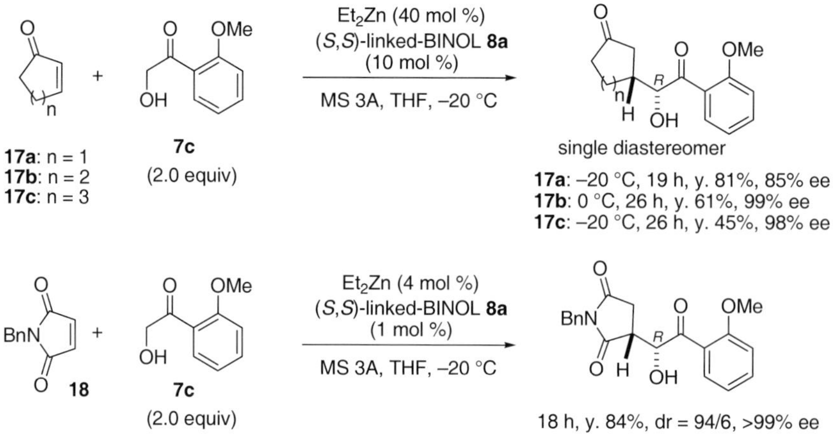

Et2Zn (40 mol %)
(S,S)-linked-BINOL 8a
(10 mol %)
MS 3A, THF, –20 °C
17a: n = 1
17b: n = 2
17c: n = 3
7c
(2.0 equiv)
single diastereomer
17a: –20 °C, 19 h, y. 81%, 85% ee
17b: 0 °C, 26 h, y. 61%, 99% ee
17c: –20 °C, 26 h, y. 45%, 98% ee
Et2Zn (4 mol %)
(S,S)-linked-BINOL 8a
(1 mol %)
MS 3A, THF, –20 °C
18
7c
(2.0 equiv)
18 h, y. 84%, dr = 94/6, >99% ee


**Scheme 2.16** Direct catalytic asymmetric Michael reaction of cyclic enones promoted by $Et_2Zn$/(*S*,*S*)-linked-BINOL = 4/1 complex with MS 3A.

10 mol % of the chiral ligand was required. In the case of maleimide **18**, the reactivity was much higher than with cyclic enones, and the product was obtained in 84% yield with excellent stereoselectivity using 1 mol % of the chiral ligand. The ketone moiety in the Michael products was successfully transformed into a synthetically more useful aryl ester through regioselective Baeyer–Villiger oxidation [24].

With the second-generation catalytic system, 4 : 1 $Et_2Zn$/(*S*,*S*)-linked-BINOL complex with 3 Å molecular sieves, the catalyst turnover was improved substantially, so that the catalyst loading could be reduced to as little as 0.01 mol % in the reaction with methyl vinyl ketone (**14e**) [24]. With 1.1 equiv. of ketone **7c**, the reaction proceeded at room temperature by using 0.01 mol % of ligand **8a** and 0.04 mol % of $Et_2Zn$ to afford the Michael adduct in 78% yield and with 89% ee (Scheme 2.17). By employing 2-hydroxy-2′-methoxypropiophenone (**10**) as the donor substrate, a chiral tetrasubstituted carbon center was constructed (Scheme 2.18). The direct Michael reaction of **10** with vinyl ketones **14** proceeded in the presence of 5 mol % of catalyst to furnish the chiral tertiary alcohol in good yield (78–95%) and with high ee (90–96%) (Table 2.13). It is worth noting that only (*S*)-**10** is consumed during the reaction, which suggests that the $Et_2Zn$/(*S*,*S*)-linked-BINOL complex differen-

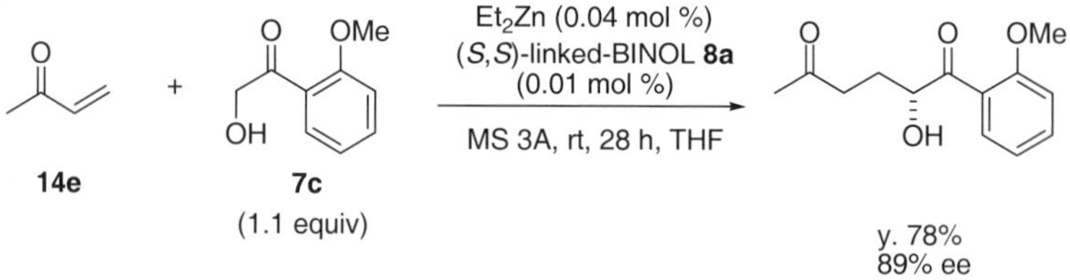

Et2Zn (0.04 mol %)
(S,S)-linked-BINOL 8a
(0.01 mol %)
MS 3A, rt, 28 h, THF
14e
7c
(1.1 equiv)
y. 78%
89% ee


**Scheme 2.17** Direct catalytic asymmetric Michael reaction of methyl vinyl ketone **14e** promoted by $Et_2Zn$/(*S*,*S*)-linked-BINOL = 4/1 complex with MS 3A.

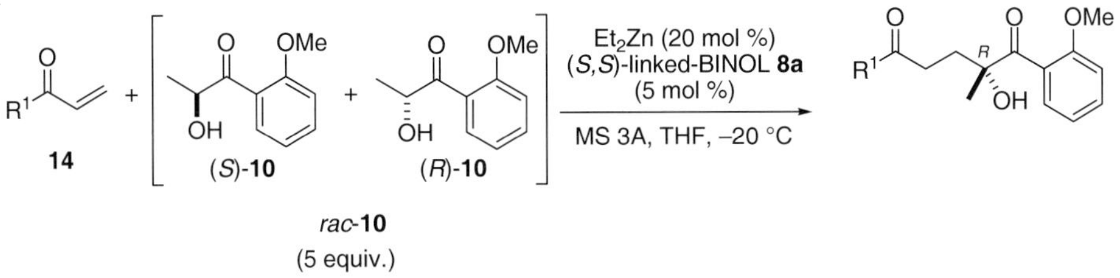

**Scheme 2.18** Direct catalytic asymmetric Michael reaction of 2-hydroxy-2′-methoxypropiophenone (10) promoted by Et$_2$Zn/(S,S)-linked-BINOL = 4/1 complex with MS 3A.

**Table 2.13** Direct catalytic asymmetric Michael reaction of 2-hydroxy-2′-methoxypropiophenone (10) promoted by Et$_2$Zn/(S,S)-linked-BINOL = 4/1 complex with MS 3A.[a]

| Entry | Vinyl ketone R$^1$ = | | Time (h) | Yield (%) | ee (%) |
|---|---|---|---|---|---|
| 1 | p-MeOC$_6$H$_4$ | 14a | 12 | 95 | 90 |
| 2 | C$_6$H$_5$ | 14b | 24 | 82 | 93 |
| 3 | o-MeOC$_6$H$_4$ | 14c | 24 | 78 | 91 |
| 4 | p-BrC$_6$H$_4$ | 14g | 24 | 82 | 93 |
| 5 | CH$_3$ | 14e | 16 | 88 | 96 |
| 6 | CH$_3$CH$_2$ | 14f | 24 | 86 | 90 |

[a] Reaction conditions: Et$_2$Zn (20 mol %), (S,S)-linked-BINOL **8a** (5 mol %), vinyl ketone/hydroxy ketone = 1/5.

tiates the configuration of ketone **10** in forming the Zn enolate. Therefore, the products from the (S,S)-catalyst, which have an (R)-OH configuration, would not be recognized by the (S,S)-catalyst and are excluded from the catalytically active Zn center to realize a high catalyst turnover.

## 2.6
### Nitroaldol (Henry) Reaction

The nitroaldol (Henry) reaction is an atom-economical approach for obtaining β-nitroalcohols from aldehydes and nitroalkanes [25]. In 1992, Shibasaki [26] reported the first successful example of a catalytic asymmetric nitroaldol reaction using heterobimetallic catalysts. In 2002, dinuclear Zn catalysis was introduced to this valuable transformation by Trost [27, 28] (Scheme 2.19). A systematic modification of chiral ligand **4a** was carried out and various derivatives were evaluated in catalytic asymmetric nitroaldol reactions. The effect of substituents on the phenol ring was examined by using benzaldehyde as a substrate (Scheme 2.20). The introduction of a 4-methoxy substituent resulted in dramatic decreases in both the reaction rate and enantioselectivity (22% ee). Placing a fluoro substituent in the *para* position, on the other hand, led to a slight increase in enantioselectivity

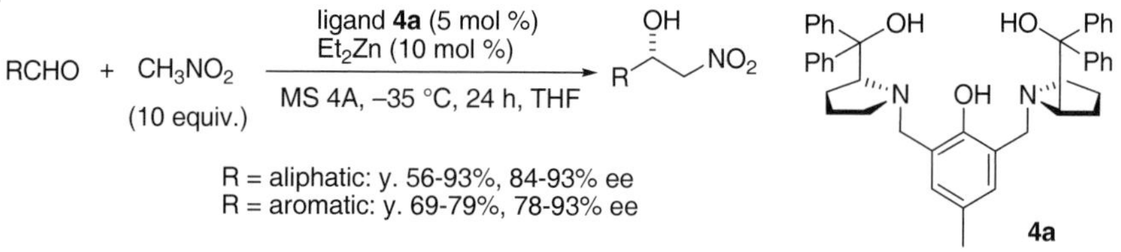
ligand 4a (5 mol %)
Et2Zn (10 mol %)
RCHO + CH3NO2
(10 equiv.)
MS 4A, –35 °C, 24 h, THF
OH
R
NO2
Ph
OH
HO
Ph
Ph
Ph
N
OH
N
4a
R = aliphatic: y. 56-93%, 84-93% ee
R = aromatic: y. 69-79%, 78-93% ee

**Scheme 2.19** Catalytic asymmetric nitroaldol reaction promoted by dinuclear $Zn_2$-**4a** complex.

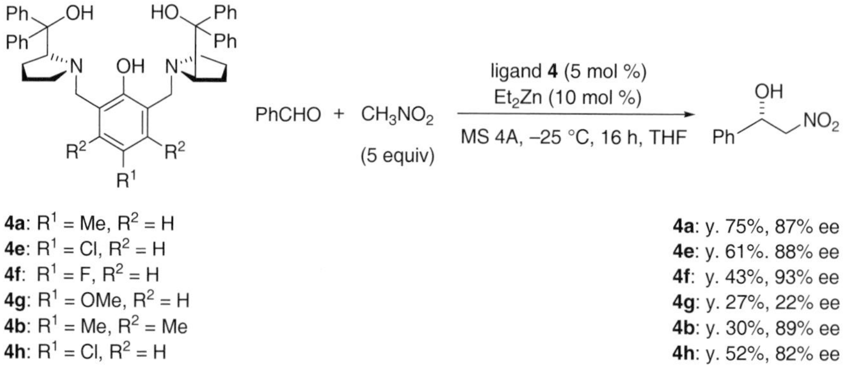
Ph
OH
HO
Ph
Ph
Ph
N
OH
N
R2
R2
R1
PhCHO + CH3NO2
(5 equiv)
ligand 4 (5 mol %)
Et2Zn (10 mol %)
MS 4A, –25 °C, 16 h, THF
OH
Ph
NO2
4a: R1 = Me, R2 = H
4e: R1 = Cl, R2 = H
4f: R1 = F, R2 = H
4g: R1 = OMe, R2 = H
4b: R1 = Me, R2 = Me
4h: R1 = Cl, R2 = H
4a: y. 75%, 87% ee
4e: y. 61%. 88% ee
4f: y. 43%, 93% ee
4g: y. 27%, 22% ee
4b: y. 30%, 89% ee
4h: y. 52%, 82% ee

**Scheme 2.20** Catalytic asymmetric nitroaldol reaction promoted by dinuclear Zn catalyst.

(93% ee). The impact of the diaryl carbinol moieties of the ligands was examined with 4-*tert*-butyldimethylsilyloxybenzaldehyde (**19**). The biphenyl ligand **4c** gave a small but real increase in ee (90%) compared to the parent ligand **4a** (Scheme 2.21). The nitro functionality in the product was easily reduced to an amine by hydrogenation over Pd/C, and (*R*)-**20** was successfully transformed into (–)-denopamine, a selective $\beta_1$-adrenoreceptor agonist (Scheme 2.22).

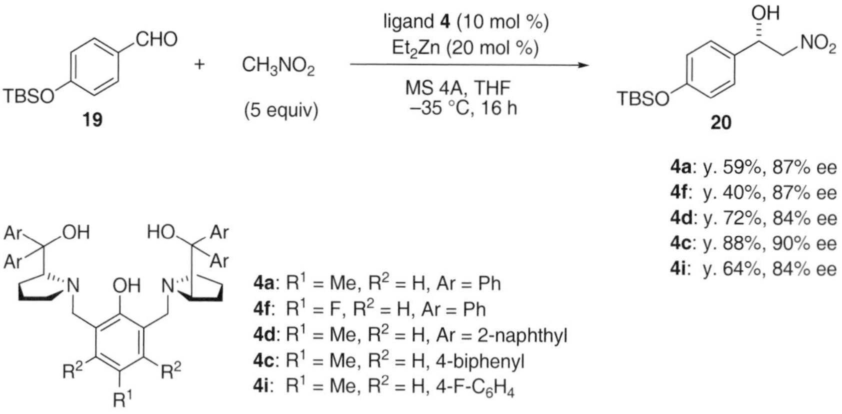
CHO
TBSO
19
+ CH3NO2
(5 equiv)
ligand 4 (10 mol %)
Et2Zn (20 mol %)
MS 4A, THF
–35 °C, 16 h
OH
NO2
TBSO
20
4a: y. 59%, 87% ee
4f: y. 40%, 87% ee
4d: y. 72%, 84% ee
4c: y. 88%, 90% ee
4i: y. 64%, 84% ee
Ar
OH
HO
Ar
Ar
Ar
N
OH
N
R2
R2
R1
4a: R1 = Me, R2 = H, Ar = Ph
4f: R1 = F, R2 = H, Ar = Ph
4d: R1 = Me, R2 = H, Ar = 2-naphthyl
4c: R1 = Me, R2 = H, 4-biphenyl
4i: R1 = Me, R2 = H, 4-F-C6H4

**Scheme 2.21** Catalytic asymmetric nitroaldol reaction promoted by dinuclear Zn catalyst.

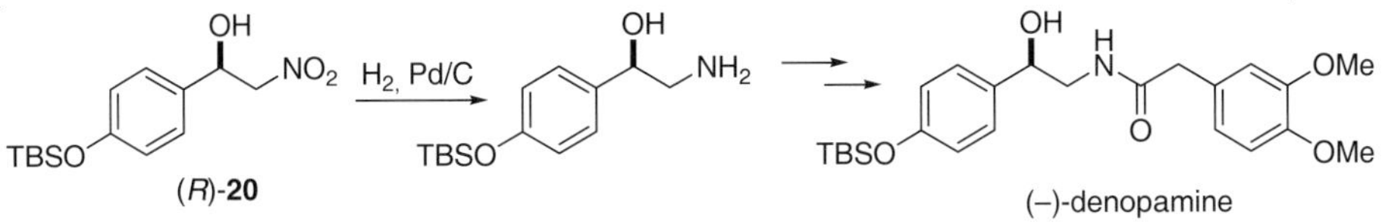

**Scheme 2.22** Transformation of nitroaldol product into (–)-denopamine.

## 2.7
## Conclusions

The past few years have witnessed an explosive growth of interest and substantial progress in Zn polymetallic asymmetric catalysis, especially in catalytic asymmetric C–C bond-forming reactions. The mild Lewis acidities and Brønsted basicities of Zn complexes make them suitable for biomimetic and atom-economical transformations such as the direct aldol reaction, which meets contemporary demands for environmentally friendly organic synthesis. The key factor for obtaining the high catalytic efficiency is the simultaneous activation of multiple reaction partners in concert, rather like in *in vivo* transformations by sophisticated enzymes. Zn polymetallic asymmetric catalysis is still in its infancy. An in-depth understanding of the reaction mechanisms and the rational design of catalysts will open up the inherent potential of the world of this asymmetric methodology.

## References

1    S. Inoue, H. Koinuma, T. Tsuruta,
     *J. Polym. Sci., Part B* **1969**, *7*, 287.
2    Reviews: (a) A. Rokicki, W. Kuran,
     *J. Macromol. Sci., Rev. Macromol. Chem.*
     **1981**, *C21*, 135; (b) S. Inoue, *Carbon
     Dioxide as a Source of Carbon* (Eds.:
     M. Aresta, G. Forti), Reidel Publishing
     Co., Dordrecht, The Netherlands, **1987**,
     p. 331; (c) D. J. Darensbourg,
     M. W. Holtcamp, *Coord. Chem. Rev.*
     **1996**, *153*, 155; (d) M. Super,
     E. J. Beckmann, *Trends Polym. Sci.* **1997**,
     *5*, 236.
3    N. Spassky, A. Momtaz, A. Kassamaly,
     M. Sepulchre, *Chirality* **1992**, *4*, 295,
     and references cited therein.
4    (a) K. Nozaki, K. Nakano, T. Hiyama,
     *J. Am. Chem. Soc.* **1999**, *121*, 11008;
     (b) K. Nakano, K. Nozaki, T. Hiyama,
     *J. Am. Chem. Soc.* **2003**, *125*, 5501.

5    Coates also reported asymmetric
     alternating copolymerization promoted
     by Zn monomer complexes with other
     ligands, see: M. Cheng, N. A. Darling,
     E. B. Lobkovsky, G. W. Coates, *Chem.
     Commun.* **2000**, 2007.
6    B. M. Trost, *Science* **1991**, *254*, 1471.
7    Review: T. D. Machajewski, C.-H. Wong,
     *Angew. Chem. Int. Ed.* **2000**, *39*, 1352.
8    (a) For a recent review of direct catalytic
     asymmetric aldol reactions, see:
     B. Alcaide, P. Almendros, *Eur. J. Org.
     Chem.* **2002**, 1595; (b) for a review of early
     important contributions on direct
     catalytic asymmetric aldol reactions using
     Au-catalysis, see: M. Sawamura, Y. Ito,
     *Chem. Rev.* **1992**, *92*, 857; (c) for a review
     of the other type of direct aldol reaction,
     that is, organocatalysis by mimics of class
     I aldolases, see: B. List, *Tetrahedron* **2002**,
     *58*, 5573.

9   Recent advances in direct catalytic
    asymmetric aldol reactions: (a) using
    unmodified aldehydes as donors, see:
    A. B. NORTHRUP, D. W. C. MACMILLAN,
    J. Am. Chem. Soc. 2002, 124, 6798;
    (b) with Ni-bis(oxazoline) complex, see:
    D. A. EVANS, C. W. DOWNEY, J. L. HUBBS,
    J. Am. Chem. Soc. 2003, 125, 8706.

10  (a) Y. M. A. YAMADA, N. YOSHIKAWA,
    H. SASAI, M. SHIBASAKI, Angew. Chem.
    Int. Ed. Engl. 1997, 36, 1871;
    (b) N. YOSHIKAWA, Y. M. A. YAMADA,
    J. DAS, H. SASAI, M. SHIBASAKI,
    J. Am. Chem. Soc. 1999, 121, 4168.

11  B. M. TROST, H. ITO, J. Am. Chem. Soc.
    2000, 122, 12003.

12  B. M. TROST, E. R. SILCOFF, H. ITO,
    Org. Lett. 2001, 3, 2497.

13  (a) W. NOTZ, B. LIST, J. Am. Chem. Soc.
    2000, 122, 7368; (b) K. SAKTHIVEL,
    W. NOTZ, T. BUI, C. F. BARBAS, III,
    J. Am. Chem. Soc. 2001, 123, 5260.

14  B. M. TROST, H. ITO, E. R. SILCOFF,
    J. Am. Chem. Soc. 2001, 123, 3367.

15  (a) N. YOSHIKAWA, N. KUMAGAI,
    S. MATSUNAGA, G. MOLL, T. OHSHIMA,
    T. SUZUKI, M. SHIBASAKI, J. Am. Chem.
    Soc. 2001, 123, 2466; (b) N. KUMAGAI,
    S. MATSUNAGA, N. YOSHIKAWA,
    T. OHSHIMA, M. SHIBASAKI, Org. Lett.
    2001, 3, 1539; (c) N. KUMAGAI,
    S. MATSUNAGA, T. KINOSHITA, S. HARADA,
    S. OKADA, S. SAKAMOTO, K. YAMAGUCHI,
    M. SHIBASAKI, J. Am. Chem. Soc. 2003,
    125, 2169; (d) for an important contri-
    bution based on the use of LLB · KOH
    catalyst, see: N. YOSHIKAWA, T. SUZUKI,
    M. SHIBASAKI, J. Org. Chem. 2002, 67,
    2556.

16  B. M. TROST, V. S. C. YEH, Org. Lett. 2002,
    4, 3513.

17  (a) For a review of linked-BINOL 8a, see:
    S. MATSUNAGA, T. OHSHIMA,
    M. SHIBASAKI, Adv. Synth. Catal. 2002,
    344, 4; (b) commercial source of linked-
    BINOL 8a: Wako Pure Chemical
    Industries Ltd.; Catalog No; (f) or
    (S,S)-8a, No. 155-02431.

18  For other representative examples of
    direct catalytic asymmetric Mannich

reactions, see: (a) with an Al catalyst:
S. YAMASAKI, T. IIDA, M. SHIBASAKI,
Tetrahedron Lett. 1999, 40, 307; (b) with a
Cu catalyst: K. JUHL, N. GATHERGOOD,
K. A. JØRGENSEN, Angew. Chem. Int. Ed.
2001, 40, 2995; with proline catalysis:
(c) B. LIST, P. POJARLIEV, W. T. BILLER,
H. J. MARTIN, J. Am. Chem. Soc. 2002,
124, 827; (d) A. CÓRDOVA, W. NOTZ,
G. ZHONG, J. M. BETANCORT,
C. F. BARBAS, III, J. Am. Chem. Soc. 2002,
124, 1842; (e) A. CÓRDOVA,
S.-I. WATANABE, F. TANAKA, W. NOTZ,
C. F. BARBAS, III, J. Am. Chem. Soc. 2002,
124, 1866.

19  B. M. TROST, L. R. TERRELL, J. Am. Chem.
    Soc. 2003, 125, 338.

20  S. MATSUNAGA, N. KUMAGAI, S. HARADA,
    M. SHIBASAKI, J. Am. Chem. Soc. 2003,
    125, 4712.

21  Reviews: (a) N. KRAUSE, A. HOFFMANN-
    RÖDER, Synthesis 2001, 171; (b) E. N.
    JACOBSEN, A. PFALTZ, H. YAMAMOTO
    (Eds.), Comprehensive Asymmetric
    Catalysis, Springer, Berlin, 1999,
    Chapter 31.

22  For several examples of direct catalytic
    asymmetric Michael reactions of
    unmodified ketones, see: (a) F.-Y. ZHANG,
    E. J. COREY, Org. Lett, 2000, 2, 1097;
    (b) J. M. BETANCORT, K. SAKTHIVEL,
    R. THAYUMANAVAN, C. F. BARBAS, III,
    Tetrahedron Lett. 2001, 42, 4441;
    (c) B. LIST, P. POJARLIEV, H. J. MARTIN,
    Org. Lett. 2001, 3, 2423.

23  N. KUMAGAI, S. MATSUNAGA,
    M. SHIBASAKI, Org. Lett. 2001, 3, 4251.

24  S. HARADA, N. KUMAGAI, T. KINOSHITA,
    S. MATSUNAGA, M. SHIBASAKI,
    J. Am. Chem. Soc. 2003, 125, 2582.

25  E. N. JACOBSEN, A. PFALTZ, H. YAMAMOTO
    (Eds.), Comprehensive Asymmetric Cata-
    lysis, Springer, Berlin, 1999, chapter 29.3.

26  H. SASAI, T. SUZUKI, S. ARAI, T. ARAI,
    M. SHIBASAKI, J. Am. Chem. Soc. 1992,
    114, 4418.

27  B. M. TROST, V. S. C. YEH, Angew. Chem.
    Int. Ed. 2002, 41, 861.

28  B. M. TROST, V. S. C. YEH, H. ITO,
    N. BREMEYER, Org. Lett. 2002, 4, 2621.

# 3
# Group 13–Alkali Metal Heterobimetallic Asymmetric Catalysis

*Takashi Ohshima and Masakatsu Shibasaki*

## 3.1
## Introduction

Group 13 metal complexes exhibit Lewis acidity because the metal center possesses a vacant orbital, and they have been utilized as Lewis acid reagents/catalysts in a wide variety of organic reactions. Thus, anionic fragments interact with the vacant orbital of the group 13 metal, resulting in the formation of 'ate' complexes. Heterobimetallic complexes that consist of a group 13 metal and an alkali metal are of particular interest because they function at the same time as both a Lewis acid and a Brønsted base, in a similar fashion as some enzymes. The alkali metal in the heterobimetallic catalyst counteracts the negative charge placed on the group 13 metal. Even tetra-coordinated complexes of group 13 metals other than boron are able to act as Lewis acids, by using relatively low energy vacant orbitals such as the 3d orbital of aluminum, and, therefore, the formation of penta- or hexa-coordinated complexes is possible. In view of the success of rare earth–alkali metal heterobimetallic asymmetric catalysis (LnMB: **1**) (Figure 3.1; see also Chapter 5), extension of the heterobimetallic concept to group 13 metal complexes is promising

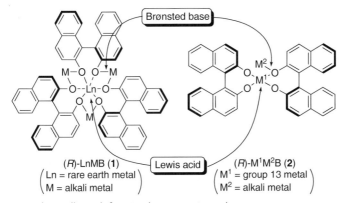

**Figure 3.1** Heterobimetallic multifunctional asymmetric catalysts.

*Multimetallic Catalysts in Organic Synthesis.* Edited by M. Shibasaki and Y. Yamamoto
Copyright © 2004 WILEY-VCH Verlag GmbH & Co. KGaA, Weinheim
ISBN: 3-527-30828-8

and their asymmetric catalysts can be expected to show multifunctionality (Lewis acidity and Brønsted basicity with control of the orientation of substrates) to effectively promote asymmetric reactions in a similar way as LnMB.

This chapter gives a survey of the relevant literature concerning group 13–alkali metal heterobimetallic asymmetric catalysis published up to the end of 2003, beginning with the development of AlLibis(binaphthoxide) complex (ALB: **2**, $M^1$ = Al, $M^2$ = Li), and including their application to enantioselective syntheses of several natural products.

## 3.2
### Catalytic Asymmetric Michael Reaction of Stabilized Carbon Nucleophiles

### 3.2.1
### Development of ALB – The First Example of a Group 13–Alkali Metal Heterobimetallic Asymmetric Catalyst

The catalytic asymmetric Michael reaction of enolates with $\alpha,\beta$-unsaturated carbonyl compounds is one of the most important carbon–carbon bond-forming reactions due to the ready availability of both substrates, the usefulness of enantiomerically enriched Michael products, and its high atom economy [1]. In particular, 1,3-di-carbonyl compounds are highly promising and desirable Michael donors for the enantioselective construction of carbon–carbon bonds. Prior to the development of ALB, efficient catalytic asymmetric Michael reactions of malonates with various enones were realized by Shibasaki et al. [2] using the rare earth metal containing heterobimetallic asymmetric catalyst LSB (L = La, S = Na), affording Michael adducts with up to 92% ee (see Section 5.5). Mechanistic studies of LSB-catalyzed Michael reactions revealed that LSB acts as a base catalyst (through the O–Na moiety) and, at the same time, shows Lewis acid character (through the La center). As an extension of the heterobimetallic concept, an amphoteric asymmetric catalyst assembled from aluminum and an alkali metal was examined by Shibasaki et al. [3], and a new class of heterobimetallic asymmetric catalysts, ALB, which consist of aluminum, lithium, and BINOL, proved to be superior to LSB in the asymmetric Michael reaction of malonates with enones. ALB was initially prepared from diisobutylaluminum hydride, 2 mol equivalents of BINOL, and 1 mol equivalent of butyllithium in THF, use of which afforded Michael product **11** in 46% yield and with 98% ee. It was subsequently found that the same catalyst could be prepared more efficiently from $LiAlH_4$ and two equivalents of BINOL in THF and showed higher reactivity (88% yield, 99% ee) (Table 3.1, entry 3). Although other alkali and alkaline earth metals (M = Na, K, and Ba) also give rather good results, lithium is the best partner for aluminum in terms of reactivity and selectivity (entries 3–6).

Shibasaki et al. [4] also examined other combinations of group 13 metals with alkali metals with the aim of developing more efficient heterobimetallic catalysts for the asymmetric Michael reaction. As expected, boron–alkali metal–BINOL complexes did not promote the reaction (Table 3.2, entries 1–3). On the other hand,

**Table 3.1** Catalytic asymmetric Michael reactions promoted by (R)-AMB complexes.

**3**: n = 1   **5**: $R^1$ = Et, $R^2$ = Me
**4**: n = 2   **6**: $R^1$ = Bn, $R^2$ = H
              **7**: $R^1$ = Me, $R^2$ = H
              **8**: $R^1$ = Et, $R^2$ = H

| Entry | Enone | Donor | Product | M | Time (h) | Yield (%) | ee (%) |
|---|---|---|---|---|---|---|---|
| 1 | 3 | 5 | 9 | Li | 72 | 84 | 91 |
| 2 | 3 | 6 | 10 | Li | 60 | 93 | 91 |
| 3 | 4 | 6 | 11 | Li | 72 | 88 | 99 |
| 4 | 4 | 6 | 11 | Na | 72 | 50 | 98 |
| 5 | 4 | 6 | 11 | K | 7 | 43 | 87 |
| 6 | 4 | 6 | 11 | Ba | 6 | 100 | 84 |
| 7 | 4 | 7 | 12 | Li | 72 | 90 | 93 |
| 8 | 4 | 8 | 13 | Li | 72 | 87 | 95 |

**Table 3.2** Catalytic asymmetric Michael reaction of **6** with **4**.

| Entry | $M^1$ | $M^2$ | Time (h) | Yield (%) | ee (%) |
|---|---|---|---|---|---|
| 1 | B | Li | 17 | 0 | – |
| 2 | B | Na | 21 | 0 | – |
| 3 | B | K | 15 | 0 | – |
| 4 | Al | Li (ALB) | 72 | 88 | 99 |
| 5 | Al | Na | 72 | 50 | 98 |
| 6 | Al | K | 72 | 43 | 87 |
| 7 | Ga | Li | 43 | 71 | 49 |
| 8 | Ga | Na (GaSB) | 143 | 45 | 98 |
| 9 | Ga | K | 44 | 50 | 86 |
| 10 | In | Li | 24 | 77 | 2 |
| 11 | In | Na | 95 | 25 | 10 |
| 12 | In | K (InPB) | 168 | 61 | 84 |

gallium–sodium–BINOL complex (GaSB) (entry 8) and indium–potassium–BINOL complex (InPB) (entry 12) were found to be rather effective catalysts for the asymmetric Michael reaction of **6** with **4**. The former was better than the latter in terms of enantioselectivity. Later, the reactivity and selectivity of ALB and GaSB were further improved (see Section 3.2.2).

Feringa et al. [5] applied ALB to the asymmetric Michael reaction of nitroesters with methyl vinyl ketone (15), in which a chiral center is constructed on the Michael donor side. When nitroester 14 was used as a Michael donor, the asymmetric Michael reaction was catalyzed by 5 mol % of ALB, which was prepared from $LiAlH_4$ and 2.45 mol equivalents of BINOL, giving the Michael product 16 in 81% yield and with 80% ee (Scheme 3.1).

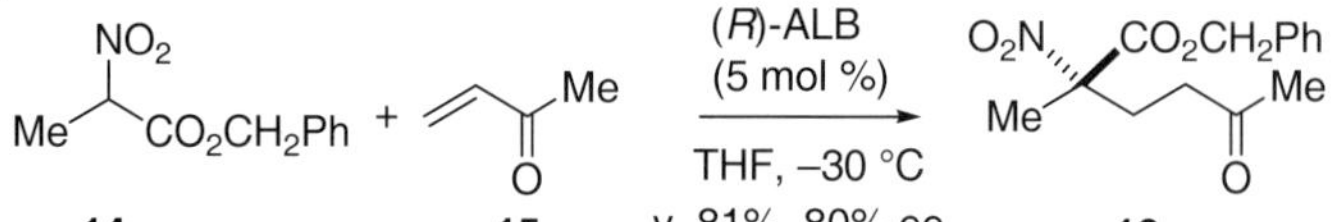

**Scheme 3.1**

ALB is also applicable to the catalytic kinetic resolution reaction of (±)-5-methylbicyclo[3.3.0]oct-1-ene-3,6-dione (17). Although LSB and SmSB, which are very efficient catalysts for the catalytic asymmetric Michael reaction of thiols with α,β-unsaturated carbonyl compounds (see Section 5.5) [6], gave unsatisfactory results (only up to 37% ee), ALB catalyzed the kinetic resolution reaction of 17 with thiol 18 to afford the Michael product 19 (48%, 78% ee) and (+)-17 (49%, 77% ee) in the presence of 4-methoxyphenol (1.2 mol equivalent with respect to ALB) (Scheme 3.2) [7].

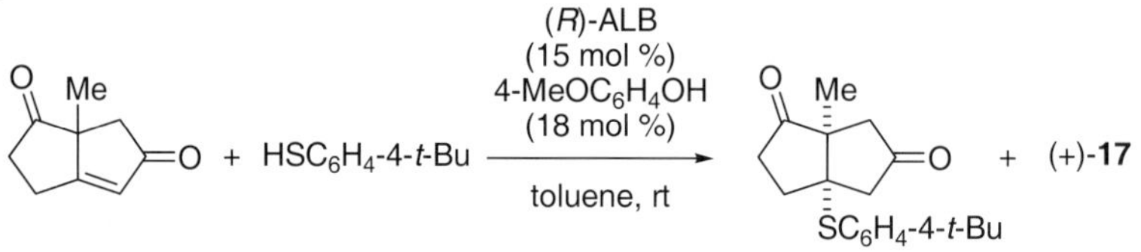

**Scheme 3.2**

The structure of ALB was unequivocally determined by X-ray crystallography (Figure 3.2) [3]. The complex has a tetrahedral geometry around the aluminum center with an average Al–O distance of 1.75 Å. The long Li–O(1) distance of 2.00 Å indicates the ionic character between $Li^+$ and $[Al(BINOL)_2]^-$. In view of the fact that the electronegativity of lithium (1.0) is lower than that of aluminum (1.5), a lithium enolate should be generated preferentially from a malonate. Moreover, $^{27}Al$ NMR studies on the Michael reaction clearly indicate that the carbonyl groups of the enones coordinate to the aluminum center. Consequently, ALB acts as a hetero-bimetallic multifunctional asymmetric catalyst. The structure of GaSB, which was prepared from $GaCl_3$, 4 mol equivalents of $NaO^tBu$, and 2 mol equivalents of BINOL in $THF/Et_2O$, was determined on the basis of its $^{13}C$ NMR spectrum and its LDI-TOF mass spectrum [4]. The complex consists of one Ga atom, one Na atom, and two molecules of BINOL in a similar arrangement as in ALB.

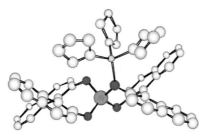

**Figure 3.2** Crystal structure of ALB: $[C_{40}H_{24}AlLiO_4] \cdot 3$ THF.

The realization of three-component coupling through tandem Michael-aldol reaction highlights a characteristic chemical property of ALB that distinguishes it from rare earth metal containing heterobimetallic complexes (LnMB). The above-mentioned mechanistic considerations suggest that the reaction of a lithium enolate derived from a malonate with an enone should lead to an aluminum enolate intermediate. On the basis of the electronegativity of aluminum, protonation of the aluminum enolate should be slower than that of the corresponding lithium, sodium, and/or lanthanoid enolates. Thus, it is possible that such an aluminum enolate could be trapped by an electrophile such as an aldehyde in preference to being protonated. The three-component coupling reaction of cyclopentenone (**3**), diethyl methylmalonate (**5**), and hydrocinnamaldehyde (**20**) was realized by Shibasaki et al. using 10 mol % of ALB to afford the product **21** in 64% yield and with 91% ee (Table 3.3, entry 1) [4]. The use of LLB, LSB, and a lithium-free La–BINOL complex (entries 2–4) gave very unsatisfactory results. This is probably because the lanthanum enolate is more reactive towards the acidic proton.

A proposed reaction pathway for the three-component coupling reactions is illustrated in Scheme 3.3. The reaction of **5** with ALB gives the corresponding lithium enolate **I**, which reacts with **3** (precoordinated to aluminum) to give the aluminum enolate **II** in an enantioselective manner. Further reaction of **II** with the aldehyde leads to alkoxide **III**. Although it is unclear as to whether the aluminum

**Table 3.3** Catalytic asymmetric tandem Michael-aldol reaction.

| Entry | Catalyst | 9 (Michael product) | | 21 | |
|---|---|---|---|---|---|
| | | Yield (%) | ee (%) | Yield (%) | ee (%) |
| 1 | ALB | 7 | 90 | 64 | 91 |
| 2 | LLB | 46 | 3 | 30 | – |
| 3 | LSB | 73 | 86 | trace | – |
| 4 | La–BINOL | 57 | 83 | trace | – |

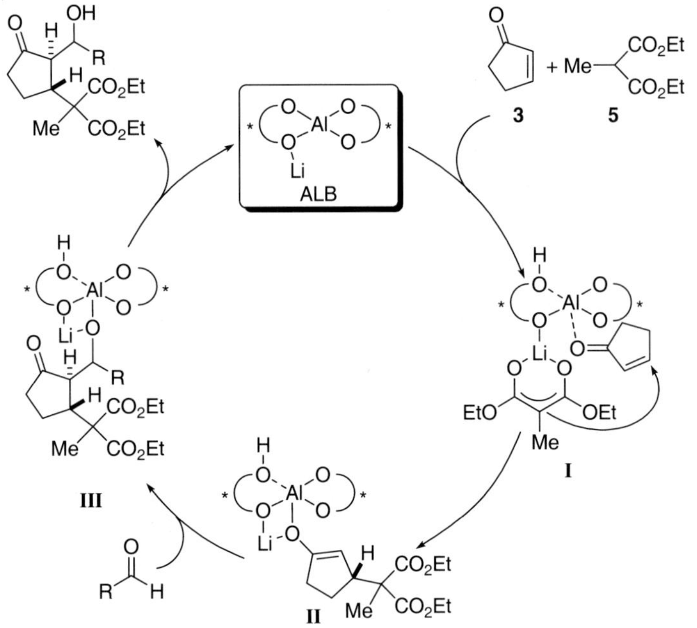

**Scheme 3.3** Proposed mechanism for the tandem Michael-aldol reaction promoted by ALB.

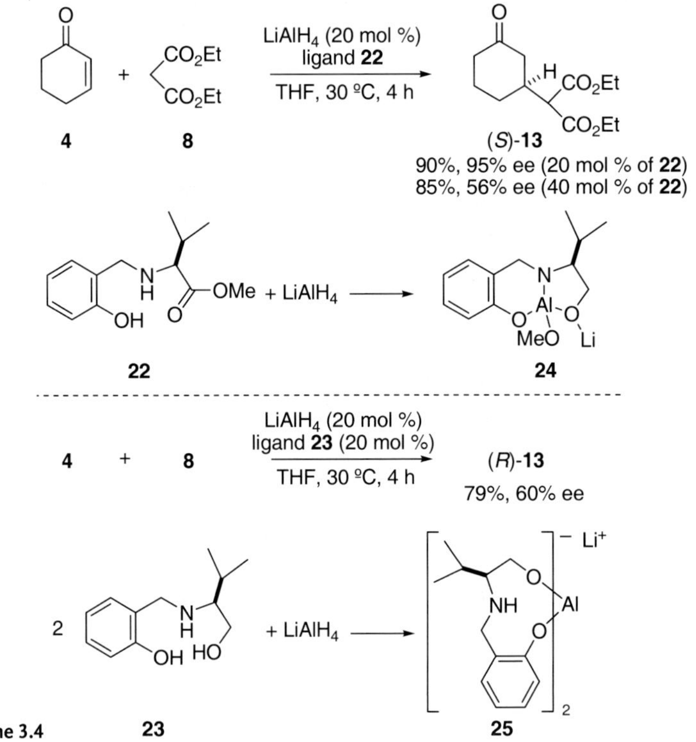

**Scheme 3.4**

or lithium alkoxide is generated, a proton from an acidic OH group may be transferred to give the three-component coupling product and regenerate ALB, thereby completing the catalytic cycle.

Recently, other types of heterobimetallic asymmetric catalysts have been reported by Narasimhan et al. [8], in which the presumably tridentate ligands **22** and **23** gave rise to interesting enantiomer-switching properties (Scheme 3.4). The heterobimetallic asymmetric catalysts prepared from LiAlH$_4$ with amino ester ligand **22** in ratios of 1 : 1 and 1 : 2 afforded the (S)-isomer of the product with 95% ee and 56% ee, respectively. On the other hand, the catalyst prepared from LiAlH$_4$ with amino alcohol ligand **23** in a ratio of 1 : 2 afforded the (R)-isomer with 60% ee. Based on measurements of the H$_2$ evolved during the catalyst preparation and $^{27}$Al NMR studies, the ALB-like $C_2$ symmetric complex **25** was proposed as being formed in the latter case, while in the former case the formation of complex **24** was proposed.

### 3.2.2
### Development of the Second-Generation Heterobimetallic Catalysts – Self-Assembly of Heterobimetallic Catalysts and Reactive Nucleophiles

Conventional bases such as NaO$^t$Bu can generate reactive nucleophiles from suitably acidic starting materials, which can then react with electrophiles to give racemic products when both of the starting materials are achiral. However, in the presence of heterobimetallic asymmetric catalysts, Shibasaki et al. found that very rapid self-assembly occurred with reactive nucleophiles, generating more efficient catalysts than the parent heterobimetallic catalysts, which were thus named "second-generation catalysts". For example, the addition of almost one molar excess of NaO$^t$Bu did not reduce the optical purity of the Michael adducts, but enhanced the reactivity of the catalyst [4]. Whereas the asymmetric Michael reaction of **6** with **4** catalyzed by 10 mol % of GaSB required 143 h at room temperature to give **11** in 45% yield with 98% ee (Table 3.4, entry 1), the reaction with 10 mol % of GaSB and 9 mol % of NaO$^t$Bu at room temperature gave **11** in 87% yield with 98% ee after just 21 h (entry 2). Addition of the sodium salt of dibenzyl malonate (9 mol %) instead of NaO$^t$Bu gave an almost identical result (quantitative yield, 96% ee, entry 5). The asymmetric Michael reaction of **6** with 2-cyclopentenone (**3**) in the presence of 10 mol % of GaSB and 9 mol % of NaO$^t$Bu also proceeded smoothly to give **10** in 96% yield with 98% ee (room temperature, 22 h, entry 7). Moreover, an efficient catalytic asymmetric synthesis of **27** was also realized for the first time (79% yield, > 99% ee, entry 9).

Ichihara et al. utilized this catalytic asymmetric Michael reaction for the enantioselective synthesis of (+)-coronatine (**35**) [9b]. As compared with their first approach starting from chiral enone **28**, the key intermediate (+)-**31** was more efficiently synthesized using the above-mentioned catalytic asymmetric Michael reaction (Scheme 3.5). In the former case (**28** to **31**), recrystallization of **30** was required to increase the optical purity, which might be decreased in a Michael reaction (chirality transfer method, 100% ee of **28** to 92% ee of **29**). In the latter case (**3** to **31**), using 10 mol % of (S)-GaSB with 7 mol % of NaO$^t$Bu, the Michael

**Table 3.4** Enhancement of catalyst efficiency using GaSB.

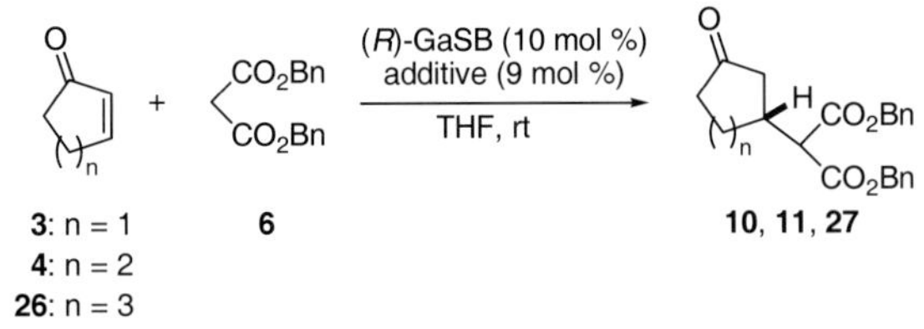

3: n = 1    6    10, 11, 27
4: n = 2
26: n = 3

| Entry | Enone | Additive | Product | Time (h) | Yield (%) | ee (%) |
|-------|-------|----------|---------|----------|-----------|--------|
| 1 | 4 | – | 11 | 143 | 45 | 98 |
| 2 | 4 | NaO-t-Bu | 11 | 21 | 87 | 98 |
| 3 | 4 | NaO-t-Bu[a] | 11 | 6 | 71 | 84 |
| 4 | 4 | NaO-t-Bu[b] | 11 | 6 | 91 | 60 |
| 5 | 4 | Na-malonate | 11 | 21 | quant. | 96 |
| 6 | 3 | – | 10 | 72 | 32 | 89 |
| 7 | 3 | NaO-t-Bu | 10 | 22 | 96 | 98 |
| 8 | 26 | – | 27 | 73 | trace | – |
| 9 | 26 | NaO-t-Bu | 27 | 73 | 79 | > 99 |

[a] 20 mol % of NaO$^t$Bu was used.
[b] 30 mol % of NaO$^t$Bu was used.

adduct **33** was obtained in 95% yield (4.6 g). The optical purity of **33** was determined to be 98% ee after two-step conversion to (+)-**31**. In this way, the preparation of optically active ester (+)-**31** was efficiently improved from 26% overall yield in six steps with 96% ee to 70% overall yield in three steps with 98% ee.

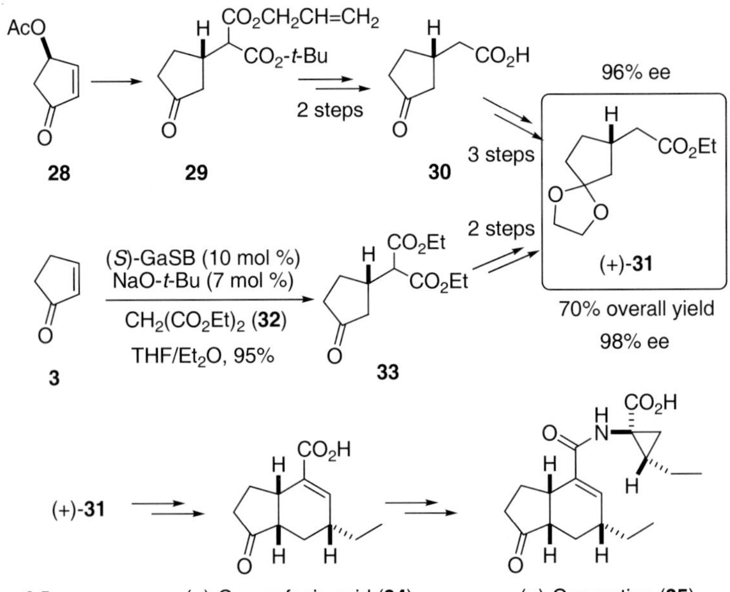

28    29    30    96% ee
2 steps    3 steps

(+)-31
70% overall yield
98% ee

(S)-GaSB (10 mol %)
NaO-t-Bu (7 mol %)
CH$_2$(CO$_2$Et)$_2$ (**32**)
THF/Et$_2$O, 95%
3    33    2 steps

**Scheme 3.5**    (+)-Coronafacic acid (**34**)    (+)-Coronatine (**35**)

**Table 3.5** Enhancement of catalyst efficiency using ALB.

| Entry | Scale (mmol) | ALB (x mol %) | KO-t-Bu | MS 4A | Time (h) | Yield (%) | ee (%) |
|---|---|---|---|---|---|---|---|
| 1 | 1 | 10 | − | − | 72 | 90 | 93 |
| 2 | 2 | 5 | + | − | 48 | 97 | 98 |
| 3 | 2 | 0.3 | + | − | 120 | 74 | 88 |
| 4 | 2 | 0.3 | + | +[a] | 120 | 94 | 99 |
| 5 | 500 | 1.0 | + | +[b] | 72 | 96 | 99 |

[a] MS 4A (8.3 g) was used for ALB (1 mmol).
[b] MS 4A (2.0 g) was used for ALB (1 mmol).

In a similar way as described above for GaSB, significant improvement in the catalytic activity of ALB was realized without any loss of enantioselectivity with the advent of "second-generation ALB", formed by self-assembly of a complex between ALB and the alkali metal malonate or alkoxide [4, 10a]. This protocol allowed the catalyst loading to be reduced from 10 mol % to 0.3 mol % [10a]. The addition of KO$^t$Bu to ALB can accelerate the catalytic asymmetric Michael reaction without lowering the high enantiomeric excess (Table 3.5, entry 2). However, 5 mol % of the catalyst is still required to obtain the product in excellent yield and with high enantiomeric excess. Addition of 4 Å molecular sieves to the reaction medium greatly improved the reaction. For example, in the presence of 4 Å molecular sieves the use of just 0.3 mol % of ALB and 0.27 mol % of KO$^t$Bu afforded the product **12** in 94% yield with 99% ee even at room temperature (entry 4). The role of the molecular sieves would seem to be the removal of a trace amount of H$_2$O that would otherwise gradually decompose the ALB–KO$^t$Bu catalyst.

A number of asymmetric catalyses have been reported, some of which have industrial applications. Most catalytic asymmetric carbon–carbon bond formations, however, are often difficult to reproduce on a manufacturing scale in terms of catalyst efficiency, enantioselectivity, or chemical yield. To address this issue, Shibasaki et al. succeeded in the development of a highly practical and efficient procedure for the large-scale synthesis of enantiomerically pure Michael product **12** using second-generation ALB under highly concentrated conditions [11]. Although a reaction can, in principle, be accelerated at higher concentration, a significant decrease in enantioselectivity is observed in many asymmetric catalyses, probably due to deactivation of the desired active species and/or the formation of an undesired species. In contrast, the second-generation ALB catalyzed asymmetric Michael reaction was successfully accelerated under highly concentrated conditions without lowering the chemical yield or the high enantiomeric excess (Table 3.6). Under almost neat conditions, 0.1 mol % of the catalyst forced the reaction to completion within 24 h (entry 4), and even with just 0.05 mol % of the catalyst, the

**Table 3.6** ALB-catalyzed Michael reaction under highly concentrated conditions.

| Entry | Scale (mmol) | ALB (x mol %) | Amount of THF (mL/mmol) | Time (h) | ee [a] (%) | Yield [b] (%) | ee [c] (%) |
|-------|--------------|---------------|-------------------------|----------|------------|---------------|------------|
| 1 | 50 | 0.25 | 0.050 | 15 | > 99 | 95 | > 99 |
| 2 | 50 | 0.20 | 0.042 | 24 | > 99 | 95 | > 99 |
| 3 | 50 | 0.10 | 0.021 | 10 | 98 | 74 | > 99 |
| 4 | 50 | 0.10 | 0.021 | 24 | 98 | 92 | > 99 |
| 5 | 50 | 0.05 | 0.010 | 48 | 98 | 94 | > 99 |
| 6 | 500 | 0.10 | 0.021 | 24 | 98 | 92 | > 99 |

[a] Enantiomeric excess of the crude product.
[b] Combined yield of **12** after two successive crystallizations.
[c] Enantiomeric excess of the crystal.

reaction was complete within 48 h (entry 5). The work-up procedure was also improved (*vide infra*), and thus the enantiomerically pure compound **12** (> 99% ee) was obtained in all instances without the need for chromatographic separation.

Moreover, pre-manufacturing scale synthesis (greater than kilogram scale) was performed using this improved procedure [11]. The detail of the reaction, including the improved work-up procedure, is shown in Figure 3.3. Using 0.1 mol % of ALB (containing only 3.4 g of BINOL) with 0.09 mol % of KO$^t$Bu and 4 Å molecular sieves, the asymmetric Michael reaction of dimethyl malonate (**7**) (6.0 mol, 686 mL) with 2-cyclohexenone (**4**) (6.0 mol, 581 mL) was complete within 24 h at ambient temperature. After a standard quenching procedure, the organic layer was concentrated to half of its original volume and treated with hexane to afford the enantiomerically pure product **12** as white crystals in 91% yield (1.24 kg) after three successive crystallizations. This method is one of the most practical and efficient catalytic asymmetric carbon–carbon bond formations showing great enantioselectivity, and its industrial application (greater than 50 kg scale) is currently under investigation.

The optically pure Michael product **12**, synthesized using the above-mentioned second-generation ALB, was successfully applied for the enantioselective synthesis of *Strychnos* alkaloids (–)-tubifolidine (**39**) [10a] and (–)-19,20-dihydroakuammicine (**40**) [10b]. These natural products were synthesized through Fischer indole synthesis and a one-pot oxidation–cyclization process promoted by DDQ (Scheme 3.6) [10c].

(–)-Strychnine (**45**) is the flagship compound of the family of *Strychnos* alkaloids and, considering its molecular weight, is one of the most complex of natural products [12]. The structural complexity of strychnine coupled with its biological activity has served as the impetus for numerous synthetic investigations. For the enantioselective synthesis of (–)-strychnine, the Michael adduct **12** was effectively utilized by

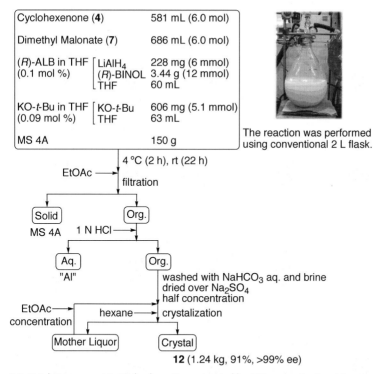

| | |
|---|---|
| Cyclohexenone (**4**) | 581 mL (6.0 mol) |
| Dimethyl Malonate (**7**) | 686 mL (6.0 mol) |

| (*R*)-ALB in THF (0.1 mol %) | LiAlH₄ | 228 mg (6 mmol) |
|---|---|---|
| | (*R*)-BINOL | 3.44 g (12 mmol) |
| | THF | 60 mL |

| KO-*t*-Bu in THF (0.09 mol %) | KO-*t*-Bu | 606 mg (5.1 mmol) |
|---|---|---|
| | THF | 63 mL |

| MS 4A | 150 g |
|---|---|

The reaction was performed using conventional 2 L flask.

```
                    | 4 °C (2 h), rt (22 h)
         EtOAc ———→  |
                     filtration
          ┌──────────┴──────────┐
       ┌─────┐              ┌─────┐
       │Solid│              │ Org.│
       └─────┘              └─────┘
       MS 4A    1 N HCl ———→  │
                ┌─────────────┴──┐
             ┌─────┐         ┌─────┐
             │ Aq. │         │ Org.│
             └─────┘         └─────┘
              "Al"
```

washed with NaHCO₃ aq. and brine
dried over Na₂SO₄
half concentration

```
   EtOAc ———→
  concentration   hexane ———→  crystalization
       ┌──────────────┐    ┌───────┐
       │Mother Liquor │    │Crystal│
       └──────────────┘    └───────┘
```

**12** (1.24 kg, 91%, >99% ee)

**Figure 3.3** Catalytic asymmetric Michael reaction promoted by ALB on greater than kilogram scale.

(*R*)-**12**

**36**

**37**

DDQ
52%
(conv. 67%)

**38**

(−)-tubifolidine (**39**)

(−)-19,20-dihydroakuammicine (**40**)

**Scheme 3.6** Catalytic asymmetric syntheses of (−)-tubifolidine (**39**) and (−)-19,20-dihydroakuammicine (**40**).

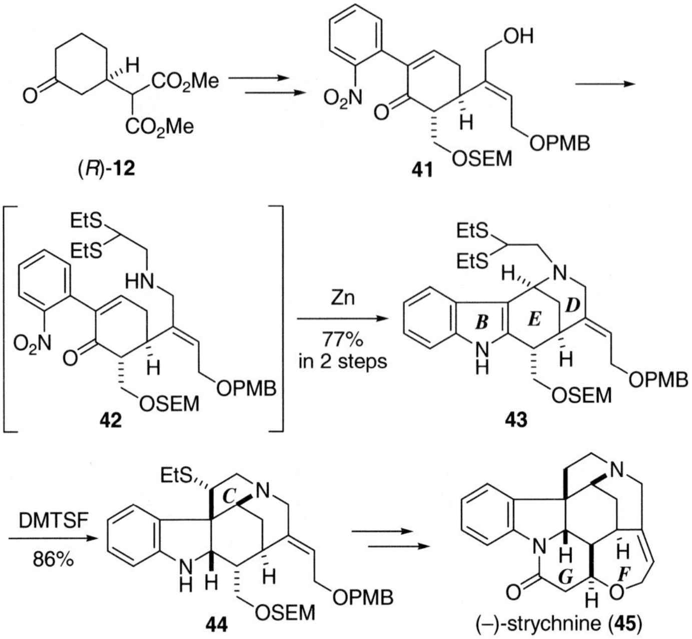

**Scheme 3.7** Catalytic asymmetric synthesis of (–)-strychnine (**45**).

Shibasaki et al. [13]. The previous method for the synthesis of **39** and **40** was not applicable to the enantioselective synthesis of strychnine due to difficulties in the construction of the FG-rings and Z-selective elaboration of the hydroxyethylidene substituent. Thus, a new ring-construction process, including a novel domino cyclization promoted by Zn (simultaneous construction of B- and D-rings) and DMTSF-promoted thionium ion cyclization (construction of C-ring), was employed for the assembly of the bridged framework (BCDE-ring system) of *Strychnos* alkaloids (Scheme 3.7).

Regioselective and enantioselective Michael reaction of the Horner–Wadsworth–Emmons reagent with enones was achieved by means of the self-assembled catalyst composed of ALB and a sodium base [14]. Depending on the structure of the starting compounds, Horner–Wadsworth–Emmons reagents generally react with α,β-unsaturated carbonyl compounds to give either 1,2-adducts (Horner–Wadsworth–Emmons products) or 1,4-adducts (Michael products). Although the attempted reaction of **46** with **4** catalyzed by ALB alone did not afford the product **47** and that catalyzed by NaO$^t$Bu alone gave only the 1,2-adduct **48** in 8–9% yield, the use of a combination of ALB (10 mol %) and NaO$^t$Bu (0.9 mol equivalent with respect to ALB) gave the Michael adduct **47** in 64% yield and with 99% ee without any formation of 1,2-adduct **48** (Scheme 3.8). The Michael adduct **50** was successfully transformed into a key intermediate in the synthesis of (+)-coronafacic acid (**34**) [9].

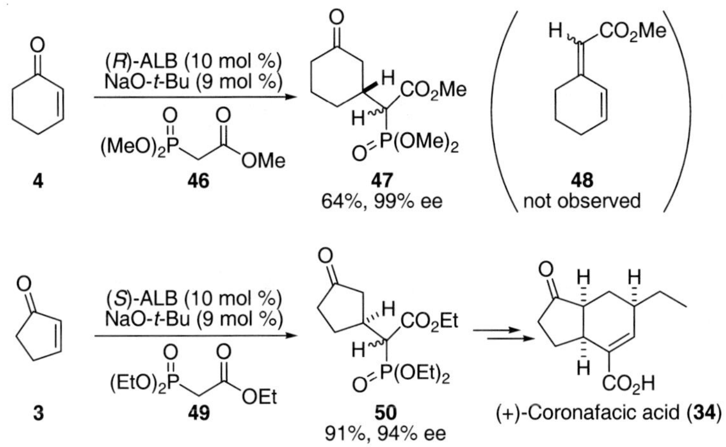

**Scheme 3.8**

The self-assembled catalyst composed of ALB and a sodium base has also been successfully applied to the above-mentioned three-component coupling reaction of an enone, an aldehyde, and a malonate [15]. A catalytic asymmetric synthesis of 11-deoxy-PGF$_{1\alpha}$ (**56**) was achieved by utilizing this reaction as the first key step (Scheme 3.9). The actual structure of the activated catalyst may be the self-assembled complex **53**, highly facially selective Michael addition at which would produce the

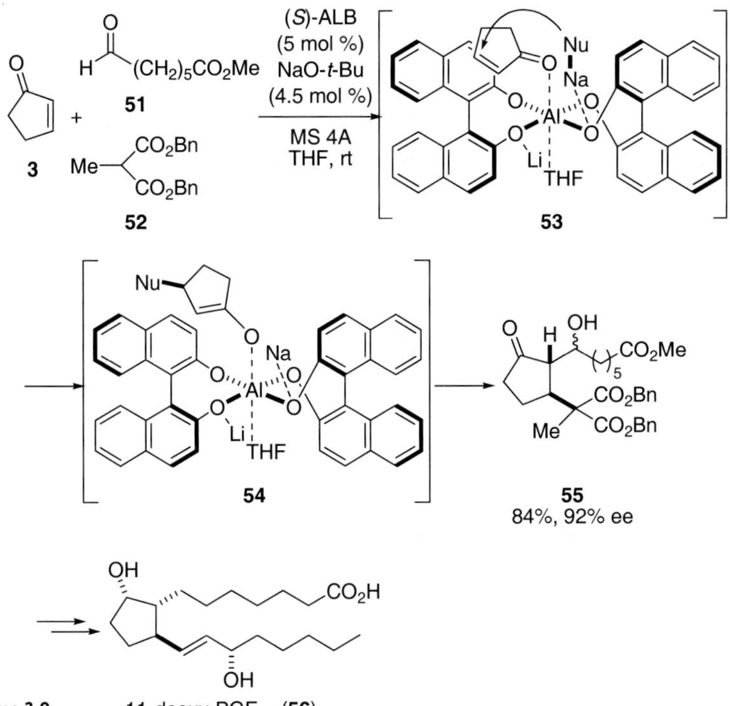

**Scheme 3.9**  11-deoxy-PGF$_{1\alpha}$ (**56**)

intermediary aluminum enolate **54**, which further reacts with the aldehyde prior to the protonation.

Immobilized catalysts have several advantages over homogeneous catalysts, such as ease of separation and reusability. The preparation of immobilized catalysts has traditionally involved the attachment of chiral ligands to a sterically disordered polymer backbone such as polystyrene. In such cases, the chiral ligands are randomly oriented along the polymer chain, making it difficult to systematically modify the heterobimetallic catalysts on the polymer. To address this issue, several methods for the immobilization of ALB-type heterobimetallic asymmetric catalysts were described by Sasai et al., such as the use of a sterically regular polymeric BINOL [16a], a soluble polymer-supported BINOL [16b], and a BINOL-containing dendrimer [16c]. Recently, these authors also reported a new approach for constructing immobilized multicomponent asymmetric catalysts based on the use of a catalyst analogue in the copolymerization [16d]. The resulting polymer **57** was found to promote the Michael reaction of **6** with **4** in 73% yield and with 91% ee (Scheme 3.10).

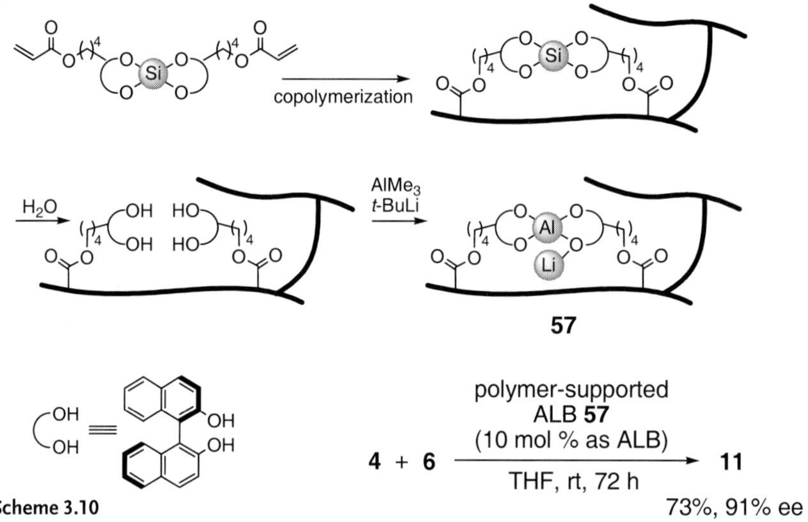

**Scheme 3.10**

## 3.3
## Catalytic Asymmetric Ring-Opening Reaction of *meso*-Epoxides

### 3.3.1
### Ring-Opening Reaction with Thiols

The asymmetric ring-opening reaction of *meso*-epoxides is an attractive and quite powerful method in asymmetric synthesis [17], since it simultaneously constructs two contiguous stereogenic centers. Although various types of stoichiometric or catalytic asymmetric epoxide ring-opening reactions have been reported, only a few practical methods have been reported to date, and these generally require the use of silylated compounds as nucleophiles [18]. Thus, a reaction using non-silylated nuclephiles such as RSH, ROH, HCN, or $HN_3$ is highly desirable. An important breakthrough in this field was achieved by Shibasaki et al. [19] with the development of the catalytic asymmetric ring-opening reaction of *meso*-epoxides with $^t$BuSH catalyzed by the heterobimetallic asymmetric catalyst GaLB. Preliminary studies using $PhCH_2SH$ and cyclohexene oxide (58) as the nucleophile revealed that LaMB (M = Li or Na) or ALB showed only low catalyst activity, affording the product of epoxide opening in yields of just 1–10%, whereas GaSB showed high catalytic activity (87% yield) in this reaction with moderate enantioselectivity (40% ee). Interestingly, the enantiomeric excess of the product gradually increased as the reaction proceeded, presumably due to the concomitant decrease in the concentration of remaining thiol. Moreover, the use of a stoichiometric amount of GaLB afforded the product in 87% yield and with 88% ee, indicating the occurrence of an undesired ligand exchange of the BINOL by the thiol. To prevent this undesired side reaction, more sterically hindered thiols were examined, and finally $^t$BuSH (65) was found to be a superior nucleophile for the asymmetric ring-opening reaction of 58. Thus, with 65 the product 66 was obtained with 98% ee even at room temperature, albeit only in 35% yield (Table 3.7, entry 1). The efficiency of this reaction was found to be

**Table 3.7** Catalytic asymmetric ring-opening reaction of *meso*-epoxides with $^t$BuSH.

| Entry | Epoxide | | Product | MS 4A[a] (g) | Time (h) | Yield (%) | ee (%) |
|---|---|---|---|---|---|---|---|
| 1 | | 58 | 66 | – | 65 | 35 | 98 |
| 2 | | 58 | 66 | 0.2 | 9 | 80 | 97 |
| 3 | | 59 | 67 | 0.2 | 36 | 74 | 95 |
| 4 | TBDPSO TBDPSO | 60 | 68 | 0.2 | 12 | 83 | 96 |
| 5 | TBDPSO TBDPSO | 61 | 69 | 0.2 | 137 | 64 | 91 |
| 6 | | 62 | 70 | 0.2 | 24 | 89 | 91 |
| 7[b] | SO₂N | 63 | 71 | 0.2 | 72 | 89 | 89 |
| 8 | TrO TrO | 64 | 72 | 2.0 | 48 | 89 | 82 |

[a] Weight per 1 mmol of epoxide.
[b] Carried out at 50 °C in the presence of 30 mol % of GaLB.

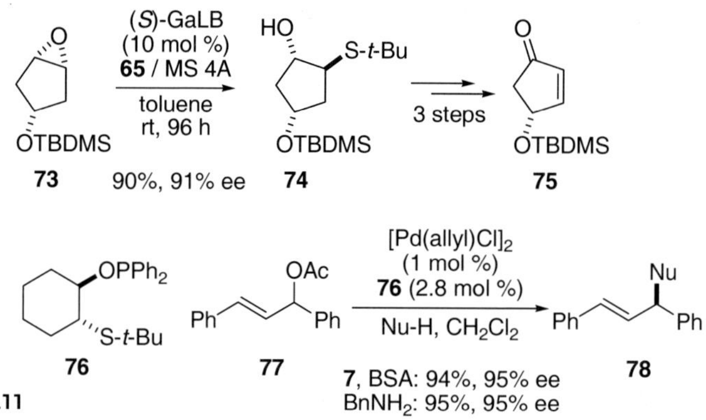

**Scheme 3.11**

greatly improved by the addition of 4 Å molecular sieves, the use of which furnished **66** in 80% yield without a decrease in enantiomeric excess (entry 2). Results illustrating the scope and limitations of this method are summarized in Table 3.7. This methodology was applied to the synthesis of **75**, an important intermediate in the synthesis of prostaglandins (Scheme 3.11) [19]. Furthermore, the almost optically pure product of epoxide opening **66** was utilized for the preparation of the attractive chiral ligand **76** by Evans et al., which, in turn, was utilized in palladium-catalyzed allylic alkylation and amination reactions, affording products with up to 95% ee.

In the present reaction, GaLB again appears to act as an asymmetric multi-functional catalyst, with a lithium binaphthoxide moiety functioning as a Brønsted base in activating the thiol **65** and controlling the orientation of the resulting lithium thiolate by chelation (Scheme 3.12). The gallium center appears to function as a

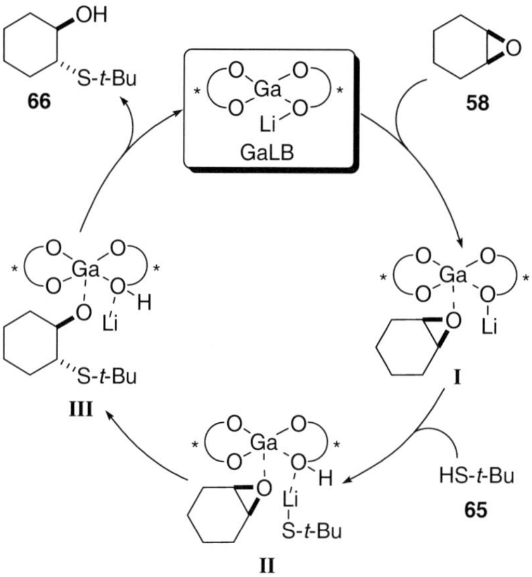

**Scheme 3.12** Proposed mechanism for the ring-opening reaction promoted by GaLB.

Lewis acid, activating and also controlling the orientation of epoxide **58**, presumably through the coordination of an axial lone pair, facilitating cleavage of a C–O bond by backside attack (**II**).

### 3.3.2
### Ring-Opening Reaction with Phenolic Oxygen – Development of a Novel Linked-BINOL Complex

Following the great success of the asymmetric ring-opening reaction of *meso*-epoxides with $^t$BuSH, oxygen nucleophiles such as alcohols, phenols, and carboxylic acids were investigated, because these provide valuable chiral building blocks such as 1,2-diol derivatives. Considering the Brønsted basicity of the alkali metal binaphthoxide moiety in the heterobimetallic complex ($pK_a$ value of BINOL ≈ 10), phenolic oxygen nucleophiles ($pK_a$ value of 4-methoxyphenol = 10.2) might also be activated to react with epoxides in a similar catalytic cycle as $^t$BuSH ($pK_a$ value of $^t$BuSH = 10.6). Indeed, the reaction of **58** with 4-methoxyphenol (**80**) in the presence of 20 mol % of GaLB and 4 Å molecular sieves (toluene, 50 °C) afforded 1,2-diol monoether **81** in 48% yield and with 93% ee (Table 3.8, entry 1) [20a]. These conditions were applicable to several unfunctionalized and functionalized *meso*-epoxides, affording products with good to high enantiomeric excesses (67–93% ee), although chemical yields were only modest (31–75%) even when using 20 mol % of the catalyst. These results can be attributed to an undesired ligand exchange between BINOL and 4-methoxyphenol (**80**), resulting in decomplexation of the

**Table 3.8** Catalytic asymmetric ring-opening reaction of *meso*-epoxides with 4-methoxyphenol (**80**).

| Entry | Epoxide | | Product | Time (h) | Yield (%) | ee (%) |
|-------|---------|---|---------|----------|-----------|--------|
| 1 | | 58 | 81 | 72 | 48 | 93 |
| 2 | | 62 | 82 | 72 | 75 | 86 |
| 3 | | 79 | 83 | 72 | 31 | 67 |
| 4 | | 59 | 84 | 72 | 74 | 87 |
| 5 | TBDPSO / TBDPSO | 60 | 85 | 96 | 34 | 80 |
| 6[a] | Mts–N | 63 | 86 | 160 | 51 | 90 |

[a] 30 mol % of GaLB was used.

GaLB and the formation of unavoidable side products derived from ring-opening of the epoxide with BINOL. Thus, the preparation of a more stable gallium heterobimetallic complex was necessary to overcome this problem. To address this issue, various modified BINOLs were examined as chiral ligands, and of these, GaLB*, which was prepared from 6,6'-bis((triethylsilyl)ethynyl)binaphthol, slightly improved the yield of the ring-opening product **81** (60% yield). The GaLB* complex, however, was not sufficiently stable to suppress the undesired ligand exchange completely, so that the use of as much as 20 mol % of the catalyst was needed to obtain the products in acceptable yields. Although another gallium complex, GaSO, prepared from 5,5',6,6',7,7',8,8'-octahydro-BINOL ($H_8$-BINOL) showed much higher catalytic activities (73% yield in only 4 h at 50 °C), the enantioselectivity of **81** was only modest (56% ee).

The next breakthrough in this field came with the development of novel linked-BINOL ligands. By linking two BINOL units in GaLB, the complex would be rendered more stable against ligand exchange without any adverse effects on the asymmetric environment. As a first trial, carbon-linked-BINOLs (Figure 3.4, **87–89**) were designed and the corresponding Ga–Li carbon-linked BINOL complexes were prepared [21a]. However, none of these complexes proved effective in the asymmetric epoxide-opening reaction of **58** with **80** (**87**, 28% yield, 27% ee; **88**, 43% yield, 10% ee; **89**, 40% yield, –1% ee). These unsatisfactory results might be attributed to the undesired oligomeric structures of these Ga–Li–carbon-linked-BINOL complexes. To overcome this problem, a novel oxygen-linked-BINOL **90** was designed by considering related work by Cram et al. regarding crown ethers incorporating chiral BINOL units [20b]. Ga–Li–linked-BINOL complex **91**, which was prepared from $GaCl_3$, 1 mol equivalent of **90**, and 4 mol equivalents of BuLi, was found to be far more stable than GaLB and very effective for the ring-opening reaction (Table 3.9). Unlike GaLB, complex **91** was found to be stable even in the presence of excess 4-methoxyphenol (**80**) and/or at higher temperatures.

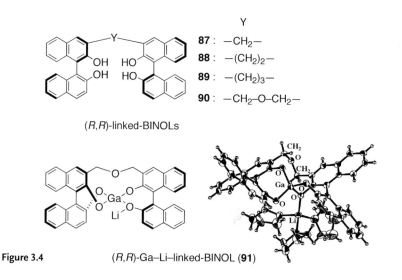

|  | Y |
|---|---|
| **87** : | $-CH_2-$ |
| **88** : | $-(CH_2)_2-$ |
| **89** : | $-(CH_2)_3-$ |
| **90** : | $-CH_2-O-CH_2-$ |

(*R,R*)-linked-BINOLs

**Figure 3.4**        (*R,R*)-Ga–Li–linked-BINOL (**91**)

**Table 3.9** Catalytic asymmetric ring-opening reaction of meso-epoxides promoted by Ga–Li–linked-BINOL complex (91).

| Entry | Epoxide | | Product | Time (h) | Yield (%) | ee (%) |
|-------|---------|---|---------|----------|-----------|--------|
| 1[a] | | 58 | 81 | 96 | 72 | 91 |
| 2 | | 62 | 82 | 63 | 88 | 85 |
| 3[a] | | 79 | 83 | 108 | 82 | 66 |
| 4[a] | | 59 | 84 | 36 | 94 | 85 |
| 5 | TBDPSO / TBDPSO | 60 | 85 | 96 | 72 | 79 |
| 6[b] | Mts–N | 63 | 86 | 160 | 77 | 78 |
| 7 | TBSO | 92 | 95 | 48 | 67 | 87 |
| 8[c] | OMe / OMe | 93 | 96 | 70 | 85 | 96 |
| 9 | Me / Me | 94 | 97 | 140 | 72 | 91 |

[a] The reaction was performed at 75 °C.
[b] 30 mol % of GaLB was used.
[c] 2.0 equiv. of **80** was used.

Based on $^{13}$C NMR as well as FAB and (–)-LDI TOF mass spectral data, the structure of Ga–Li–linked-BINOL catalyst **91** was suggested to the assumed monomeric one. Although all attempts to obtain X-ray quality crystals of the complex **91** prepared from GaCl$_3$ were unsuccessful, an X-ray quality crystal of LiCl-free Ga–Li–linked-BINOL complex revealed a monomeric tetracoordinated structure, similar to the structure of the ALB · (thf)$_3$ complex (Figure 3.4). The LiCl-free Ga–Li–linked-BINOL complex was prepared from Ga(O$^i$Pr)$_3$, 1 mol equivalent of linked-BINOL **90**, and 1 mol equivalent of BuLi. The sample obtained by treating this crystal with 3 mol equivalents of LiCl in THF was also found to catalyze the ring-opening reaction of epoxide **58** to give **81** with 90% ee. Further application of linked-BINOLs, including related S-linked-BINOLs [22] and NR-linked-BINOLs [10c], to other metals led to the development of highly efficient multifunctional catalysts such as La/linked-BINOL complex (see Chapter 5) and Et$_2$Zn/linked-BINOL complex (see Chapter 2).

## 3.4
## Catalytic Asymmetric Mannich Reactions

### 3.4.1
### Direct Catalytic Asymmetric Mannich-Type Reaction of Unmodified Ketones

The asymmetric Mannich reaction is one of the most powerful C–C bond-forming reactions in organic chemistry [23]. It provides optically active β-amino ketones and esters (Mannich bases), which are versatile building blocks for the synthesis of numerous pharmaceuticals and natural products. Although various methods have been developed for the enantioselective synthesis of Mannich bases, actual asymmetric Mannich-type reactions using achiral substrates are quite scarce. The first successful catalytic asymmetric additions of enolates to imines have only recently been reported by the groups of Kobayashi, Sodeoka, Lectka, and Jacobsen [23b, 24]. All of these Mannich reactions, however, required pre-conversion of the ketone or the ester moiety to a more reactive species such as a metal enolate or a ketene silyl acetal. The most effective and atom-economic asymmetric Mannich reaction would be a direct reaction of unmodified ketones. Shibasaki et al. [25] disclosed the first report of a direct catalytic asymmetric Mannich reaction using a heterobimetallic asymmetric catalyst. For the one-pot, three-component coupling reaction of propiophenone (98), $(CH_2O)_n$, and pyrrolidine, in which $H_2O$ is unavoidably generated as a side product, LLB was utilized and afforded the corresponding Mannich product with 64% ee. The yield, however, was only 16%, probably due to decomposition of LLB by the excess $H_2O$ generated, despite the presence of 3 Å molecular sieves, and competing formation of $C_4H_8NCH_2C_4H_8$. Significant improvement of this direct Mannich reaction was achieved by in situ generation of iminium ions $[R_2N=CH_2]^+$ from aminomethyl ethers such as 103. Among the various catalyst systems examined, a combination of ALB and La(OTf)$_3$ · n H$_2$O (1 : 1) in the presence of 3 Å molecular sieves was found to catalyze direct asymmetric Mannich-type reactions, providing β-amino aryl ketones in good yields (65–76%) and with moderate selectivities (31–44% ee) (Table 3.10).

**Table 3.10** Direct catalytic asymmetric Mannich-type reaction.

| Entry | Ar | R$^1$ | Substrate | Product | Yield (%) | ee (%) |
|-------|-----|-------|-----------|---------|-----------|--------|
| 1 | Ph | Me | 98 | 104 | 65 | 40 |
| 2[a] | Ph | Et | 99 | 105 | 69 | 34 |
| 3 | 4-MeO-C$_6$H$_4$ | Me | 100 | 106 | 76 | 31 |
| 4 | 2-naphthyl | Me | 101 | 107 | 61 | 44 |
| 5 | 6-MeO-2-naphthyl | Me | 102 | 108 | 69 | 44 |

[a] 103 was added over 36 h.

3.4.2
**Enantio- and Diastereoselective Catalytic Nitro-Mannich Reactions**

The nitro-Mannich reaction gives rise to β-nitroamines, and the nitro group can be easily converted into the amino function, thereby providing the useful vicinal diamines. The first example of an enantioselective nitro-Mannich reaction was reported by Shibasaki et al. [26], in which $YbKH_2(binaphthoxide)_3$ (109) was utilized as an efficient catalyst. Although the catalyst 109 gave the product with up to 91% ee, the reaction suffered from the limitation that only nitromethane could be used as the nucleophile; for example, catalyst 109 did not promote the reaction of imine 110 with nitroethane (114) at all. In contrast, the second-generation ALB generated from ALB and $KO^tBu$ promoted the reaction of 110 with 114 to afford the product 118 in 77% yield with a *dr* of 6 : 1 and with 83% ee [27a]. Results illustrating the scope and limitations of this reaction are summarized in Table 3.11.

**Table 3.11** Diastereoselective catalytic nitro-Mannich reactions.

110: Ar = Ph
111: Ar = 4-MeOC$_6$H$_4$
112: Ar = 4-MeC$_6$H$_4$
113: Ar = 4-ClC$_6$H$_4$

114: R = Me
115: R = Et
116: R = (CH$_2$)$_2$OBn
117: R = (CH$_2$)$_3$OBn

118-124

| Entry | Imine | Nitroalkane | Product | x mol % | Yield (%) | dr (anti/syn) | ee (anti, %) |
|-------|-------|-------------|---------|---------|-----------|---------------|--------------|
| 1 | 110 | 114 | 118 | 20 | 77 | 6 : 1 | 83 |
| 2 | 110 | 115 | 119 | 20 | 98 | 6 : 1 | 74 |
| 3 | 110 | 116 | 120 | 20 | 95 | 7 : 1 | 82 |
| 4 | 110 | 117 | 121 | 20 | 75 | 6 : 1 | 77 |
| 5 | 111 | 115 | 122 | 20 | 77 | 6 : 1 | 78 |
| 6 | 112 | 115 | 123 | 20 | 68 | 6 : 1 | 77 |
| 7 | 113 | 115 | 124 | 20 | 89 | 3 : 1 | 71 |
| 8 | 110 | 114 | 118 | 10 | 97 | 5 : 1 | 75 |
| 9 | 110 | 115 | 119 | 10 | 87 | 5 : 1 | 71 |
| 10 | 110 | 116 | 120 | 10 | 87 | 5 : 1 | 60 |
| 11 | 110 | 117 | 121 | 10 | 93 | 5 : 1 | 63 |
| 12 | 111 | 115 | 122 | 10 | 96 | 5 : 1 | 63 |
| 13 | 112 | 115 | 123 | 10 | 98 | 5 : 1 | 81 |
| 14 | 113 | 115 | 124 | 10 | 97 | 3 : 1 | 74 |

This enantio- and diastereoselective catalytic nitro-Mannich reaction was successfully applied in the enantioselective synthesis of CP-99994 (129, competent antagonist of substance P) (Scheme 3.13) [27b].

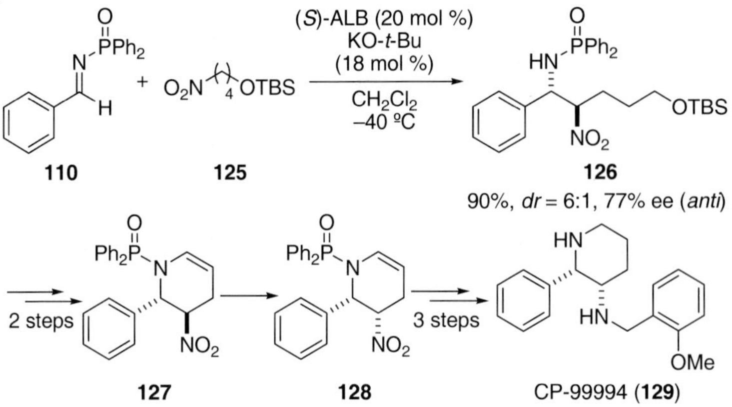

Scheme 3.13

## 3.5
## Catalytic Asymmetric Hydrophosphonylation and Hydrophosphinylation of Aldehydes

### 3.5.1
### Catalytic Asymmetric Hydrophosphonylation

α-Hydroxy phosphonates and phosphonic acids are interesting compounds in the design of enzyme inhibitors as some of them are known to inhibit enzymes such as rennin, EPSP synthase, and HIV protease [28]. Although several methods for the synthesis of optically active α-hydroxy phosphonates and phosphonic acids have been reported, catalytic asymmetric hydrophosphonylation of aldehydes would be the most desirable method in terms of atom economy and diversity of the products. Shibuya et al. [29a] and Spilling et al. [29b] independently reported the first example of a catalytic asymmetric hydrophosphonylation of aldehydes using an LLB complex. However, there was still room for improvement, particularly with regard to the enantioselectivity and substrate generality. Later, Shibasaki et al. reported the first example of an efficient catalytic asymmetric hydrophosphonylation of imines promoted by LPB [29c]. Then, the catalytic asymmetric hydrophosphonylation of aldehydes was greatly improved by Shibasaki et al. [30] using ALB in toluene. As shown in Table 3.12, several aromatic aldehydes as well as α,β-unsaturated aldehydes were transformed to the corresponding α-hydroxy phosphonates in up to 95% yield and with up to 90 % ee (entry 3).

### 3.5.2
### Catalytic Asymmetric Hydrophosphinylation

ALB was further applied to the catalytic asymmetric hydrophosphinylation of aldehydes by Shibuya et al. [31a]. This was the first example of a catalytic asymmetric synthesis of α-hydroxy-H-phosphinates that allowed conversion to chiral α-hydroxy-phosphinate derivatives through reaction with an electrophile such as an aldehyde

**Table 3.12** Catalytic asymmetric hydrophosphonylation of aldehydes.

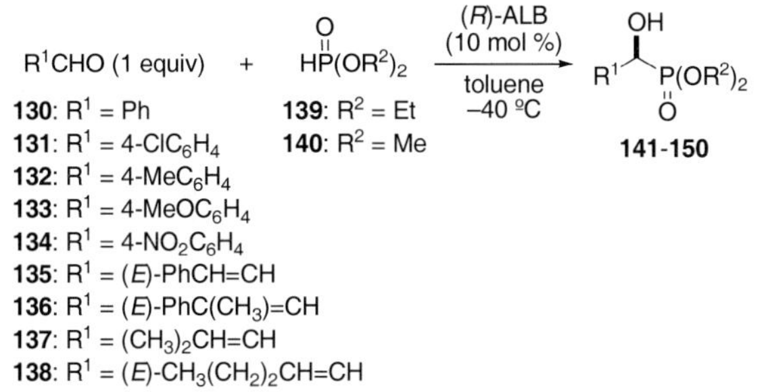

R$^1$CHO (1 equiv)  +  HP(OR$^2$)$_2$  $\xrightarrow[\substack{\text{toluene} \\ -40\ ^\circ\text{C}}]{\substack{(R)\text{-ALB} \\ (10\ \text{mol}\ \%)}}$  R$^1$$\overset{\text{OH}}{\underset{\underset{\text{O}}{\overset{\|}{\text{P}}}(\text{OR}^2)_2}{|}}$

130: R$^1$ = Ph                139: R$^2$ = Et
131: R$^1$ = 4-ClC$_6$H$_4$        140: R$^2$ = Me              141-150
132: R$^1$ = 4-MeC$_6$H$_4$
133: R$^1$ = 4-MeOC$_6$H$_4$
134: R$^1$ = 4-NO$_2$C$_6$H$_4$
135: R$^1$ = (E)-PhCH=CH
136: R$^1$ = (E)-PhC(CH$_3$)=CH
137: R$^1$ = (CH$_3$)$_2$CH=CH
138: R$^1$ = (E)-CH$_3$(CH$_2$)$_2$CH=CH

| Entry | Aldehyde | Phosphite | Product | Time (h) | Yield (%) | ee (%) |
|-------|----------|-----------|---------|----------|-----------|--------|
| 1 | 130 | 139 | 141 | 90 | 39 | 73 |
| 2 | 130 | 140 | 142 | 51 | 90 | 85 |
| 3[a] | 130 | 140 | 142 | 90 | 95 | 90 |
| 4 | 131 | 140 | 143 | 38 | 80 | 83 |
| 5 | 132 | 140 | 144 | 92 | 82 | 86 |
| 6 | 133 | 140 | 145 | 115 | 88 | 78 |
| 7 | 134 | 140 | 146 | 66 | 85 | 71 |
| 8 | 135 | 140 | 147 | 83 | 85 | 82 |
| 9 | 136 | 140 | 148 | 83 | 93 | 89 |
| 10 | 137 | 140 | 149 | 94 | 72 | 68 |
| 11 | 138 | 140 | 150 | 39 | 53 | 55 |

[a] 1.2 equiv. of **130** and 9 mol % of ALB were used.

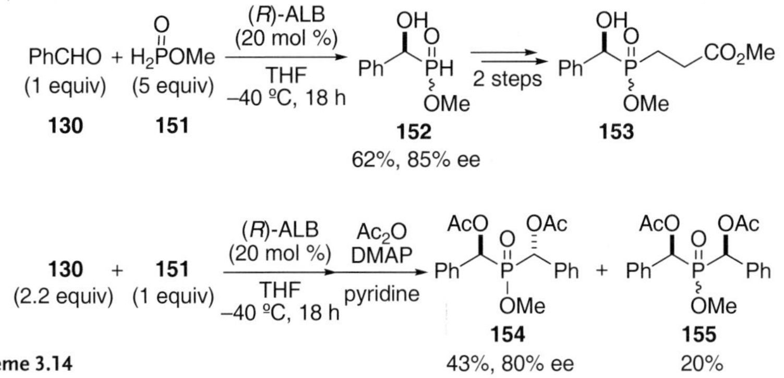

PhCHO + H$_2$POMe  $\xrightarrow[\substack{\text{THF} \\ -40\ ^\circ\text{C},\ 18\ \text{h}}]{\substack{(R)\text{-ALB} \\ (20\ \text{mol}\ \%)}}$  Ph$\overset{\text{OH}}{\underset{\underset{\text{OMe}}{\overset{\|}{\text{PH}}}}{|}}$  $\xrightarrow{\text{2 steps}}$  Ph$\overset{\text{OH}}{\underset{\underset{\text{OMe}}{\overset{\|}{\text{P}}}}{|}}$CO$_2$Me

(1 equiv)  (5 equiv)

**130**      **151**                            **152**                      **153**
                                          62%, 85% ee

**130**  +  **151**  $\xrightarrow[\substack{\text{THF} \\ -40\ ^\circ\text{C},\ 18\ \text{h}}]{\substack{(R)\text{-ALB} \quad \text{Ac}_2\text{O} \\ (20\ \text{mol}\ \%) \quad \text{DMAP} \\ \text{pyridine}}}$  Ph$\overset{\text{AcO} \quad \text{OAc}}{\underset{\underset{\text{OMe}}{\overset{\|}{\text{P}}}}{|}}$Ph  +  Ph$\overset{\text{AcO} \quad \text{OAc}}{\underset{\underset{\text{OMe}}{\overset{\|}{\text{P}}}}{|}}$Ph

(2.2 equiv)  (1 equiv)

**Scheme 3.14**                                     **154**                      **155**
                                          43%, 80% ee              20%

or acrylate (Scheme 3.14). Moreover, optically active α,α'-dihydroxyphosphinates were directly synthesized by the reaction of methyl phosphinate with excess amounts of aldehydes through double-hydrophosphinylation, albeit with low diastereoselectivities.

Yokomatsu and Shibuya et al. [31b, c] also achieved the diastereoselective hydrophosphinylation of N,N-dibenzyl-α-amino aldehydes to afford the corresponding β-amino-α-hydroxyphosphinates. When ethyl phosphinate was employed in

**Table 3.13** Diastereoselective hydrophosphinylation.

| Entry | Aldehyde | Product | Catalyst | Yield (%) | syn/anti |
|-------|----------|---------|----------|-----------|----------|
| 1 | 156 | 159 | (R)-ALB | 66 | 84 : 13 |
| 2 | 156 | 159 | (S)-ALB | 56 | 6 : 94 |
| 3 | 157 | 160 | (R)-ALB | 54 | 94 : 6 |
| 4 | 157 | 160 | (S)-ALB | 71 | 2 : 98 |

the hydrophosphinylation, both *syn*- and *anti*-β-amino-α-hydroxy-*H*-phosphinates could be selectively synthesized by tuning the chirality of the ALB, indicating that the reactions proceeded in a highly catalyst-controlled manner (Table 3.13).

## 3.6
## Conclusion

As reviewed in this chapter, group 13–alkali metal heterobimetallic asymmetric catalysts, such as ALB, GaLB, and GaSB, are able to promote a variety of asymmetric reactions in a highly enantioselective manner (up to > 99% ee) with high substrate generality. In some cases, as little as 0.1 mol % of the catalyst is sufficient for completion of the reaction in high yield and high ee (ALB-catalyzed Michael reactions of malonate with enones, for example), although in many cases more than 10 mol % of the catalyst is still required. The group 13–alkali metal hetero-bimetallic asymmetric catalysts can be prepared from commercially available and relatively inexpensive reagents, and the preparation methods are very simple. For example, ALB is prepared by simply adding a solution of BINOL (2 mol equivalent to Al) in THF to LiAlH$_4$ at 4 °C (ice bath). Therefore, the asymmetric reactions catalyzed by these catalysts are highly desirable for both enantioselective syntheses of complex molecules and for industrial-scale syntheses. The notable abilities of these group 13–alkali metal heterobimetallic asymmetric catalysts can be ascribed to their multifunctionality (Lewis acidity and Brønsted basicity, with control of the orientation of the substrates). The concept of multifunctionality, as realized in the rare earth–alkali metal and the group 13–alkali metal heterobimetallic asymmetric catalysts, has opened up a new field in asymmetric catalysis, and further progress along these lines can be expected to be highly fruitful.

# References

1   For recent reviews of catalytic
    asymmetric Michael reactions, see:
    (a) M. KANAI, M. SHIBASAKI, in *Catalytic
    Asymmetric Synthesis* (Ed.: I. OJIMA), 2nd
    ed., Wiley, New York, 2000, p. 569–592;
    (b) K. TOMIOKA, Y. NAGAOKA, in
    *Comprehensive Asymmetric Catalysis*
    (Eds.: E. N. JACOBSEN, A. PFALTZ,
    H. YAMAMOTO), Springer, Berlin, 1999,
    Vol. 3, Chapter 31.1; (c) M. SIBI,
    S. MANYEM, *Tetrahedron* 2000, 56,
    8033–8061; (d) N. KRAUSE,
    A. HOFFMANN-RÖDER, *Synthesis* 2001,
    171–196; (e) A. ALEXAKIS, C. BENHAIM,
    *Eur. J. Org. Chem.* 2002, 3221–3226;
    (f) J. CHRISTOFFERS, A. BARO, *Angew.
    Chem. Int. Ed.* 2003, 42, 1688–1690.

2   H. SASAI, T. ARAI, Y. SATOW, K. N. HOUK,
    M. SHIBASAKI, *J. Am. Chem. Soc.* 1995,
    117, 6194–6198.

3   T. ARAI, H. SASAI, K. AOE, K. OKAMURA,
    T. DATE, M. SHIBASAKI, *Angew. Chem. Int.
    Ed. Engl.* 1996, 35, 104–106.

4   T. ARAI, Y. M. A. YAMADA, N. YAMAMOTO,
    H. SASAI, M. SHIBASAKI, *Chem. Eur. J.*
    1996, 2, 1368–1372.

5   E. KELLER, N. VELDMAN, A. L. SPEK,
    B. L. FERINGA, *Tetrahedron: Asymmetry*
    1997, 8, 3403–3413.

6   E. EMORI, T. ARAI, H. SASAI, M. SHIBASAKI,
    *J. Am. Chem. Soc.* 1998, 120, 4043–4044.

7   E. EMORI, T. IIDA, M. SHIBASAKI,
    *J. Org. Chem.* 1999, 64, 5318–5320.

8   S. VELMATHI, S. SWARNALAKSHMI,
    S. NARASIMHAN, *Tetrahedron: Asymmetry*
    2003, 14, 113–117.

9   (a) S. NARA, H. TOSHIMA, A. ICHIHARA,
    *Tetrahedron Lett.* 1996, 37, 6745–6748;
    (b) S. NARA, H. TOSHIMA, A. ICHIHARA,
    *Tetrahedron* 1997, 53, 9509–9524.

10  (a) S. SHIMIZU, K. OHORI, T. ARAI,
    H. SASAI, M. SHIBASAKI, *J. Org. Chem.*
    1998, 63, 7547–7551; (b) K. OHORI,
    S. SHIMIZU, T. OHSHIMA, M. SHIBASAKI,
    *Chirality* 2000, 12, 400–403. Recently, this
    synthetic process was further improved
    using La–NMe-linked BINOL complex
    catalyzed asymmetric Michael reaction
    of β-keto esters; see: (c) K. MAJIMA,
    R. TAKITA, A. OKADA, T. OHSHIMA,
    M. SHIBASAKI, *J. Am. Chem. Soc.* 2003,
    125, 15837–15845.

11  Y. XU, K. OHORI, T. OHSHIMA,
    M. SHIBASAKI, *Tetrahedron* 2002, 58,
    2585–2588.

12  For a representative review, see:
    J. BONJOCH, D. SOLÉ, *Chem. Rev.* 2000,
    100, 3455–3482.

13  T. OHSHIMA, Y. XU, R. TAKITA,
    S. SHIMIZU, D. ZHONG, M. SHIBASAKI,
    *J. Am. Chem. Soc.* 2002, 124, 14546–
    14547 (Additions and Corrections 2003,
    125, 2014).

14  T. ARAI, H. SASAI, K. YAMAGUCHI,
    M. SHIBASAKI, *J. Am. Chem. Soc.* 1998,
    120, 441–442.

15  K. YAMADA, T. ARAI, H. SASAI, M. SHIBA-
    SAKI, *J. Org. Chem.* 1998, 63, 3666–3672.

16  (a) T. ARAI, G.-S. HU, X.-F. ZHENG, L. PU,
    H. SASAI, *Org. Lett.* 2000, 2, 4261–4263;
    (b) D. JAYAPRAKASH, H. SASAI, *Tetra-
    hedron: Asymmetry* 2001, 12, 2589–2595;
    (c) T. ARAI, T. SEKIGUTI, Y. IIZUKA,
    S. TAKIZAWA, S. SAKAMOTO,
    K. YAMAGUCHI, H. SASAI, *Tetrahedron:
    Asymmetry* 2002, 13, 2083–2087;
    (d) T. ARAI, T. SEKIGUTI, K. OTSUKI,
    S. TAKIZAWA, H. SASAI, *Angew. Chem. Int.
    Ed.* 2003, 42, 2144–2147.

17  For reviews, see: (a) D. M. HODGSON,
    A. R. GIBBS, G. P. LEE, *Tetrahedron* 1996,
    52, 14361–14384; (b) M. C. WILLIS,
    *J. Chem. Soc., Perkin Trans 1* 1999, 1765–
    1784.

18  E. N. JACOBSEN, H. W. MICHAEL, in
    *Comprehensive Asymmetric Catalysis*
    (Eds.: E. N. JACOBSEN, A. PFALTZ,
    H. YAMAMOTO), Springer, Berlin, 1999;
    Vol. 3, Chapter 35.

19  T. IIDA, N. YAMAMOTO, H. SASAI,
    M. SHIBASAKI, *J. Am. Chem. Soc.* 1997,
    119, 4783–4784.

20  (a) T. IIDA, N. YAMAMOTO, S. MATSUNAGA,
    H.-G. WOO, M. SHIBASAKI, *Angew. Chem.
    Int. Ed.* 1998, 37, 2223–2226;
    (b) S. MATSUNAGA, J. DAS, J. ROEL,
    E. M. VOGL, N. YAMAMOTO, T. IIDA,
    K. YAMAGUCHI, M. SHIBASAKI, *J. Am.
    Chem. Soc.* 2000, 122, 2252–2260.

21  (a) E. M. VOGL, S. MATSUNAGA, M. KANAI,
    T. IIDA, M. SHIBASAKI, *Tetrahedron Lett.*
    1998, 39, 7917–7920; (b) H. ISHITANI,
    T. KITAZAWA, S. KOBAYASHI, *Tetrahedron
    Lett.* 1999, 40, 2161–2164. For a review of

linked-BINOLs, see: (c) S. Matsunaga, T. Ohshima, M. Shibasaki, *Adv. Synth. Catal.* **2002**, *344*, 3–15.

22 N. Kumagai, S. Matsunaga, T. Kinoshita, S. Harada, S. Okada, S. Sakamoto, K. Yamaguchi, M. Shibasaki, *J. Am. Chem. Soc.* **2003**, *125*, 2169–2178.

23 For recent excellent reviews, see: (a) M. Arend, B. Westermann, N. Risch, *Angew. Chem. Int. Ed.* **1998**, *37*, 1044–1070; (b) A. Córdova, *Acc. Chem. Res.* ASAP article.

24 (a) H. Ishitani, M. Ueno, S. Kobayashi, *J. Am. Chem. Soc.* **1997**, *119*, 7153–7154; (b) E. Hagiwara, A. Fujii, M. Sodeoka, *J. Am. Chem. Soc.* **1998**, *120*, 2474–22475; (c) D. Ferraris, B. Young, T. Dudding, T. Lectka, *J. Am. Chem. Soc.* **1998**, *120*, 4548–4549; (d) A. G. Wenzel, E. N. Jacobsen, *J. Am. Chem. Soc.* **2002**, *124*, 12964–12965.

25 (a) S. Yamasaki, T. Iida, M. Shibasaki, *Tetrahedron Lett.* **1999**, *40*, 307–310; (b) S. Yamasaki, T. Iida, M. Shibasaki, *Tetrahedron* **1999**, *55*, 8857–8867.

26 K. Yamada, S. J. Harwood, H. Gröger, M. Shibasaki, *Angew. Chem. Int. Ed.* **1999**, *38*, 3504–3506.

27 (a) K. Yamada, G. Moll, M. Shibasaki, *Synlett* **2001**, 980–982; (b) N. Tsuritani, K. Yamada, N. Yoshikawa, M. Shibasaki, *Chem. Lett.* **2002**, 276–277.

28 For comprehensive reviews, see: (a) P. Kafarski, B. Lejczak, *Phosphorus, Sulfur Silicon Relat. Elem.* **1991**, *63*, 193; (b) P. A. Bartlett, C. K. Marlowe, P. P. Giannousis, J. E. Hanson, *Cold Spring Harbor Symp. Quant. Biol.* **1987**, *LII*, 83.

29 (a) T. Yokomatsu, T. Yamagishi, S. Shibuya, *Tetrahedron: Asymmetry* **1993**, *4*, 1783–1784; (b) N. P. Rath, S. D. Spilling, *Tetrahedron Lett.* **1994**, *35*, 227–230; (c) H. Sasai, S. Arai, Y. Tahara, M. Shibasaki, *J. Org. Chem.* **1995**, *60*, 6656–6657.

30 T. Arai, M. Bougauchi, H. Sasai, M. Shibasaki, *J. Org. Chem.* **1996**, *61*, 2926–2927.

31 (a) T. Yamagishi, T. Yokomatsu, K. Suemune, S. Shibuya, *Tetrahedron* **1999**, *55*, 12125–12136; (b) T. Yamagishi, K. Suemune, T. Yokomatsu, S. Shibuya, *Tetrahedron* **2002**, *58*, 2577–2583; (c) T. Yamagishi, T. Kusano, T. Yokomatsu, S. Shibuya, *Synlett* **2002**, 1471–1474.

# 4
# Rare Earth Bimetallic Asymmetric Catalysis

*Motomu Kanai and Masakatsu Shibasaki*

## 4.1
## Introduction

A reasonable design for a rare earth bimetallic asymmetric catalyst is to use one rare earth metal as a Lewis acid to activate a substrate and the other as a nucleophile generator (Figure 4.1). As described in the previous chapters, if dual generation of an activated substrate and an activated nucleophile takes place at positions defined by asymmetric catalysts, the direction of the bond formation should be controlled with high enantioselectivity (face- or position-selectivity) [1].

It is now well recognized that a rare earth metal can function as a Lewis acid, and thus activation of a substrate through coordination can be easily understood [2, 3]. For rare earth metal-mediated nucleophile generation, two possible mechanisms can be considered. One involves deprotonation of a *pre*-nucleophile by a rare earth metal alkoxide acting as a Brønsted base, and the other involves transmetalation from a fairly reactive silicon-conjugated nucleophile (such as TMSCN) to a highly reactive rare earth metal-conjugated nucleophile. Although the former deprotonation mechanism appears to be simple [4], and there are several examples of the use of rare earth metal naphthoxides as Brønsted bases for nucleophile generation in asymmetric catalysis [5], there is as yet no example of this mechanism being operative in an asymmetric bimetallic system. On the other hand, the methodology of nucleophile generation through transmetalation is utilized in bimetallic asymmetric catalysts. This method is currently limited to the use of cyanide as the nucleophile. Transmetalation from TMSCN to a rare earth metal cyanide was initially reported by Utimoto et al. in their studies aimed at developing epoxide-opening, aziridine-opening, and cyanosilylation reactions using ytterbium cyanide as catalyst in the presence of TMSCN [6, 7]. Three catalytic asymmetric reactions using a combination of Lewis acid activation of substrates and transmetalation activation of TMSCN are reviewed in this chapter: catalytic asymmetric cyanosilylation of ketones [8] and Strecker reaction of ketoimines [9] reported by our group, and a catalytic asymmetric ring-opening of *meso*-epoxides with TMSCN reported by Jacobsen's group [10].

*Multimetallic Catalysts in Organic Synthesis.* Edited by M. Shibasaki and Y. Yamamoto
Copyright © 2004 WILEY-VCH Verlag GmbH & Co. KGaA, Weinheim
ISBN: 3-527-30828-8

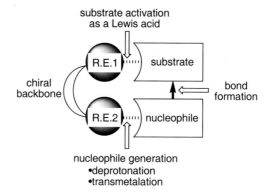

**Figure 4.1** General concept of rare earth bimetallic asymmetric catalysis.

## 4.2
## Catalytic Asymmetric Cyanosilylation of Ketones

### 4.2.1
### Catalytic Asymmetric Synthesis of a Camptothecin Intermediate: Discovery of an (S)-Selective Lanthanide Bimetallic Catalyst for the Cyanosilylation of Ketones

We reported the first example of catalytic enantioselective cyanosilylation of ketones with broad substrate generality using a titanium catalyst generated from D-glucose-derived ligands **1** and **2** (Figure 4.2) [11–13]. (*R*)-*Tertiary* cyanohydrins are obtained from both aromatic and aliphatic ketones with high enantioselectivity using 1–10 mol % of **1** or **2** (see Table 4.1). Mechanistic studies indicated that the titanium-**1** complex functions as a two-center catalyst, with the titanium activating the substrate ketone as a Lewis acid and the phosphine oxide activating TMSCN as a Lewis base (see **15** in Scheme 4.3). In collaboration with Curran's group at the University of Pittsburgh, we applied this reaction to the catalytic enantioselective synthesis of (20*S*)-camptothecin **5** (Scheme 4.1) and its analogues, which are among the most effective agents for the treatment of solid tumors [14]. For this purpose, however, the more expensive L-glucose was needed as the chiral source. The synthetic utility of the reaction would be greatly improved if both enantiomers of the product could be obtained with similar efficiency using a readily available chiral source.

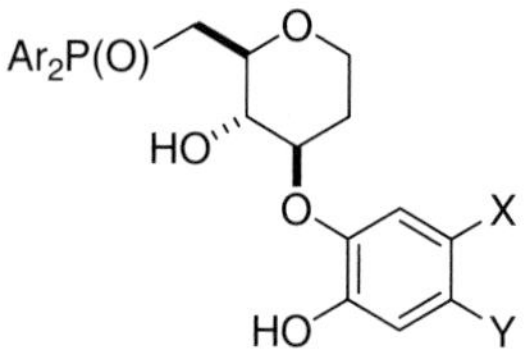

Ar₂P(O) ...

**1**: Ar = Ph, X, Y = H
**2**: Ar = Ph, X = COPh, Y = H
**3**: Ar = Ph, X, Y = F
**4**: Ar = *p*-tol, X, Y = F

**Figure 4.2** D-Glucose-derived ligands for catalytic asymmetric cyanosilylation of ketones.

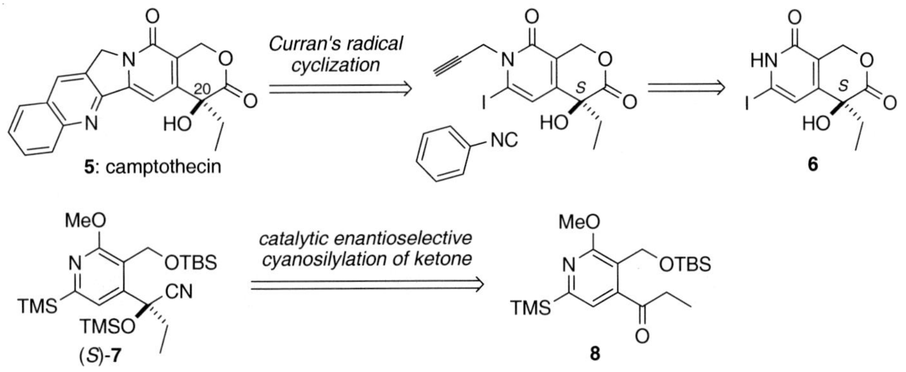

**Scheme 4.1** Synthetic plan for camptothecin.

Based on Curran's established synthetic route, α-hydroxy lactone **6** offers a general synthetic intermediate for the synthesis of the camptothecin family. Therefore, we selected ethyl ketone **8** as a substrate for the catalytic enantioselective cyanosilylation (Scheme 4.1). Although the Ti-**1** complex (20 mol %) produced the undesired (R)-**7** in low yield with low enantioselectivity, a samarium catalyst (5 mol %) prepared from Sm(O$^i$Pr)$_3$ [15] and **1** in a ratio of 1 : 1.8 gave the desired (S)-**7** in 92% yield with 72% ee after 24 h at –40 °C with THF as solvent. When propionitrile was used as the solvent, both the reactivity and enantioselectivity were improved, and **7** was obtained in 98% yield with 84% ee after 18 h.

Although the initial results were promising, neither the enantioselectivity nor the catalyst loading were satisfactory for application to plant-scale synthesis of camptothecin analogues. Therefore, we continued to optimize the catalyst. We anticipated that the catalyst could be improved by increasing the Lewis acidity of the metal and Lewis basicity of the phosphine oxide, by introducing electron-withdrawing groups on the catechol moiety and electron-donating groups on the phosphine oxide, respectively. Among the many ligands investigated [16], ligand **4**,

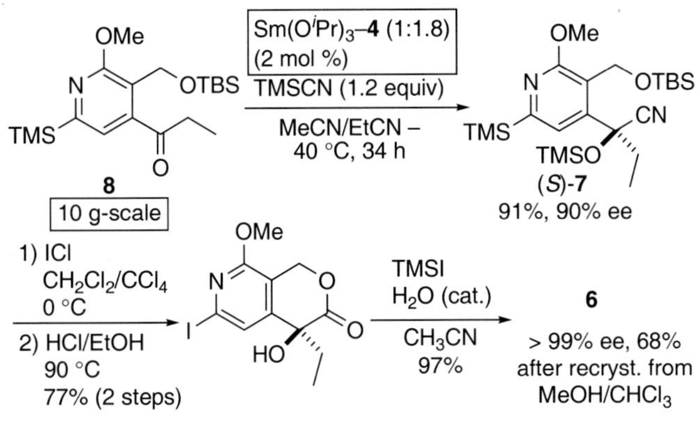

**Scheme 4.2** Catalytic enantioselective synthesis of key intermediate of camptothecin.

incorporating difluorocatechol and di-*p*-tolylphosphine oxide moieties, produced significantly improved results [17]. Thus, 2 mol % of the catalyst derived from **4** gave **7** in 95% yield with 89% ee in 19 h. Under similar conditions, the original catalyst derived from **1** gave the product in 100% yield with 82% ee in 44 h. Using a mixed solvent of acetonitrile and propionitrile (1 : 1), the enantiomeric excess achieved using **4** was further improved to 90%. This reaction was performed on a 10 g-scale without any difficulties. The key intermediate **6** for the camptothecin family was synthesized in three steps from (*S*)-cyanohydrin **7** (Scheme 4.2). Enantiomerically pure **6** was obtained by recrystallization from MeOH/CHCl$_3$ (8 : 1).

### 4.2.2
### Generality of Catalytic Asymmetric Cyanosilylation of Ketones Using Lanthanide Bimetallic Complexes

Next, to demonstrate the generality of this catalytic (*S*)-selective cyanosilylation, we used acetophenone (**9a**) as the substrate in conjunction with ligand **1**. With this substrate, the catalyst derived from **4** (a better catalyst for camptothecin synthesis as described above) gave comparable results to those achieved with the **1**-derived catalyst. In the presence of 5 mol % of Sm-**1** complex (1 : 1.8), the reaction in THF was completed within 2 h at –40 °C, giving (*S*)-cyanohydrin **10a** in 85% yield with 82% ee. When gadolinium was used instead of samarium, the enantiomeric excess was further improved to 89%. The optimal ratio of gadolinium to **1** was found to be 1 : 2 (92% ee), as shown in Figure 4.3, although the enantiomeric excess reached a plateau at a **1**/Gd ratio of 1.5 : 1. Under the optimized conditions in propionitrile solution, **10a** was obtained in 89% yield with 95% ee (Table 4.1, entry 1). The results of catalytic enantioselective cyanosilylations of ketones using Gd-**1** and Ti-**1** (or **2**) complexes are summarized in Table 4.1. Both enantiomers of the ketone cyano-

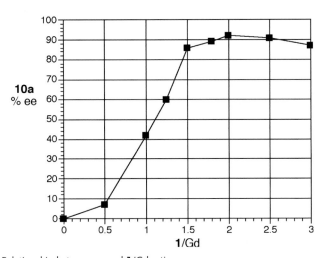

**Figure 4.3** Relationship between ee and 1/Gd ratio.

**Table 4.1** Catalytic enantioselective cyanosilylation of ketones.

| Entry | Ketone | | Metal | Ligand | Loading (x mol %) | Solvent | Temp (°C) | Time (h) | Yield (%) | ee (%) | Absolute configuration |
|---|---|---|---|---|---|---|---|---|---|---|---|
| 1 | | R = H | 9a Gd(O$^i$Pr)$_3$ | 1 | 5 | EtCN | −40 | 2 | 89 | 95 | (S)[a] |
| 2 | | | 9a Gd(O$^i$Pr)$_3$ | 1 | 1 | EtCN | −40 | 16 | 93 | 91 | (S)[a] |
| 3 | | CH$_3$ | 9a Ti(O$^i$Pr)$_4$ | 2 | 1 | THF | −20 | 88 | 92 | 94 | (R)[a] |
| 4 | | R = Cl | 9b Gd(O$^i$Pr)$_3$ | 1 | 5 | THF | −60 | 55 | 89 | 89 | (S) |
| 5 | | | 9b Ti(O$^i$Pr)$_4$ | 1 | 1 | THF | −25 | 92 | 72 | 90 | (R) |
| 6 | | | 9c Gd(O$^i$Pr)$_3$ | 1 | 5 | EtCN | −60 | 18 | 93 | 96 | (S) |
| 7 | | | 9c Ti(O$^i$Pr)$_4$ | 1 | 10 | THF | −40 | 80 | 82 | 95 | (R) |
| 8 | | | 9d Gd(O$^i$Pr)$_3$ | 1 | 5 | THF | −60 | 14 | 93 | 97 | (S) |
| 9 | | | 9d Ti(O$^i$Pr)$_4$ | 2 | 1 | THF | −10 | 92 | 90 | 92 | (R) |
| 10 | | | 9e Gd(O$^i$Pr)$_3$ | 1 | 5 | EtCN | −60 | 6.5 | 94 | 87 | (S) |
| 11 | | | 9e Ti(O$^i$Pr)$_4$ | 1 | 10 | THF | −50 | 88 | 72 | 91 | (R) |
| 12 | | | 9f Gd(O$^i$Pr)$_3$ | 1 | 5 | EtCN | −60 | 19 | 96 | 76 | (S) |
| 13 | | | 9f Ti(O$^i$Pr)$_4$ | 2 | 2.5 | THF | −30 | 92 | 72 | 90 | (R) |
| 14 | | | 9g Gd(O$^i$Pr)$_3$ | 1 | 5 | EtCN | −60 | 9 | 92 | 94 | (S) |
| 15 | | | 9h Gd(O$^i$Pr)$_3$ | 1 | 5 | EtCN | −60 | 1 | 97 | 66 | (S) |
| 16 | | | 9h Ti(O$^i$Pr)$_4$ | 1 | 10 | THF | −50 | 36 | 92 | 85 | (R) |
| 17 | | | 9i Gd(O$^i$Pr)$_3$ | 1 | 5 | EtCN | −60 | 0.5 | 79 | 47 | (S) |
| 18 | | | 9i Ti(O$^i$Pr)$_4$ | 2 | 2.5 | THF | −45 | 92 | 80 | 82 | (R) |

[a] The absolute configurations were determined by comparison with the reported value of optical rotation. Others were assigned tentatively.

hydrins can be synthesized from a broad range of ketones using catalysts derived from one chiral source by switching the metal from titanium to gadolinium. It is noteworthy that enzymes (oxynitrilase) cannot promote enantioselective cyanation of ketones with such a broad substrate generality as our catalyst does: in enzyme-catalyzed reactions, high enantioselectivity is obtained with alkyl methyl ketones, but only low chemical yields and enantioselectivities are obtained with other ketones such as aryl-, ethyl-, and propyl-substituted derivatives [12f].

4.2.3
**Reaction Mechanism**

A preliminary catalyst structure, the proposed reaction mechanism, and comparison with the reaction mechanism of Ti-1 are postulated in Scheme 4.3; these postulates were based on the following experiments [8]. When Pr(O$^i$Pr)$_3$ and 1 were mixed in a 1 : 2 ratio [18], complete ligand exchange was observed by $^1$H NMR analysis, indicating the generation of the presumed complex 11 and free 1. After evaporating the $^i$PrOH, excess TMSCN (> 4 mol equiv.) was added at –40 °C. Thereafter, peaks corresponding to monosilylated 1 (δ = 0.34 ppm) and disilylated 1 (δ = 0.38, 0.50 ppm) were observed at an intensity corresponding to ca. 90% yield from 11 and free 1, respectively. These observations indicated that 11 was converted to praseodymium cyanide **12** by monosilylation of 1, and that free 1 was disilylated to give 13. Furthermore, the molecular formula of complex **12** (Ln = Gd) was confirmed by ESI-MS. The mass value and the isotope distribution completely matched the

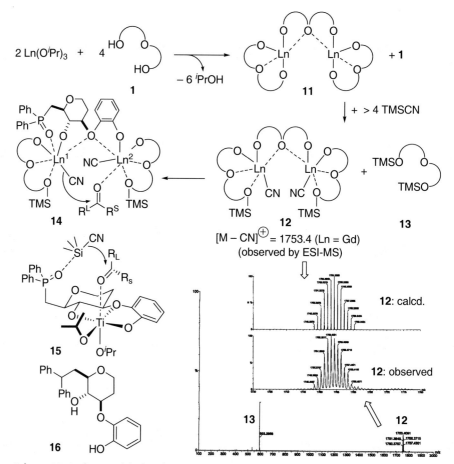

**Scheme 4.3** Working model of catalyst structure and reaction mechanism.

calculated values (Scheme 4.3). The relationship between the enantiomeric excess of the product and the 1/Gd ratio (Figure 4.3) was also consistent with these results. Therefore, the active catalyst was determined to be the 2 : 3 complex **12**.

The higher activity of the lanthanide-1 catalyst compared to Ti-1, as well as the structure of catalyst **12** and Utimoto's studies [6], suggest that the active nucleophile is the lanthanide metal cyanide, not TMSCN. This was confirmed by the following results. First, a facile CN scrambling between the gadolinium cyanide of the catalyst and TMSCN was found. Thus, after complex **12** (1 equiv.) containing $^{13}$CN had been prepared from TMS$^{13}$CN, acetophenone (**9a**) (1 equiv.) and a variable amount of TMS$^{12}$CN (1, 2, or 3 equiv.) were added; $^{13}$C NMR analysis of the product cyanohydrin indicated that the incorporation of $^{13}$CN was dependent on the ratio of added TMS$^{13}$CN and TMS$^{12}$CN. Moreover, only one signal corresponding to a cyanide ($\delta = 117$ ppm) was observed upon $^{13}$C NMR analysis of **12** (Ln = Pr) labeled with $^{13}$CN in the presence of a variable amount of TMS$^{13}$CN (0 or 2 equiv.) at $-60$ °C. With confirmation of the *pre*-equilibrium, kinetic studies were performed and the reaction order with respect to TMSCN was determined to be zero. Therefore, in the reaction catalyzed by Ln-1, the lanthanide metal cyanide acted as the active nucleophile. In contrast, the Ti-1-catalyzed reaction, in which TMSCN activated by the phosphine oxide acts as the active nucleophile, had a reaction order dependency of 0.7 with respect to TMSCN.

These findings point to the bimetallic transition state **14** in Scheme 4.3 as a working model. The Ln$^1$ cyanide is more electron-rich, and therefore more active as a nucleophile, because Ln$^1$ is bound to two alkoxides and coordinated by the phosphine oxide of the linker ligand. On the other hand, Ln$^2$ is more Lewis acidic. Intramolecular cyanide transfer from the nucleophilic Ln$^1$ cyanide to a ketone, activated by the more Lewis acidic Ln$^2$, should control the direction of the cyanide entry, giving products with high enantioselectivity. Consistent with this model, the order dependency of the reaction rate with regard to the catalyst was determined to be 0.8. The reversal of the enantioselectivity seen on switching between the Ti complex and the Ln complex is due to the different catalyst structures and the reaction mechanism. In the case of the Ti catalyst, TMSCN activated by the phosphine oxide functions as the nucleophile (**15**, Scheme 4.3). The essential contribution of the Lewis basic phosphine oxide in the reaction of the Ln catalyst was highlighted by results obtained with the catalyst prepared from control ligand **16**. Thus, Gd-**16** (1 : 2) promoted the reactions of **9a** and **9h** much more slowly in THF, giving **10a** and **10h** with only 7% ee (98% yield after 10 h at $-40$ °C) and 2% ee (97% yield after 18 h at $-40$ °C), respectively. NMR studies revealed that less TMSCN was consumed during the formation of catalyst Pr-**16** (1 : 2) than during that of catalyst Pr-**1**, which indicated the formation of less lanthanide metal cyanide in the case of Pr-**16** compared with Pr-**1**. No ESI-MS peaks corresponding to the Gd-**16** complex were seen under any conditions, which suggests a lack of defined structure in this case. Therefore, the phosphine oxide facilitates the lanthanide metal cyanide formation and stabilizes the active 2 : 3 complex **12**, as well as activates the lanthanide metal cyanide. Further studies aimed at determining the three-dimensional catalyst structure by X-ray crystallography are currently ongoing.

4.2.4
**Application to Catalytic Enantioselective Synthesis of an Oxybutynin Intermediate**

Oxybutynin (Ditropan, **18**) is a widely utilized muscarinic receptor antagonist for the treatment of urinary urgency, frequency, and incontinence [19]. A number of derivatives have been synthesized, mainly by modifying the ester (or the amide) moiety and/or the cycloalkyl moiety. Some of these analogues have improved $M_3$-receptor subtype selectivity and thus the side effects caused by antagonizing other subtypes, such as $M_1$- and $M_2$-receptors, are minimized. A common structural feature of oxybutynin and related compounds is the presence of a chiral *tertiary* α-hydroxy carbonyl moiety. Although oxybutynin is currently prescribed in racemic form, the (S)-enantiomer is proposed to have an improved therapeutic profile. Therefore, there is a high demand for the development of an efficient enantio-selective synthetic route. Previously reported methods utilized diastereoselective reactions that required a stoichiometric amount of chiral auxiliary or a chiral starting material to construct the chiral stereocenter of **18** [20]. We thus planned to apply our gadolinium-catalyzed enantioselective cyanosilylation of ketones to a practical synthesis of oxybutynin intermediate **17** [21].

Using 1 mol % of Gd-**1** complex, the reaction of ketone **9j** proceeded at –40 °C, and the product was obtained in 100% yield with 94% ee. This reaction proved to be practical, and could be performed on a 100 g-scale (Scheme 4.4). After the usual aqueous work-up, the crude oil was purified by passage through a short column of silica gel. Pure **10j** was obtained by eluting with AcOEt/hexane (1 : 20), and further elution with MeOH/CHCl₃ (1 : 15) allowed recovery of a ligand-containing fraction (a mixture of ligand **1** and the silylated ligand). The pure ligand **1** was recovered in 98% yield after acidic desilylation (1 M aq. HCl in THF) and recrystallization.

Having established a practical method for catalytic enantioselective cyanosilylation of the ketone, the next task was to convert the cyanide to the carboxylic acid. Careful reduction of **10j** with DIBAL-H in toluene, desilylation with 4 N aq. HCl in THF, followed by oxidation with NaClO₂ gave the known α-hydroxy carboxylic acid **17** in 80% overall yield (3 steps). Chemically pure **17** was obtained through a base/acid aqueous work-up without silica gel column chromatography. Recrystallization from CH₂Cl₂/hexane gave enantiomerically pure intermediate **17** for oxybutynin synthesis [22].

4.2.5
**Catalytic Enantioselective Cyanosilylation of Ketones**
**Containing Sterically Similar Substituents**

The successful attainment of excellent enantioselectivity in oxybutynin synthesis indicated that the Gd-**1** complex can discriminate between phenyl and cyclohexyl groups on the prochiral carbonyl carbon [23]. The unusually high enantio-differentiation ability of the catalyst towards the apparently difficult substrate **9j** led us to study the applicability of the catalytic enantioselective cyanosilylation to other aryl ketones bearing sterically similar substituents. As shown in Table 4.2,

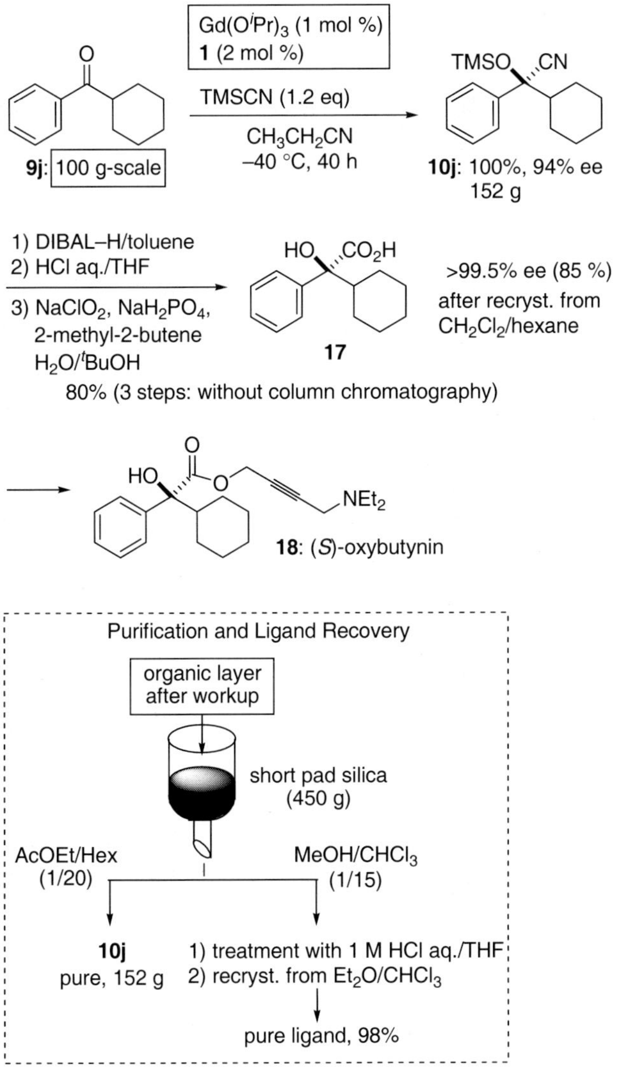

**Scheme 4.4** Application to oxybutynin synthesis.

high to excellent enantioselectivities were obtained, except in the case of cyclopentyl phenyl ketone (**9n-H**) and phenyl isopropyl ketone (**9q-H**). Products obtained with high enantioselectivity should be very useful for synthesizing new chiral oxybutynin analogues.

The sharp contrast in the enantiomeric excesses of **10n-H** and **10q-H** and of products from other ketones is intriguing from a mechanistic point of view. The results could not be attributed to differences in the ground-state structures between **9n-H** and **9q-H** and the other ketones, based on a comparison of the most stable conformations of these substrates. The reactivity–enantioselectivity relationship

**Table 4.2** Catalytic enantioselective cyanosilylation of ketones containing sterically similar substituents.

| Entry | Ketone | | Time (h) | Yield (%) | ee (%) |
|---|---|---|---|---|---|
| 1 | (MeO-) | 9k | 22 | 99 | 94 |
| 2 | (F$_3$C-) | 9l | 1 | 96 | 83 |
| 3 | | 9m | 5 | 99 | 94 |
| 4 | | 9n-H | 64 | 87 | 22 |
| 5 | | 9n-D[b] | 1 | 99 | 95 |
| 6 | | 9o | 2 | 99 | 97 |
| 7 | | 9p | 48 | 97 | 82 |
| 8 | | 9q-H | 20 | 99 | 38 |
| 9 | | 9q-D[b] | 1 | 81 | 96 |
| 10 | | 9r | 1 | 87 | 85 |

[a] The absolute configuration was tentatively assigned based on the analogy to 10j.
[b] 87%-D for 9n-D and 80%-D for 9q-D.

indicated a general trend in that high enantioselectivity was obtained when the reactions proceeded smoothly. Our explanation for this trend is that the asymmetric catalyst might act as a Brønsted base and deprotonate the starting ketones, if the α-proton is relatively acidic. This competitive pathway might cause a ligand exchange to produce a less enantioselective catalyst containing the substrates. This hypothesis is supported by a dramatic deuterium kinetic isotope effect, as a result of which the undesired deprotonation is suppressed [24]. Reactions using **9n-D** and **9q-D** proceeded rapidly, and the product was obtained with 95–96% ee (Table 4.2; entry 5 *vs.* entry 4; entry 9 *vs.* entry 8) [21]. As far as we are aware, this is the first example of a dramatic advantage being gained through an isotope effect in catalytic enantioselective reactions. The results also provide a very important insight into the reaction mechanism, in which the Brønsted basicity of the Gd-**1** catalyst can cause undesired decomposition of the highly enantioselective catalytic species.

## 4.3
## Catalytic Enantioselective Strecker Reaction of Ketoimines

The catalytic enantioselective Strecker reaction (cyanation of imines) is one of the most direct and efficient methods for the asymmetric synthesis of natural and unnatural α-amino acids. Recent intensive studies in this field have led to the ability to synthesize chiral monosubstituted α-amino nitriles from a wide range of aldoimines [25]. In the case of the catalytic enantioselective Strecker reaction using ketoimines as substrates, however, only two asymmetric catalyses had been reported when we started this project: Jacobsen's Schiff-base catalysis [26] and Vallée's heterobimetallic catalysis [27]. Of these, Jacobsen's catalysis gave high enantio-selectivity in the reactions of aryl methyl ketoimines and *tert*-butyl methyl ketoimine. Considering the biological importance of disubstituted α-amino acids [28, 29], however, there was room for improvement in terms of substrate generality. The reaction analogy between the cyanosilylation of ketones and the Strecker reaction of ketoimines led us to extend the application of our chiral gadolinium complex to its use as an asymmetric catalyst for the latter reaction.

We began with optimization using acetophenone-derived imines as substrates, TMSCN as the nucleophile, and Gd-**1** as the catalyst (10 mol %). When using an *N*-benzyl-protected imine as the substrate, however, the product was obtained with only 35% ee, even after extensive screening of reaction conditions such as the lanthanide metal used and the metal/ligand ratio. Because of the oxophilic character of lanthanide metals, the use of an *N*-diphenylphosphinoylimine **19a** [30] as substrate was investigated, and the product was obtained with 82% ee. Finally, a product with 96% ee was obtained when electronically tuned Gd-**3**, containing difluorocatechol, was used as the catalyst.

Having optimized the reaction conditions, substrate generality was investigated (Table 4.3, method A). High enantioselectivity was obtained from a wide range of ketoimines using 2.5–10 mol % of the catalyst. This is the most general catalytic enantioselective Strecker reaction of ketoimines reported to date [9].

**Table 4.3** Catalytic enantioselective Strecker reaction.

| Entry | Substrate | | Method[a] | Cat. (x) | Time (h) | Yield (%) | ee (%) |
|---|---|---|---|---|---|---|---|
| 1 | | R = H: **19a** | A | 2.5 | 24 | 94 | 95[b] |
| 2 | | **19a** | B | 2.5 | 2 | 98 | 97[b] |
| 3 | | **19a** | B | 1 | 30 | 94 | 92[b] |
| 4 | | R = Cl: **19b** | A | 2.5 | 67 | 84 | 89 |
| 5 | | **19b** | B | 2.5 | 0.7 | 90 | 97 |
| 6 | | R = Me: **19c** | A | 2.5 | 52 | 93 | 98[b] |
| 7 | | **19c** | B | 2.5 | 0.7 | 96 | 99[b] |
| 8 | | **19d** | A | 2.5 | 72 | 67 | 94 |
| 9 | | **19d** | B | 2.5 | 3.5 | 93 | 98 |
| 10 | | **19e** | A | 10 | 14 | 72 | 85[b] |
| 11 | | **19e** | B | 2.5 | 2 | 97 | 93[b] |
| 12 | | **19f** | A | 5 | 68 | 79 | 83[b] |
| 13 | | **19f** | B | 5 | 0.2 | 95 | 98[b] |
| 14 | | **19g** | A | 5 | 52 | 99 | 88[b] |
| 15 | | **19g** | B | 5 | 2 | 94 | 94[b] |
| 16 | | **19h** | A | 5 | 67 | 58 | 90 |
| 17 | | **19h** | B | 2.5 | 2 | 95 | 97 |
| 18 | | **19i** | A | 2.5 | 134 | 70 | 52[b] |
| 19 | | **19i** | B | 2.5 | 0.3 | 94 | 98[b] |
| 20 | | **19j** | A | 2.5 | 158 | 67 | 48[b] |
| 21 | | **19j** | B | 2.5 | 48 | 98 | 97[b] |
| 22 | | **19k** | A | 5 | 65 | 73 | 72 |
| 23 | | **19k** | B | 5 | 2 | 96 | 93 |
| 24 | | **19l** | A | 5 | 48 | 74 | 51 |
| 25 | | **19l** | B | 2.5 | 2.5 | 91 | 80 |

**Table 4.3** (continued)

| Entry | Substrate | | Method [a] | Cat. (x) | Time (h) | Yield (%) | ee (%) |
|-------|-----------|---|-----------|----------|----------|-----------|--------|
| 26 | | 19m | B | 1 | 16 | 90 | 95 |
| 27 | | 19n | B | 1 | 21 | 93 | 93 |
| 28 | | 19o | B | 2.5 | 10 | 94 | 96 |
| 29 | | 19p | B | 2.5 | 6 | 98 | 98 |

[a] Method A = Without additive. Method B = In the presence of additive.
[b] The absolute configuration was determined to be (S).

Recently, we found that the efficiency (catalyst loading and catalyst turnover frequency) and substrate generality of our catalytic asymmetric Strecker reaction are dramatically improved by the addition of a proton source, such as 2,6-dimethylphenol (Table 4.3, method B) [31]. This improvement stems from a change in the catalyst structure from TMS-bearing catalyst **12** to proton-bearing catalyst **21** induced by the protic additive, as demonstrated by ESI-MS studies (Scheme 4.5). This new catalyst made it possible to use heteroaromatic and cyclic ketoimines as substrates for the first time (Table 4.3, entries 19, 21, and 26–29). Postulated catalytic cycles for methods A and B are compared in Scheme 4.5.

Logical consideration of the reaction mechanism for method B led us to develop more atom-economical reaction conditions using a catalytic amount of TMSCN (2.5–5 mol %) and a stoichiometric amount of HCN. The catalyst turnover number reached 1000, while maintaining the excellent enantioselectivity [31]. Enantiomerically pure α,α-disubstituted amino acids, including an important pharmaceutical lead, were easily obtained through simple acid hydrolysis of the Strecker products followed by ion-exchange column chromatography.

## 4.4
## Catalytic Enantioselective Ring-Opening of *meso*-Epoxides with TMSCN

The catalytic asymmetric ring-opening of *meso*-epoxides constitutes an important strategy for the preparation of valuable chiral building blocks [32]. Although a variety of nucleophiles have been employed, such as phenols, an azide, amines, and halides, there are quite a few examples of the use of carbon nucleophiles. This is partly due

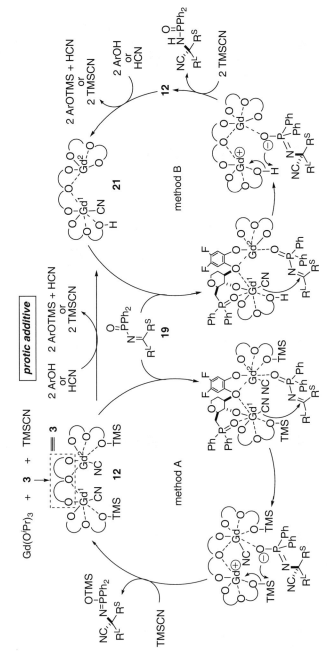

**Scheme 4.5** Proposed catalytic cycle in the absence (left) and presence (right) of a protic additive.

to the low reactivity of epoxides as electrophiles. Snapper and Hoveyda [33] reported the first example belonging to this category, using TMSCN as a nucleophile and chiral titanium complexes of Schiff base-containing peptides as catalysts. Oguni [34] reported a chiral Schiff base catalyzed epoxide-opening reaction with phenyl-lithium. The enantioselectivity and substrate generality, however, were by no means optimized in these pioneering works.

In 2000, Jacobsen reported a highly enantioselective epoxide-opening reaction with TMSCN. His group investigated chiral rare earth metal complexes based on the seminal studies by Kagan [7] and Utimoto [6], which revealed that lanthanide catalysts are capable of selectively forming a nitrile product, with minimal production of the isomeric isonitrile. After screening lanthanide metals and nitrogen-containing chiral ligands, they found that the ytterbium complex of Ph-pybox (22) gave good results in the reactions with cyclohexene oxide and *cis*-2-butene oxide, while $^t$Bu-pybox (23) gave good results with epoxides fused to five-membered rings (Table 4.4). Although 22 and 23 possess the same chirality, these catalysts gave the opposite product enantiomers. On the basis of the observed positive nonlinear effect and the second-order initial-rate kinetics with respect to the catalyst, the authors proposed a bimetallic mechanism in which the catalyst plays a dual role of cyanide delivery

**Table 4.4** Catalytic enantioselective *meso*-epoxide opening with TMSCN.

| Entry | Substrate | Ligand | Temp (°C) | Time (d) | Product | Yield (%) | ee (%) |
|-------|-----------|--------|-----------|----------|---------|-----------|--------|
| 1 | | 22 | −45 | 4 | | 90 | 91 |
| 2 | | 23 | −10 | 7 | | 83 | 92 |
| 3 | Me, Me | 22 | −40 | 7 | Me, Me | 80 | 90 |
| 4 | CO₂Et | 23 | 0 | 7 | CO₂Et | 86 | 83 |
| 5 | N-COCF₃ | 23 | −10 | 7 | N-COCF₃ | 72 | 87 |

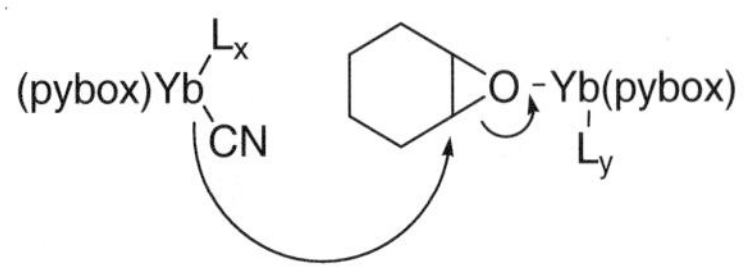

**Figure 4.4** Proposed bimetallic mechanism.

agent and Lewis acid (Figure 4.4) [10]. Although this example is a most significant achievement in the catalytic enantioselective epoxide-opening reaction with carbon nucleophiles, further studies are necessary to improve the efficiency (substrate generality and catalyst loading), considering the high importance of this reaction.

## 4.5
## Conclusion

Catalytic asymmetric reactions using chiral rare earth bimetallic complexes have been reviewed. These studies, however, are just the first steps, and studies aimed at the development of efficient asymmetric catalysis based on this strategy need to be continued. For example, there is little doubt that chiral rare earth bimetallic complexes capable of promoting reactions through dual Brønsted base/Lewis acid activation with high atom economy will be created in the future. The development of catalytic enantioselective reactions that are both highly efficient and environmentally friendly is a major target of contemporary synthetic organic chemistry.

## References and Notes

1 For recent reviews on bifunctional asymmetric catalysis, see: (a) G. J. ROWLANDS, *Tetrahedron* **2001**, *57*, 1865–1882; (b) M. SHIBASAKI, M. KANAI, K. FUNABASHI, *Chem. Commun.* **2002**, 1989–1999.

2 (a) For the initial use of a lanthanide as a Lewis acid catalyst, see: H. ADKINS, B. H. NISSEN, *J. Am. Chem. Soc.* **1922**, *44*, 2749 (acetal formation); (b) For the initial use of a lanthanide as a Lewis acid catalyst for a C–C bond formation, see: M. BEDNARSKI, S. DANISHEFSKY, *J. Am. Chem. Soc.* **1983**, *105*, 3716–3717 (Diels–Alder reaction); (c) For the initial use of a lanthanide as a stoichiometric Lewis acid for a C–C bond formation, see: B. M. TROST, M. J. BOGDANOWICZ, *J. Am. Chem. Soc.* **1973**, *95*, 2038–2040 (addition of an S-ylide to a ketone).

3 (a) For a recent review, see: S. KOBAYASHI (Chapter 19) and M. SHIBASAKI, K.-I. YAMADA, N. YOSHIKAWA (Chapter 20), in *Lewis Acids in Organic Synthesis* (Ed.: H. YAMAMOTO), Wiley-VCH, Weinheim, Germany, 2000; (b) For a review of early work, see: H. B. KAGAN, J. L. NAMY, *Tetrahedron* **1986**, *42*, 6573–6614.

4 For the initial use of a rare earth metal alkoxide as a Brønsted base catalyst for C–C bond formations, see: T. SASAI, T. SUZUKI, S. ARAI, T. ARAI, M. SHIBASAKI, *J. Am. Chem. Soc.* **1992**, *114*, 4418–4420.

5 (a) K. MAJIMA, R. TAKITA, A. OKADA, T. OHSHIMA, M. SHIBASAKI, *J. Am. Chem. Soc.* **2003**, *125*, 15837–15845; (b) Y. S. KIM, S. MATSUNAGA, J. DAS, A. SEKINE, T. OHSHIMA, M. SHIBASAKI,

*J. Am. Chem. Soc.* **2000**, *122*, 6506–6507; (c) T. Nemoto, T. Ohshima, K. Yamaguchi, M. Shibasaki, *J. Am. Chem. Soc.* **2001**, *123*, 2725–2732; (d) T. Nemoto, T. Ohshima, M. Shibasaki, *J. Am. Chem. Soc.* **2001**, *123*, 9474–9475; (e) T. Nemoto, H. Kakei, V. Gnanadesikan, S. Tosaki, T. Ohshima, M. Shibasaki, *J. Am. Chem. Soc.* **2002**, *124*, 14544–14545.

6 (a) S. Matsubara, H. Onishi, K. Utimoto, *Tetrahedron Lett.* **1990**, *31*, 6209–6212; (b) S. Matsubara, T. Kodama, K. Utimoto, *Tetrahedron Lett.* **1990**, *31*, 6379–6380; (c) S. Matsubara, T. Takai, K. Utimoto, *Chem. Lett.* **1991**, 1447–1450.

7 Kagan et al. reported lanthanide chloride-catalyzed aldol reaction, cyanosilylation, and epoxide-opening reactions prior to Utimoto's work: A. E. Vougioukas, H. B. Kagan, *Tetrahedron Lett.* **1987**, *28*, 5513–5516.

8 K. Yabu, S. Masumoto, S. Yamasaki, Y. Hamashima, M. Kanai, W. Du, D. P. Curran, M. Shibasaki, *J. Am. Chem. Soc.* **2001**, *123*, 9908–9909.

9 S. Masumoto, H. Usuda, M. Suzuki, M. Kanai, M. Shibasaki, *J. Am. Chem. Soc.* **2003**, *125*, 5634–5635.

10 S. E. Schaus, E. N. Jacobsen, *Org. Lett.* **2000**, *2*, 1001–1004.

11 (a) Y. Hamashima, M. Kanai, M. Shibasaki, *J. Am. Chem. Soc.* **2000**, *122*, 7412–7413; (b) Y. Hamashima, M. Kanai, M. Shibasaki, *Tetrahedron Lett.* **2001**, *42*, 691–694.

12 For other examples, see: (a) Y. N. Belokon', B. Green, N. S. Ikonnikov, M. North, V. I. Tararov, *Tetrahedron Lett.* **1999**, *40*, 8147–8150; (b) S.-K. Tian, L. Deng, *J. Am. Chem. Soc.* **2001**, *123*, 6195–6196; (c) Y. N. Belokon', B. Green, N. S. Ikonnikov, M. North, T. Parsons, V. I. Tararov, *Tetrahedron* **2001**, *57*, 771–779; (d) H. Deng, M. P. Isler, M. L. Snapper, A. H. Hoveyda, *Angew. Chem. Int. Ed.* **2002**, *41*, 1009–1012; (e) F. Chen, X. Feng, B. Qin, G. Zhang, Y. Jiang, *Org. Lett.* **2003**, *5*, 949–952; (f) For enzyme-catalyzed reactions, see: F. Effenberger, S. Heid, *Tetrahedron: Asymmetry* **1995**, *6*, 2945–2952.

13 Ligands **1**, **2**, and **3** are commercially available from Junsei Chemical Co., Ltd. (Fax: +81-3-3270-5461).

14 For recent enantioselective synthesis of the (20*S*)-camptothecin family, see: (a) H. Josien, S.-B. Ko, D. Bom, D. P. Curran, *Chem. Eur. J.* **1998**, *4*, 67–83; (b) D. L. Comins, J. M. Nolan, *Org. Lett.* **2001**, *3*, 4255–4257.

15 Purchased from Kojundo Chemical Laboratory Co., Ltd. (Fax: +81-492-84-1351).

16 For a flexible ligand synthesis, see: S. Masumoto, K. Yabu, M. Kanai, M. Shibasaki, *Tetrahedron Lett.* **2002**, *43*, 2919–2922.

17 (a) K. Yabu, S. Masumoto, M. Kanai, D. P. Curran, M. Shibasaki, *Tetrahedron Lett.* **2002**, *43*, 2923–2926; (b) K. Yabu, S. Masumoto, M. Kanai, W. Du, D. P. Curran, M. Shibasaki, *Heterocycles* **2003**, *59*, 369–385.

18 The praseodymium complex was used for NMR studies, because gadolinium is strongly paramagnetic. Pr-**1** (1 : 2) complex gave **10a** in 96% yield with 77% ee after 2 h at −40 °C (THF).

19 I. M. Thompson, R. Lauvetz, *Urology* **1976**, *8*, 452–454.

20 (a) P. T. Grover, N. N. Bhongle, S. A. Wald, C. H. Senanayake, *J. Org. Chem.* **2000**, *65*, 6283–6287; (b) M. Mitsuya, Y. Ogino, N. Ohtake, T. Mase, *Tetrahedron* **2000**, *56*, 9901–9907; (c) C. H. Senanayake, Q. K. Fang, P. Grover, R. P. Bakale, C. P. Vandenbossche, S. A. Wald, *Tetrahedron Lett.* **1999**, *40*, 819–822.

21 S. Masumoto, M. Suzuki, M. Kanai, M. Shibasaki, *Tetrahedron Lett.* **2002**, *43*, 8647–8651.

22 For the conversion of **17** to **18**, see: R. P. Bakale, J. L. Lopez, F. X. McConville, C. P. Vandenbossche, C. H. Senanayake, U.S. Patent no. 6140529.

23 Corey et al. reported that electronic effects can determine the effective size of the carbonyl substituents: E. J. Corey, C. J. Helal, *Tetrahedron Lett.* **1995**, *36*, 9153–9156.

24 For the use of an isotope effect in organic synthesis to prevent the undesired deprotonation, see: (a) G. B. Dudley, S. J. Danishefsky, G. Sukenick,

*Tetrahedron Lett.* **2002**, *43*, 5605–5606;
(b) E. VEDEJS, J. LITTLE, *J. Am. Chem. Soc.*
**2002**, *124*, 748–749.

**25** For reviews, see: (a) H. GRÖGER, *Chem.
Rev.* **2003**, *103*, 2795–2828; (b) L. YET,
*Angew. Chem. Int. Ed.* **2001**, *40*, 875–877.

**26** (a) P. VACHAL, E. N. JACOBSEN, *Org. Lett.*
**2000**, *2*, 867–870; (b) P. VACHAL,
E. N. JACOBSEN, *J. Am. Chem. Soc.* **2002**,
*124*, 10012–10014.

**27** M. CHAVAROT, J. J. BYRNE, P. Y. CHAVANT,
Y. VALLÉE, *Tetrahedron: Asymmetry* **2001**,
*12*, 1147–1150.

**28** For reviews, see: (a) Y. OHFUNE,
M. HARIKAWA, *J. Synth. Org. Chem.* **1997**,
*55*, 982–993; (b) C. CATIVIELA,
M. D. DIAZ-DE-VILLEGAS, *Tetrahedron:
Asymmetry* **1998**, *9*, 3517–3599.

**29** Catalytic asymmetric alkylation is another
powerful methodology for disubstituted
α-amino acid synthesis. For recent
examples, see: (a) T. OOI, M. TAKEUCHI,

M. KAMEDA, K. MARUOKA, *J. Am. Chem.
Soc.* **2000**, *122*, 5228–5229;
(b) B. M. TROST, K. DOGRA, *J. Am. Chem.
Soc.* **2002**, *124*, 7256–7257.

**30** B. KRZYZANOWSKA, W. J. STEC, *Synthesis*
**1982**, 270–273.

**31** (a) N. KATO, M. SUZUKI, M. KANAI,
M. SHIBASAKI, *Tetrahedron Lett.* **2004**, *45*,
3147–3151; (b) N. KATO, M. SUZUKI,
M. KANAI, M. SHIBASAKI, *Tetrahedron
Lett.* **2004**, *45*, 3153–3155.

**32** For a review, see: E. N. JACOBSEN,
M. H. WU, in *Comprehensive Asymmetric
Catalysis* (Eds.: E. N. JACOBSEN, A. PFALTZ,
H. YAMAMOTO), Springer, New York,
1999, Chapter 35.

**33** K. D. SHIMIZU, B. M. COLE,
C. A. KRUEGER, K. W. KUNTZ,
M. L. SNAPPER, A. H. HOVEYDA, *Angew.
Chem. Int. Ed. Engl.* **1997**, *36*, 1704–1707.

**34** N. OGUNI, Y. MIYAGI, K. ITOH,
*Tetrahedron Lett.* **1998**, *39*, 9023–9026.

# 5

# Rare Earth–Alkali Metal Heterobimetallic Asymmetric Catalysis

*Shigeki Matsunaga and Masakatsu Shibasaki*

## 5.1
## Introduction

Asymmetric catalysis has received considerable attention over the past few decades, and its contribution in organic synthesis has become increasingly important [1]. Various enantioselective chemical transformations are now performed with only catalytic amounts of chiral promoters. Some of these enantioselective transformations are applied in industrial production processes. The performance of most artificial catalysts, however, is still far from satisfactory in terms of generality and reactivity. On the other hand, enzymes catalyze various organic transformations under mild conditions, although they often lack substrate generality. One of the advantages of enzymes over most artificial catalysts is that they often contain two or more active sites for the catalysis. The synergistic function of the active sites renders substrates more reactive in the transition state and controls the positions of adjacent functional groups. This concept of multifunctional catalysis is the key to increasing the scope of natural and artificial catalysts.

Since the first report of a catalytic asymmetric nitroaldol reaction using rare earth metal complexes in 1992, Shibasaki et al. [2] continued to develop the concept of multifunctional catalysis wherein the catalysts exhibit both Lewis acidity and Brønsted basicity (Figure 5.1). The synergistic effects of the two functions enabled various transformations which were difficult to achieve using conventional

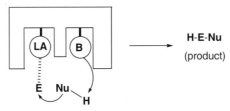

**Figure 5.1** Concept of multifunctional catalysts employing the synergistic function of a Lewis acid and a Brønsted base.
LA = Lewis acid; B = Brønsted base; E = electrophile; Nu-H = nucleophile.

*Multimetallic Catalysts in Organic Synthesis.* Edited by M. Shibasaki and Y. Yamamoto
Copyright © 2004 WILEY-VCH Verlag GmbH & Co. KGaA, Weinheim
ISBN: 3-527-30828-8

monometallic catalysts with only Lewis acidity. A variety of enantioselective transformations have been realized by choosing a combination of metals according to the type of reaction. In particular, heterobimetallic complexes that contain a rare earth metal, three alkali metals, and three 1,1'-bi-2-naphthols (BINOLs) have offered a versatile framework for the development of asymmetric catalysts. The properties of the catalyst are modified by choosing the alkali metals and further refined by choosing the appropriate rare earth metal. In this chapter, representative hetero-bimetallic rare earth–alkali metal multifunctional asymmetric catalysts and asymmetric reactions promoted by these catalysts are introduced.

## 5.2
### Development and Structural Analysis of Rare Earth–Alkali Metal Heterobimetallic Complexes

Shibasaki et al. initially reported that enantiomerically-enriched nitroaldol adducts **3** were obtained with up to 90% ee with the catalyst prepared from La(O$^t$Bu)$_3$ (1 mol equiv.), (*S*)-BINOL (1.5 mol equiv.), LiCl (2 mol equiv.), and H$_2$O (10 mol equiv.). After optimization, the method for catalyst preparation was improved to employ LaCl$_3$ · 7 H$_2$O (1 mol equiv.), Li$_2$(*S*-binol)$_2$ (1 mol equiv.), NaO$^t$Bu (1 mol equiv.), and H$_2$O (4 mol equiv.) [3]. During efforts to elucidate the structure of the catalyst, a series of complexes containing a rare earth metal and an alkali metal were synthesized. X-ray crystallographic analyses, elemental analyses, and mass spectro-metric studies of these complexes revealed that they consisted of a framework represented as M$_3$[Ln(*S*-binol)$_3$] (abbreviated as LnMB; Ln: lanthanide, M: alkali metal, B: BINOL; Figure 5.2) with one molecule of H$_2$O coordinating to Ln [4]. Each complex has a center of asymmetry at the central rare earth metal, and can exist as a mixture of diastereomers (Figure 5.3, Λ-form and Δ-form). Nevertheless, every crystal was found to have the Λ-configuration rather than the Δ-form when the complex was prepared from (*S*)-BINOL, indicating that the configuration at the central metal is affected by the configuration of the BINOL. On the basis of the elucidated structure, different procedures for catalyst preparation were developed.

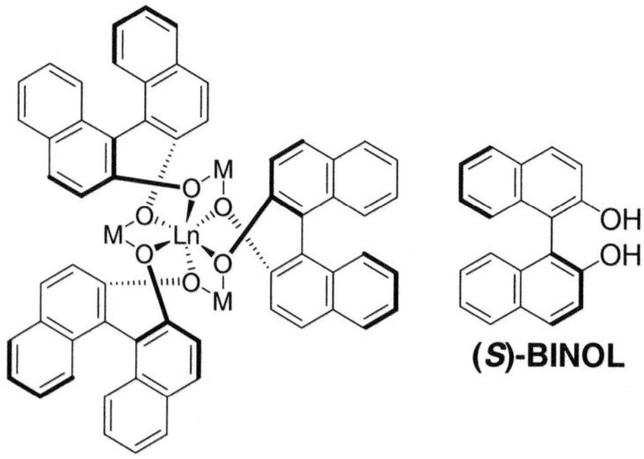

**Figure 5.2** Structure of M$_3$[Ln(binol)$_3$]-type heterobimetallic complexes (LnMB).

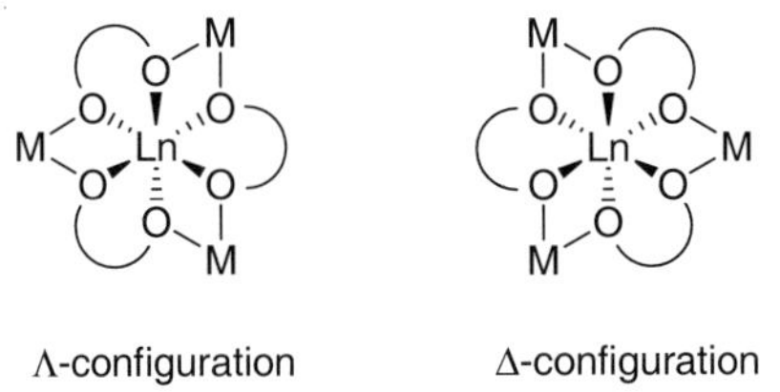

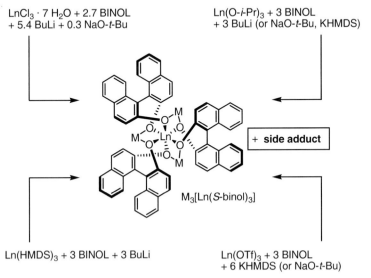

Λ-configuration        Δ-configuration

**Figure 5.3** Two possible configurations of $M_3[Ln(binol)_3]$-type (LnMB) complexes.

The catalyst **4a** ($Li_3[La(binol)_3(H_2O)]$; LLB · $H_2O$) was prepared by treatment of $La(O^iPr)_3$ [5] with Li(Hbinol) (3 mol equiv. to La) and $H_2O$ (1 mol equiv. to La) in THF. This method produces the catalytic species in high yield. An alternative method for preparing the catalyst from inexpensive hydrated $LnCl_3$ was also investigated. An equally active catalyst was thus prepared from $LaCl_3$ · 7 $H_2O$, $Li_2(binol)$ (2.7 mol equiv.), and $NaO^tBu$ (0.3 mol equiv.) in THF [6]. The catalytic activities of these complexes are maintained for several months under argon at ambient temperature.

Aspinall et al. [7] reported detailed structural studies of rare earth–alkali metal heterobimetallic complexes. Whereas the crystal structures provided by Shibasaki et al. included one molecule of water coordinated to the central metal, Aspinall et al. [7] succeeded in preparing anhydrous crystals of $M_3[Ln(binol)_3]$ (LnMB). $Li_3[Ln(binol)_3]$ complexes were obtained from $Ln[N(SiMe_3)_2]_3$ and Li(Hbinol) in THF or $Et_2O$. The $HN(SiMe_3)_2$ by-product was removed under reduced pressure, and the crystals were obtained from THF/petroleum ether. Aspinall et al. [7] reported X-ray crystal structures of a series of heterobimetallic complexes. The differences between the anhydrous and aqua crystal structures were also reported.

$LnCl_3$ · 7 $H_2O$ + 2.7 BINOL
+ 5.4 BuLi + 0.3 NaO-*t*-Bu

$Ln(O$-*i*-Pr$)_3$ + 3 BINOL
+ 3 BuLi (or NaO-*t*-Bu, KHMDS)

+ **side adduct**

$M_3[Ln(S$-binol$)_3]$

Ln(HMDS)$_3$ + 3 BINOL + 3 BuLi

$Ln(OTf)_3$ + 3 BINOL
+ 6 KHMDS (or NaO-*t*-Bu)

**Scheme 5.1** Preparation methods of rare earth–alkali metal hetero-bimetallic complexes from various rare earth metal sources.

Salvadori et al. [8] reported a solution-phase structure of $M_3[Yb(binol)_3]$ (YbMB). YbSB and YbPB catalysts were prepared from $Yb(OTf)_3$ as the metal source. Differences between the solid-phase and solution-phase structures were studied in detail by means of UV, CD, and NMR analyses. These authors also discussed the effects of the alkali metals on the charge distribution in the Yb–alkali metal heterobimetallic complexes.

All the preparative methods described above (Scheme 5.1) are considered to generate an essentially identical catalytic species. However, the amounts of contaminating side products, such as alkali metal halides, vary depending on the procedure used. Because achiral additives such as LiCl often have a strong effect on enantioselectivity and reactivity, which may be either positive or negative, the most suitable preparative method should be selected for each asymmetric reaction.

## 5.3
## Nitroaldol Reaction

The nitroaldol reaction is useful for the synthesis of various natural products and other biologically interesting compounds. The LLB complex was identified as the first highly enantioselective catalyst for the nitroaldol reaction [3]. A plausible catalytic cycle for the enantioselective nitroaldol reaction is depicted in Scheme 5.2. In this reaction, the lanthanum metal in LLB (**4a**) acts as a Lewis acid to activate the aldehyde, and the lithium binaphthoxide moiety functions as a Brønsted base to deprotonate nitromethane to give a lithium nitronate. The reaction proceeds through simple proton transfer, thus generating no waste after the catalytic cycle. This catalytic cycle features some characteristic aspects of the heterobimetallic complexes.

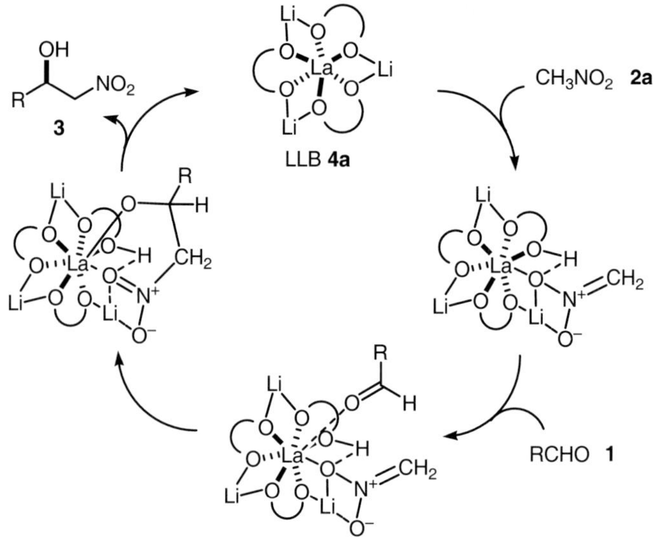

**Scheme 5.2** Plausible mechanism for catalytic asymmetric nitroaldol reaction.

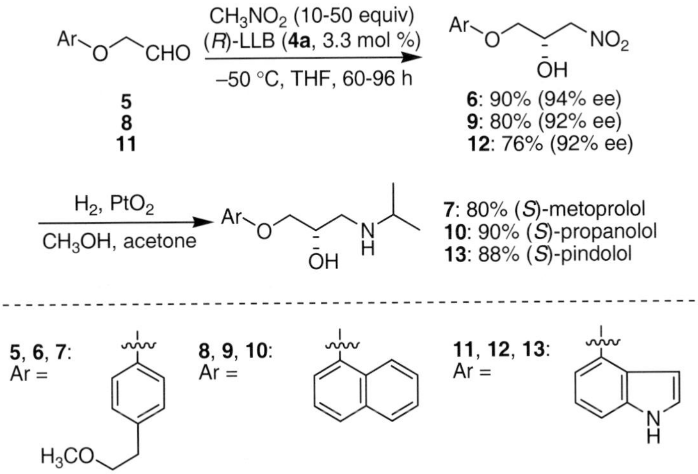

**Scheme 5.3** Catalytic asymmetric synthesis of β-blockers using (R)-LLB (**4a**) as a catalyst.

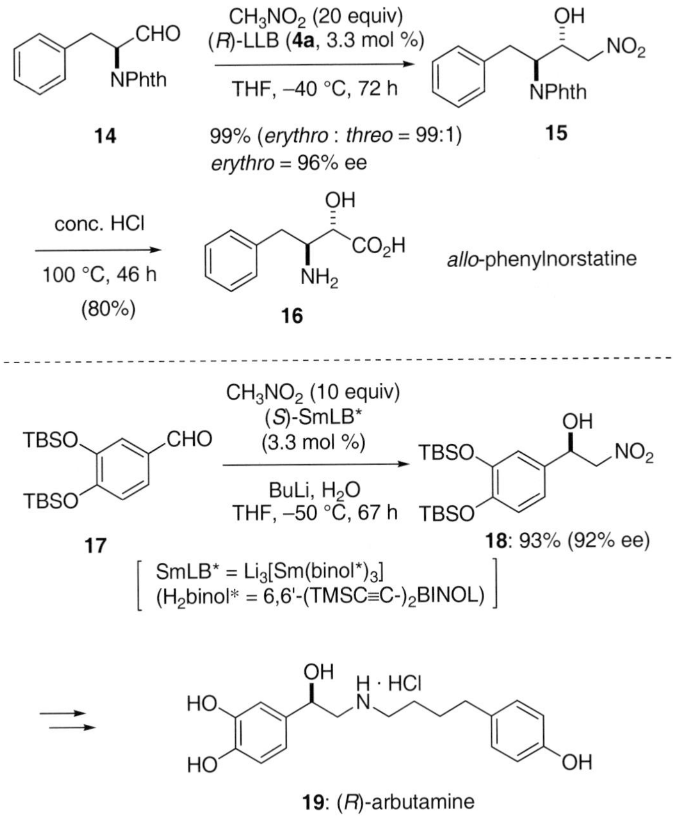

**Scheme 5.4** Catalytic asymmetric syntheses of *allo*-phenylnorstatin and arbutamine.

The catalytic asymmetric nitroaldol reaction has been applied to the synthesis of various optically active β-hydroxy nitroalkanes and to catalytic asymmetric syntheses of β-blockers (Scheme 5.3), *allo*-phenylnorstatine (Scheme 5.4), and (*R*)-arbutamine (Scheme 5.4) [9]. A tandem inter-intramolecular catalytic asymmetric nitroaldol reaction has also been reported [10].

Diastereo- and enantioselective reaction was achieved by employing a modified complex (**4e**) prepared from 6,6′-bis(triethylsilylethynyl)BINOL (**20e**) (Figure 5.4) [11]. Various nitroalkanes and nitroethanol were applicable, giving the corresponding *syn*-adducts with good diastereo- and enantioselectivities. The results are summarized in Table 5.1. The observed *syn*-selectivity was explained in terms of the Newman projections shown in Figure 5.5. This method was applied to the synthesis of *threo*-dihydrosphingosine (**24**) (Scheme 5.5).

Lanthanides exhibit a distinctive feature, known as "lanthanide contraction". The ionic radius of an octacoordinated trivalent lanthanide element decreases with increasing atomic number from 1.16 Å in La to 0.98 Å in Lu [12]. The effects of lanthanide contraction on catalyst selectivity in nitroaldol reactions are shown in Figure 5.6. The enantiomeric excesses of nitroaldols **3e–3g** obtained with the various heterobimetallic catalysts (LnLB) showed a dependence on the ionic radii of the rare earth metals [13]. Moreover, the best lanthanide differed depending on the particular substrate. These results indicate that reaction conditions can, in principle, be optimized simply by selecting the most suitable rare earth metal for the substrate.

**Table 5.1** Diastereo- and enantioselective nitroaldol reactions.

| | (*S*)-LLB* (**4a–4e**) (3.3 mol %) | | | | | | | | |
| RCHO **1** | R′CH$_2$NO$_2$ **2** ——→ THF | | | | | | | | |

**1a**: R = PhCH$_2$CH$_2$
**1b**: R = CH$_3$(CH$_2$)$_4$

**2b**: R′ = Me
**2c**: R′= Et
**2d**: R′ = CH$_2$OH

**3a**: R = PhCH$_2$CH$_2$, R′ = Me
**3b**: R = PhCH$_2$CH$_2$, R′ = Et
**3c**: R = PhCH$_2$CH$_2$, R′ = CH$_2$OH
**3d**: R = CH$_3$(CH$_2$)$_4$, R′ = CH$_2$OH

| Entry | Alde-hyde R | Nitro-alkane R′ | Cat. | Time (h) | Temp (°C) | Product | Yield (%) | syn/anti | ee (%) |
|-------|------|------|------|------|------|------|------|------|------|
| 1 | 1a | 2b | 4a | 75 | −20 | 3a | 79 | 74/26 | 66 |
| 2 | 1a | 2b | 4b | 75 | −20 | 3a | 80 | 74/26 | 65 |
| 3 | 1a | 2b | 4c | 75 | −20 | 3a | 77 | 84/16 | 90 |
| 4 | 1a | 2b | 4d | 75 | −20 | 3a | 72 | 85/15 | 92 |
| 5 | 1a | 2b | 4e | 57 | −20 | 3a | 70 | 89/11 | 93 |
| 6 | 1a | 2b | 4e | 115 | −40 | 3a | 21 | 94/6 | 97 |
| 7 | 1a | 2c | 4a | 138 | −40 | 3b | 89 | 85/15 | 87 |
| 8 | 1a | 2c | 4e | 138 | −40 | 3b | 85 | 93/7 | 95 |
| 9 | 1a | 2d | 4a | 111 | −40 | 3c | 62 | 84/16 | 66 |
| 10 | 1a | 2d | 4e | 111 | −40 | 3c | 97 | 92/8 | 97 |
| 11 | 1b | 2d | 4a | 93 | −40 | 3d | 70 | 87/13 | 78 |
| 12 | 1b | 2d | 4e | 93 | −40 | 3d | 96 | 92/8 | 95 |

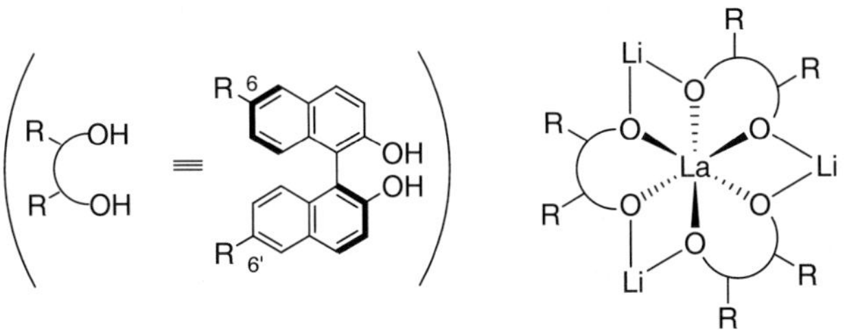

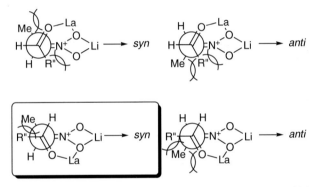

**20a**: R = H
**20b**: R = Me
**20c**: R = C≡C-H
**20d**: R = C≡CSiMe₃
**20e**: R = C≡CSiEt₃

**4a**: R = H
**4b**: R = Me
**4c**: R = C≡C-H
**4d**: R = C≡CSiMe₃
**4e**: R = C≡CSiEt₃

**Figure 5.4** Heterobimetallic complexes **4** prepared from 6,6′-disubstituted BINOLs **20**.

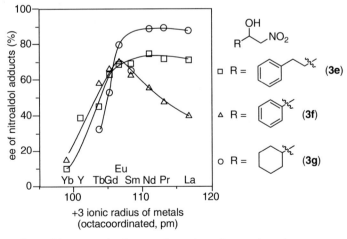

**Figure 5.5** Newman projections of intermediates in the diastereoselective nitroaldol reaction.

**Figure 5.6** Effects of the ionic radii of rare earth metals on the enantiomeric excess of nitroaldol adducts.

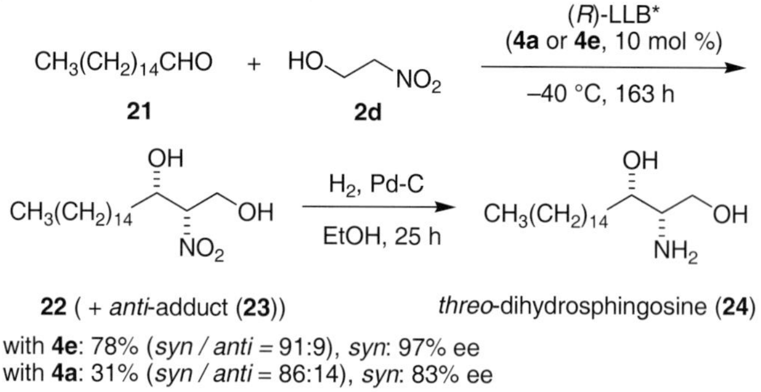

(R)-LLB*
(4a or 4e, 10 mol %)
CH3(CH2)14CHO
HO
NO2
–40 °C, 163 h
21
2d
OH
CH3(CH2)14
OH
NO2
H2, Pd-C
EtOH, 25 h
OH
CH3(CH2)14
OH
NH2
22 ( + anti-adduct (23))
threo-dihydrosphingosine (24)
with 4e: 78% (syn / anti = 91:9), syn: 97% ee
with 4a: 31% (syn / anti = 86:14), syn: 83% ee


**Scheme 5.5** Catalytic asymmetric synthesis of *threo*-dihydrosphingosine.

Although the heterobimetallic complexes afforded nitroaldol adducts with good stereoselectivities, most reactions required a long reaction time, even with relatively high catalyst loadings (3 to 10 mol %). To achieve more efficient catalysis, a strategy for accelerating the reaction was investigated. A plausible mechanism for the catalytic asymmetric nitroaldol reaction is shown in Scheme 5.6. The concentration of intermediate (A) was considered to be rather low in the reaction mixture because of the presence of an acidic OH group in the proximity of the lithium nitronate. The nitronate could be protonated by the OH group to give the nitroalkane and LLB (**4a**). To avoid this undesirable pathway and to accelerate the desired carbon–carbon bond formation, a catalytic amount of base (1 mol equiv. to La) was added to remove the proton from (A). In fact, the second-generation LLB (LLB-II;

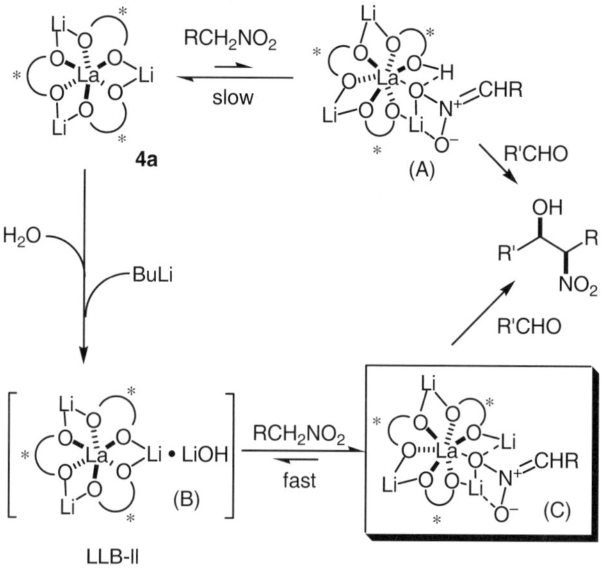

RCH2NO2
slow
4a
(A)
R'CHO
H2O
BuLi
R'CHO
Li • LiOH
(B)
RCH2NO2
fast
CHR
(C)
LLB-II


**Scheme 5.6** Proposed mechanism for the catalytic asymmetric nitroaldol reaction promoted by LLB or LLB-II (LLB · LiOH).

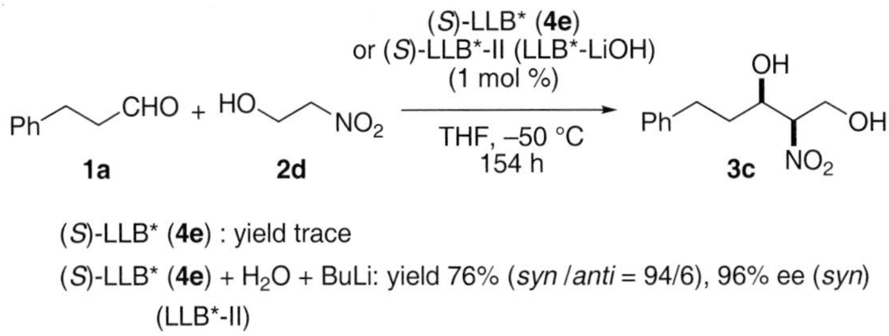

Scheme 5.7 shows the reaction scheme with the following conditions:

(S)-LLB* (4e)
or (S)-LLB*-II (LLB*-LiOH)
(1 mol %)

Ph ~ CHO + HO ~ NO₂ → (THF, –50 °C, 154 h) → Ph ~ OH, OH, NO₂

1a    2d    3c

(S)-LLB* (4e) : yield trace

(S)-LLB* (4e) + H₂O + BuLi: yield 76% (syn /anti = 94/6), 96% ee (syn)
(LLB*-II)

**Scheme 5.7** Acceleration of the nitroaldol reaction by second-generation LLB* (4e) (LLB* · LiOH).

LLB · LiOH), prepared from LLB, $H_2O$ (1 mol equiv. to La), and BuLi (0.9 mol equiv. to La), was found to efficiently accelerate catalytic asymmetric nitroaldol reactions. The nitroaldol reaction proceeded smoothly even with a reduced catalyst loading (1 mol %) as shown in Scheme 5.7 [14]. LLB (4a) and lithium nitronates evidently self-assemble to form complex (C), because the optical purity of the nitroaldols was the same as that obtained with LLB alone. Racemic reaction catalyzed by lithium nitronate alone was negligible.

## 5.4
## Direct Aldol Reaction with LLB · KOH Complex

The aldol reaction is one of the most useful carbon–carbon bond-forming reactions in which one or two stereogenic centers are concomitantly constructed. Diastereo- and enantioselective aldol reactions have been performed with excellent chemical yields and stereoselectivities using chiral catalysts [15]. Most cases, however, required the pre-conversion of donor substrates into more reactive species (26, Scheme 5.8), such as enol silyl ethers or ketene silyl acetals (Mukaiyama-type aldol addition reaction), using no less than stoichiometric amounts of silicon atoms and bases (Scheme 5.8 a). From an atom-economic point of view [16], such stoichiometric amounts of reagents, which afford wastes such as salts, should be excluded from the process. Thus, direct catalytic asymmetric aldol reaction, utilizing an unmodified

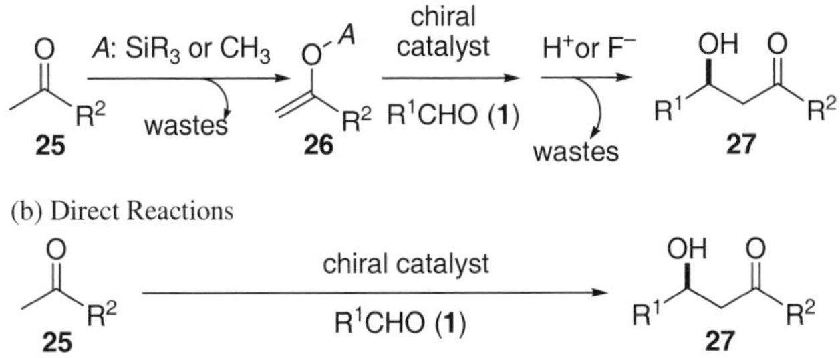

(a) Aldol Reactions with Latent Enolates

25 → (A: SiR₃ or CH₃, wastes) → 26 → (chiral catalyst, R¹CHO (1)) → (H⁺ or F⁻, wastes) → 27

(b) Direct Reactions

25 → (chiral catalyst, R¹CHO (1)) → 27

**Scheme 5.8** Aldol-type addition of latent enolates and direct aldol reaction of unmodified donors.

ketone or ester as a nucleophile (Scheme 5.8 b), is desirable. Many researchers have directed considerable attention to this field, which is reflected in the increasing number of publications [17]. In the early stages of the investigation, the subject appeared to be formidable, because the *in situ* formation of metal enolates from ketones is generally much less favorable than that from nitroalkanes due to the high $pK_a$ values of the former (nitroalkanes ca. 10, ketones ca. 17 in $H_2O$).

Various rare earth–alkali metal heterobimetallic catalysts, such as $Li_3[La(binol)_3]$ (LLB, **4a**), $Na_3[La(binol)_3]$ (LSB), and $K_3[La(binol)_3]$ (LPB), were screened for their efficacy in promoting the reaction of pivalaldehyde (**1c**) and acetophenone (**25a**). LLB afforded the aldol adducts with up to 94% ee [18]. Other complexes, such as $Na_3[La(binol)_3]$ and $K_3[La(binol)_3]$, showed much lower reactivity and selectivity.

**Table 5.2** Direct catalytic asymmetric aldol reactions of methyl ketone **25** promoted by (*S*)-LLB · KOH.

$$R^1CHO + \underset{\substack{25 \\ (3-15\ equiv)}}{\overset{O}{\underset{}{\|}}R^2} \xrightarrow[\text{THF, -60 to -20 °C}]{\substack{(S)\text{-LLB (3-8 mol \%)} \\ \text{KHMDS (7.2 mol \%)} \\ H_2O\ (16\ mol\ \%)}} \underset{27}{\overset{OH\quad O}{R^1 \underset{}{\phantom{x}} R^2}}$$

| Entry | Aldehyde | | Ketone (equiv.) | | Time (h) | Yield (%) | ee (%) |
|---|---|---|---|---|---|---|---|
| 1 | CHO (isopropyl-CHO) | 1c | -Ph | 25a (5) | 15 | 75 | 88 |
| 2 | | 1c | -Ph | 25a (5) | 28 | 85 | 89 |
| 3 | | 1c | -CH$_3$ | 25b (10) | 20 | 62 | 76 |
| 4 | | 1c | -CH$_2$CH$_3$ | 25c (15) | 95 | 72 | 88 |
| 5 | Ph⌒CHO | 1d | -Ph | 25a (5) | 18 | 83 | 85 |
| 6[a)] | | 1d | -Ph | 25a (5) | 33 | 71 | 85 |
| 7 | BnO⌒CHO | 1e | -Ph | 25a (5) | 36 | 91 | 90 |
| 8 | | 1e | -Ph | 25a (5) | 24 | 70 | 93 |
| 9 | ⌒CHO | 1f | -Ph | 25a (5) | 15 | 90 | 33 |
| 10 | | 1f | -*m*-NO$_2$-C$_6$H$_4$ | 25d (3) | 70 | 68 | 70 |
| 11 | ⌒CHO | 1g | -*m*-NO$_2$-C$_6$H$_4$ | 25d (3) | 96 | 60 | 80 |
| 12 | ⌒⌒CHO | 1h | -*m*-NO$_2$-C$_6$H$_4$ | 25d (5) | 96 | 55 | 42 |
| 13 | Ph⌒⌒CHO | 1a | -*m*-NO$_2$-C$_6$H$_4$ | 25d (3) | 31 | 50 | 30 |

[a)] (*S*)-LLB (3 mol %), KHMDS (2.7 mol %), $H_2O$ (6 mol %) were used.

Although enantioselectivity was relatively good in an initial trial (up to 94% ee), the catalytic activity of LLB itself was rather low, requiring a catalyst loading of at least 20 mol % and anhydrous reaction conditions. After optimization, the addition of catalytic amounts of bases (KOH) proved effective in enhancing the catalytic activity [19]. For example, the reaction between **1d** and **25a** reached completion after 18 h to afford the product in 83% yield and with 85% ee, when potassium bis(trimethyl-silyl)amide (KHMDS) (0.9 mol equiv. to La) and $H_2O$ (2 mol equiv. to La) were added (Table 5.2, entry 5). Because the catalytic species is likely to consist of three different metals (La, Li, and K), the catalyst was called a "heteropolymetallic catalyst, LLB · KOH". Table 5.2 summarizes the results of aldol reactions using the LLB · KOH catalyst. The aldol products were obtained from α,α-disubstituted aldehydes with ee values ranging from 76 to 93% (entries 1–8).

Because KOH, generated from KHMDS and $H_2O$, dissolves in THF upon addition of LLB ($Li_3[La(binol)_3]$), it was postulated that the KOH interacts with LLB. A plausible reaction mechanism is outlined in Figure 5.7, in which the ketone is deprotonated by KOH, and the aldehyde is activated and fixed by the lanthanum metal. As was also mentioned in the catalytic cycle for the nitroaldol reaction, the reaction proceeds through simple proton transfer without any waste, enabling atom-economical transformation. Initial rate kinetic studies and an observed isotope effect in a kinetic study using [$D_3$]acetophenone ($k_H/k_D = 5$) indicated that the rate-determining step is enolate formation. The coordination of aldehydes to the central metal of the catalyst was confirmed by a $^1H$ NMR study. An upfield shift of the formyl hydrogen signal in pivalaldehyde was observed upon addition of 20 mol % of PrLB ($Li_3[Pr(binol)_3]$), whereas there was no shift after the addition of $Li_2(binol)$.

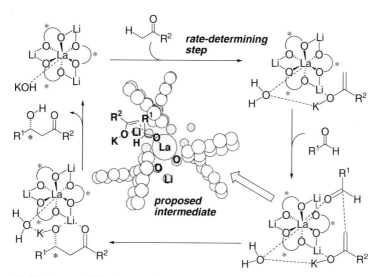

**Figure 5.7** Working model for direct catalytic asymmetric aldol reactions promoted by (S)-LLB · KOH complex.

The direct aldol reaction using LLB · KOH catalyst has been applied to the resolution of racemic aldehyde **28** in an enantioselective total synthesis of epothilones (**30**) (Scheme 5.9) [20]. The aldol reaction has also been applied in the formal total synthesis of fostriecin, in which anhydrous LLB itself was used as the catalyst (Scheme 5.10) [21].

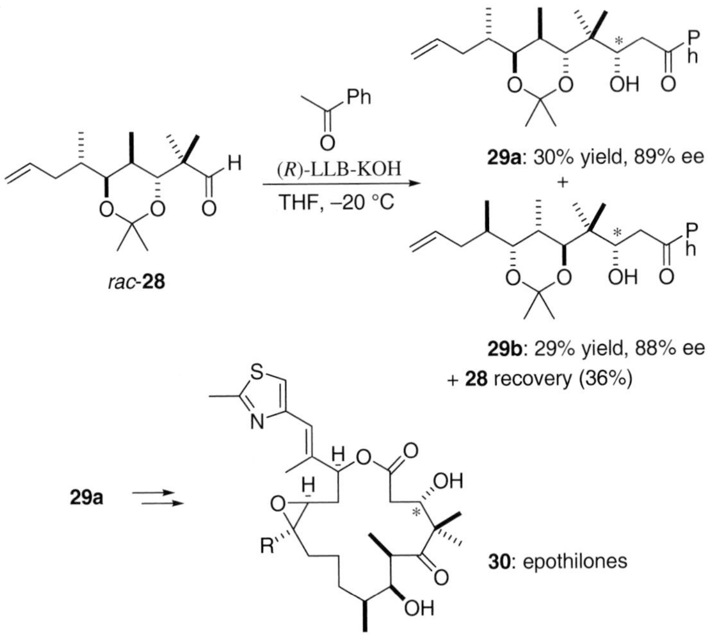

**Scheme 5.9** Catalytic resolution of racemic aldehyde **28** by LLB · KOH in the enantioselective total synthesis of epothilones.

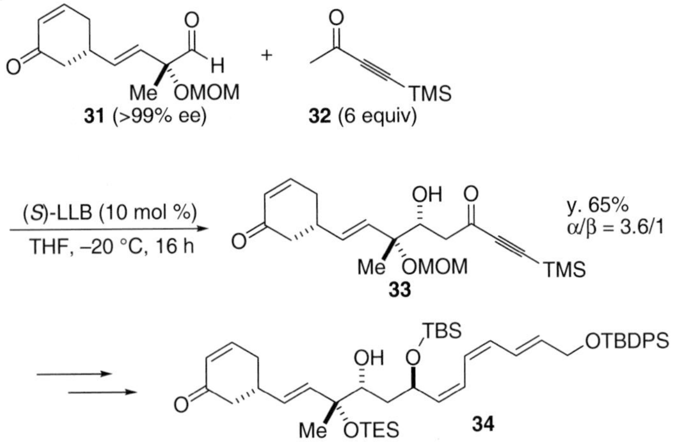

**Scheme 5.10** Application to the formal total synthesis of fostriecin.

**Table 5.3** Direct catalytic asymmetric aldol reaction of hydroxy ketone **35** promoted by (S)-LLB · KOH.

| Entry | Aldehyde | | Ketone (R-) | | Time (h) | Yield (%) | dr (anti/syn) | ee (%) (anti/syn) |
|---|---|---|---|---|---|---|---|---|
| 1 | Ph⌒⌒CHO | 1i | H- | 35a | 24 | 84 | 84/16 | 95/74 |
| 2 | | 1i | H- | 35a | 40 | 78 | 78/22 | 92/70 |
| 3 | | 1i | 4-MeO- | 35b | 35 | 50 | 81/19 | 98/79 |
| 4 | | 1i | 2-Me- | 35c | 35 | 90 | 77/23 | 84/57 |
| 5 | | 1i | 4-Me- | 35d | 35 | 90 | 83/17 | 97/85 |
| 6 | ⌒⌒⌒CHO | 1b | 4-Me- | 35d | 12 | 96 | 75/25 | 96/89 |
| 7 | ⌒⌒⌒CHO | 1j | H- | 35a | 28 | 90 | 72/28 | 94/83 |
| 8 | ⌒CHO | 1k | H- | 35a | 24 | 86 | 65/35 | 90/83 |

The LLB · KOH catalyst also proved applicable to the aldol reaction with hydroxy ketones as donors, affording *anti*-diols with excellent enantiomeric excesses and moderate to good diastereoselectivities (Table 5.3) [22].

## 5.5
## Application to Catalytic Asymmetric 1,4-Addition Reactions

While investigating the structure of Li$_3$[La(binol)$_3$] (LLB, **4a**), a heterobimetallic complex consisting of La, Na, and BINOL (Na$_3$[La(binol)$_3$], LSB) was also synthesized according to a procedure similar to that for LLB. The structure of LSB was determined by X-ray crystallographic analysis [2]. As mentioned above in Section 5.3, LLB proved effective for the nitroaldol reaction; however, it was not at all effective for the Michael reaction, as shown in Table 5.4, entry 2. With LSB as catalyst, this tendency was reversed. LSB was effective for the Michael reaction, but not for the nitroaldol reaction. These contrasting results illustrate how a subtle change in the structure and charge distribution of a heterobimetallic catalyst can affect the enantioselection of the targeted reaction. Substituted or unsubstituted malonates reacted with α,β-unsaturated ketones to give the Michael adducts in good yields and with high enantiomeric excesses. The results are summarized in Tables 5.4 and 5.5 [23, 24].

**Table 5.4** Catalytic asymmetric Michael reaction: enantioselection on acceptors.

37a: n = 2  
37b: n = 1  

38a: R$^1$ = Bn, R$^2$ = H  
38b: R$^1$ = Bn, R$^2$ = Me  
38c: R$^1$ = Me, R$^2$ = H  

39a: n = 2, R$^1$ = Bn, R$^2$ = H  
39b: n = 2, R$^1$ = Bn, R$^2$ = Me  
39c: n = 2, R$^1$ = Me, R$^2$ = H  
39d: n = 1, R$^1$ = Bn, R$^2$ = Me  

| Entry | Enone | Malonate | Product | Cat. | Temp. (°C) | Time (h) | Yield (%) | ee (%) |
|---|---|---|---|---|---|---|---|---|
| 1 | 37a | 38a | 39a | LSB | rt | 12 | 98 | 85 |
| 2 | 37a | 38a | 39a | LLB | rt | 12 | 78 | 2 |
| 3 | 37a | 38b | 39b | LSB | 0 | 24 | 91 | 92 |
| 4 | 37a | 38c | 39c | LSB | rt | 12 | 98 | 83 |
| 5 | 37b | 38b | 39d | LSB | −40 | 36 | 89 | 72 |

**Table 5.5** Catalytic asymmetric Michael reaction: enantioselection on donors.

| Michael donor | Michael acceptor | Product | Cat. ((R)-LSB) Amount (mol %) | Time (h) | Yield (%) | ee (%) |
|---|---|---|---|---|---|---|
| 40a | 41a | 42a | 5 | 19 | 89 | 91 |
| 40b | 41a | 42b | 5 | 16 | 93 | 83 |
| 40c | 41a | 42c | 5 | 16 | 98 | 89 |
| 40d | 41b | 42d | 20 | 18 | 97 | 84 |
| 40e | 41c | 42e | 20 | 93 | 69 | 89 |

Heterobimetallic complexes containing sodium (Na$_3$[Ln(binol$_3$)], LnSB) were found to efficiently promote the addition of thiols **43a–c** to α,β-unsaturated thioesters and α,β-unsaturated ketones [25]. LSB (Na$_3$[La(binol$_3$)]) and SmSB (Na$_3$[Sm(binol$_3$)]) afforded the products with up to 93% ee (Tables 5.6 and 5.7). The chirality of the products in Table 5.7 was generated in the protonation step. In other words, catalytic

**Table 5.6** Catalytic asymmetric conjugate addition of thiols to enones.

| Entry | Enone | | | R² | | Product | Time | Yield (%) | ee (%) |
|---|---|---|---|---|---|---|---|---|---|
| 1 | n = 2, R¹ = H | 37a | | 4-t-BuPh | 43a | 44a | 20 min | 93 | 84 |
| 2 | n = 2, R¹ = H | 37a | | Ph | 43b | 44b | 20 min | 87 | 68 |
| 3 | n = 2, R¹ = H | 37a | | PhCH₂ | 43c | 44c | 14 h | 86 | 90 |
| 4 | n = 1, R¹ = H | 37b | | PhCH₂ | 43c | 44d | 4 h | 94 | 56 |
| 5[a] | n = 3, R¹ = H | 37c | | PhCH₂ | 43c | 44e | 41 h | 87 | 83 |
| 6[a], [b] | n = 2, R¹ = Me | 37d | | PhCH₂ | 43c | 44f | 43 h | 56 | 85 |

[a] 20 mol % of catalyst was used, and toluene was used as solvent.
[b] At –20 °C.

**Table 5.7** Catalytic asymmetric protonations in conjugate additions of thiols.

| Enone | | Product | Ln | Cat. | Temp. | Time | Yield | ee |
|---|---|---|---|---|---|---|---|---|
| R¹ | R² | | | (mol %) | (°C) | (h) | (%) | (%) |
| EtO | Me | 45a 46a | La | 20 | –20 | 48 | 50 | 82 |
| EtS | Me | 45b 46b | La | 20 | –78 | 2 | 93 | 90 |
| EtS | Me | 45b 46b | La | 10 | –78 | 8 | 90 | 88 |
| EtS | Me | 45b 46b | Sm | 10 | –78 | 7 | 86 | 93 |
| EtS | Me | 45b 46b | Sm | 2 | –78 | 6 | 89 | 88 |
| EtS | i-Pr | 45c 46c | Sm | 10 | –78 | 7 | 78 | 90 |
| EtS | PhCH₂ | 45d 46d | Sm | 10 | –78 | 7 | 89 | 87 |
| EtS | Ph | 45e 46e | Sm | 10 | –93 | 1 | 98 | 84 |

[a] Reaction was performed in toluene.

asymmetric protonation using a heterobimetallic complex was achieved. The proposed mechanism is shown in Scheme 5.11. The reaction was applied to the total synthesis of epothilones, together with a direct aldol reaction [20].

For the 1,4-addition of nitromethane to enones, the catalyst LPB ($K_3[La(binol_3)]$) gave the best results, as shown in Scheme 5.12. The addition of a proton source was crucial in promoting this reaction with high selectivity [26].

Recently, the catalyst YLB ($Li_3[Y(binol_3)]$) has been found to promote a few asymmetric transformations through a different reaction mechanism to those of the above-mentioned reactions, opening a new possibility for heterobimetallic catalysis. Thus, anhydrous YLB proved effective in promoting catalytic asymmetric 1,4-addition of O-alkylhydroxylamines [27]. As summarized in Table 5.8, the

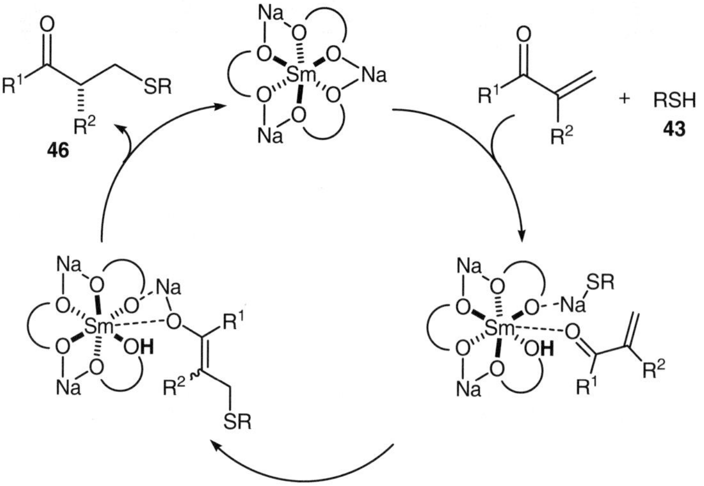

**Scheme 5.11** Proposed mechanism for catalytic asymmetric protonation in the conjugate addition of thiols.

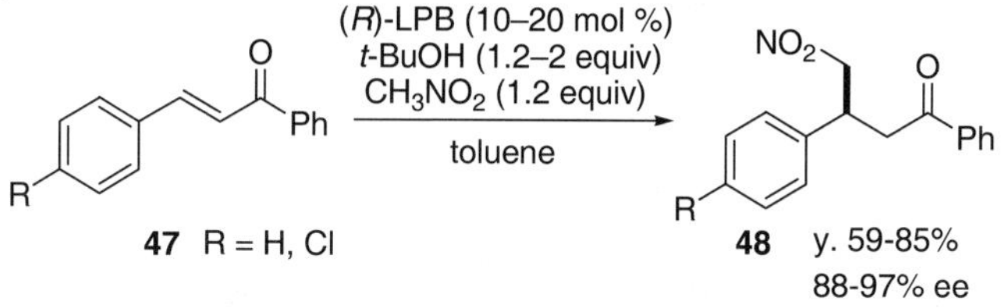

**Scheme 5.12** Catalytic asymmetric Michael addition of nitromethane to enones.

reactions proceeded smoothly with catalyst loadings of as little as 0.5–3 mol %. The results indicated that neither the amine nor the product inhibited the catalytic activity of YLB, even with a reduced catalyst loading, unlike in simple Lewis acid catalysis. Preliminary mechanistic studies revealed that the heterobimetallic cooperative function of the Y and Li metal centers played a key role in achieving high catalyst turnover. Neither Li-BINOL, nor Y-BINOL, nor YPB ($K_3[Y(binol_3)]$) was effective for the reaction at all (Table 5.9). The mechanistic aspects of this reaction are also noteworthy. On the basis of mechanistic studies on heterobimetallic catalysis, the Y metal center might function as the Lewis acid that activates the enone. Considering the $pK_a$ value of the amine proton, however, it is unlikely that Li-binaphthoxide functions as a Brønsted base and activates the amine moiety. Thus, the role of the heterobimetallic catalysis appears to be different from that in other examples. On the basis of the different results obtained with Li (YLB) and K (YPB) (Table 5.9), the oxygen atom of **49** might coordinate to Li; **49** would then be positioned close to enone **47**, and the addition reaction would be accelerated (Figure 5.8). The interaction between the oxygen atom of **49** and K in YPB should be weaker than that with Li in YLB.

**Table 5.8** Catalytic asymmetric 1,4-addition of O-methylhydroxylamine (**49**) promoted by YLB.

| Entry | Enone R¹ | R² | | Product | | YLB (mol %) | Time (h) | Yield (%) | ee (%) |
|---|---|---|---|---|---|---|---|---|---|
| 1 | Ph | Ph | | 47a | 50a | 3 | 42 | 97 | 95 |
| 2 | Ph | Ph | | 47a | 50a | 1 | 48 | 95 | 96 |
| 3 | Ph | Ph | | 47a | 50a | 0.5 | 80 | 96 | 96 |
| 4 | 4-Cl-C₆H₄ | Ph | | 47b | 50b | 3 | 42 | 96 | 96 |
| 5 | 4-Cl-C₆H₄ | Ph | | 47b | 50b | 1 | 46 | 92 | 96 |
| 6 | 4-F-C₆H₄ | Ph | | 47c | 50c | 3 | 54 | 97 | 96 |
| 7[a] | 4-F-C₆H₄ | Ph | | 47c | 50c | 1 | 65 | 91 | 96 |
| 8 | 4-Me-C₆H₄ | Ph | | 47d | 50d | 3 | 48 | 96 | 94 |
| 9[a] | 4-MeO-C₆H₄ | Ph | | 47e | 50e | 3 | 74 | 91 | 96 |
| 10 | 3-Me-C₆H₄ | Ph | | 47f | 50f | 3 | 48 | 96 | 92 |
| 11[a] | 2-furyl | Ph | | 47g | 50g | 3 | 48 | 95 | 94 |
| 12[a] | 2-thienyl | Ph | | 47h | 50h | 3 | 78 | 96 | 93 |
| 13 | Ph | 4-Cl-C₆H₄ | | 47i | 50i | 3 | 48 | 92 | 92 |
| 14 | Ph | 4-Cl-C₆H₄ | | 47i | 50i | 1 | 78 | 97 | 93 |
| 15 | Ph | 4-Me-C₆H₄ | | 47j | 50j | 3 | 48 | 96 | 96 |
| 16[a] | Ph | 4-MeO-C₆H₄ | | 47k | 50k | 3 | 82 | 85 | 95 |
| 17[a] | Ph | 4-MeO-C₆H₄ | | 47k | 50k | 1 | 74 | 85 | 95 |
| 18 | Ph | 3-NO₂-C₆H₄ | | 47l | 50l | 3 | 42 | 98 | 81 |
| 19 | Ph | 3-Cl-C₆H₄ | | 47m | 50m | 3 | 48 | 95 | 92 |
| 20[a] | Ph | 2-Cl-C₆H₄ | | 47n | 50n | 3 | 122 | 92 | 82 |
| 21[a] | Ph | 2-furyl | | 47o | 50o | 3 | 84 | 80 | 92 |
| 22 | Ph | 2-thienyl | | 47p | 50p | 3 | 48 | 96 | 95 |
| 23[a] | Ph | 4-pyridyl | | 47q | 50q | 3 | 60 | 91 | 85 |
| 24[a] | Ph | n-C₅H₁₁ | | 47r | 50r | 3 | 84 | 96 | 84 |
| 25[a] | Ph | i-PrCH₂ | | 47s | 50s | 3 | 48 | 95 | 93 |
| 26 | Ph | i-Pr | | 47t | 50t | 3 | 78 | 97 | 86 |
| 27 | Ph | cyclo-hexyl | | 47u | 50u | 3 | 48 | 98 | 82 |
| 28[a] | Ph | t-Bu | | 47v | 50v | 3 | 96 | 57 | 82 |
| 29[b] | Ph | trans-PhCH=CH | | 47w | 50w | 3 | 84 | 91 | 95 |

**Table 5.9** Catalytic asymmetric 1,4-addition of O-methylhydroxylamine (**49**) using various metal complexes.

| Entry | Catalyst (x mol %) | Time (h) | Yield (%) | ee (%) | Config. |
|---|---|---|---|---|---|
| 1 | none | 42 | trace | – | – |
| 2 | BuLi/BINOL (9/9) | 42 | 11 | 12 | R |
| 3 | Y(HMDS)₃/BINOL (3/9) | 42 | 29 | 16 | R |
| 4 | YPB (3) | 42 | 19 | 12 | R |
| 5 | YLB (3) | 42 | 97 | 95 | S |

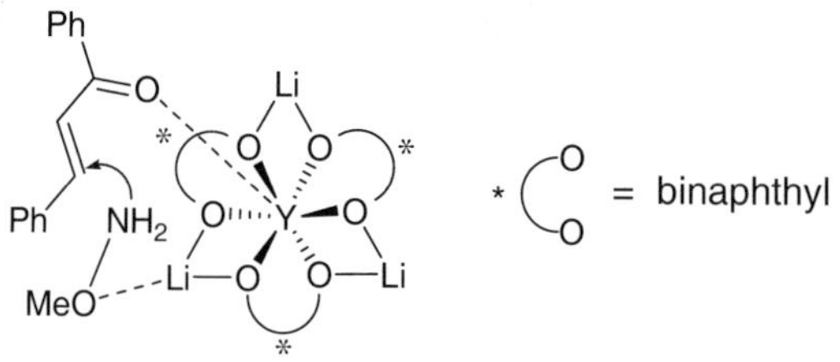

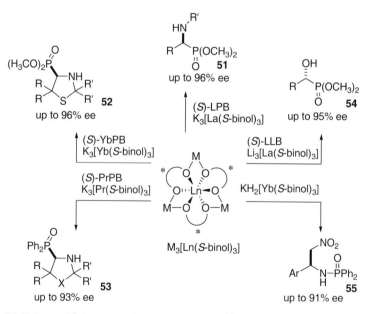

**Figure 5.8** Postulated model for 1,4-addition of **49** catalyzed by YLB.

## 5.6
## Other Examples

Rare earth–alkali metal heterobimetallic complexes have also been shown to promote other reactions with good selectivity (> 90% ee) through dual Lewis acid/Brønsted base activation, as summarized in Figure 5.9 [28–32]. In all cases, the reaction proceeded through simple proton transfer from a donor substrate to an acceptor substrate with concomitant formation of a new carbon–carbon or carbon–heteroatom bond. Vallée et al. [33] reported the hydrocyanation of aldimines and ketoimines as shown in Scheme 5.13. A novel heterobimetallic Li[Sc(binol)$_2$] complex afforded the best results in their Strecker-type reaction.

**Figure 5.9** Various catalytic asymmetric reactions promoted by rare earth–alkali metal heterobimetallic asymmetric catalysts.

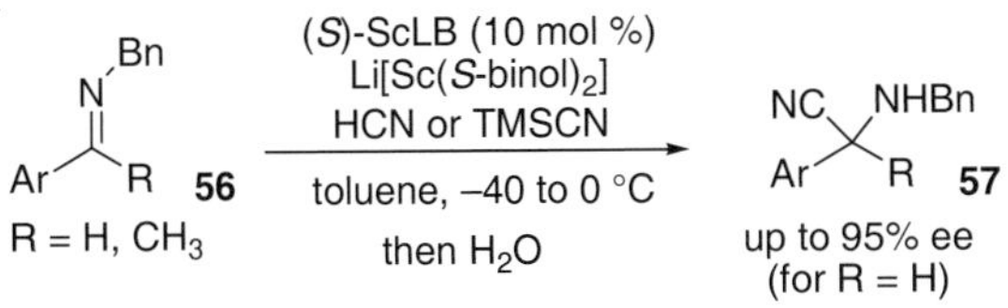

**Scheme 5.13** Strecker-type reaction catalyzed by Sc-Li complex.

## 5.7
## Miscellaneous Examples

As summarized in Table 5.10, the YLB · $H_2O$ complex was found to promote a catalytic asymmetric cyanoethoxycarbonylation reaction [34]. In this reaction, $H_2O$, BuLi, and $Ar_3P(O)$ all had to be added in order to achieve high enantiomeric excess and good reactivity. A total synthesis of patulolide C was achieved using catalytic asymmetric cyanoethoxycarbonylation reaction and chiral transfer by sigmatropic rearrangement as key steps (Scheme 5.14) [35]. The reaction mechanism is expected to be completely different from those of the other reactions described in this chapter, although detailed mechanistic studies are still under investigation.

By utilizing the multi-interaction activation of YLB catalysis, a sequential catalytic asymmetric cyanation–nitroaldol reaction with a single catalyst component was demonstrated, as summarized in Scheme 5.15 [35]. For this sequential reaction,

**Table 5.10** Catalytic asymmetric cyanoethoxycarbonylation reaction.

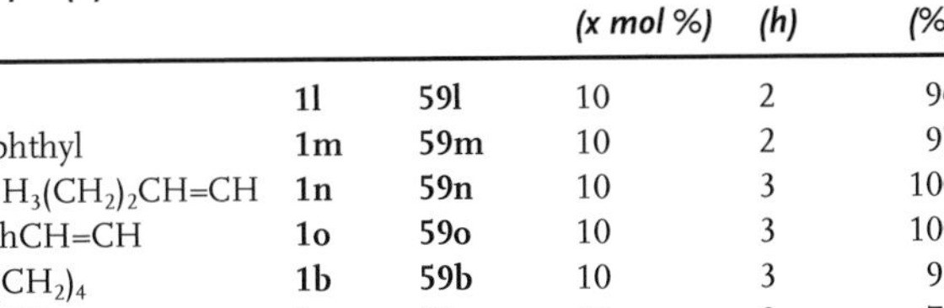

| Entry | Aldehyde (R) | | Product | Cat. (x mol %) | Time (h) | Yield (%) | ee (%) |
|-------|--------------|---|---------|----------------|----------|-----------|--------|
| 1 | Ph | 1l | 59l | 10 | 2 | 96 | 94 |
| 2 | 1-naphthyl | 1m | 59m | 10 | 2 | 97 | 90 |
| 3 | $(E)$-$CH_3(CH_2)_2CH=CH$ | 1n | 59n | 10 | 3 | 100 | 92 |
| 4 | $(E)$-PhCH=CH | 1o | 59o | 10 | 3 | 100 | 91 |
| 5 | $CH_3(CH_2)_4$ | 1b | 59b | 10 | 3 | 93 | 94 |
| 6[a] | $CH_3CH_2$ | 1p | 59p | 10 | 2 | 79 | 92 |
| 7[a] | $(CH_3)_2CH$ | 1f | 59f | 10 | 2 | 88 | 98 |
| 8[a] | $(CH_3)_2CH$ | 1f | 59f | 5 | 2 | 82 | 96 |
| 9[b] | $(CH_3)_2CH$ | 1f | 59f | 1 | 9 | 96 | 90 |
| 10 | $c$-$C_6H_{11}$ | 1q | 59q | 10 | 2 | 97 | 96 |
| 11 | $(CH_3)_3C$ | 1c | 59c | 10 | 3 | 93 | 87 |

[a] Isolated yield became lower because product was volatile.
[b] Reaction was performed on 6.0 mmol scale.

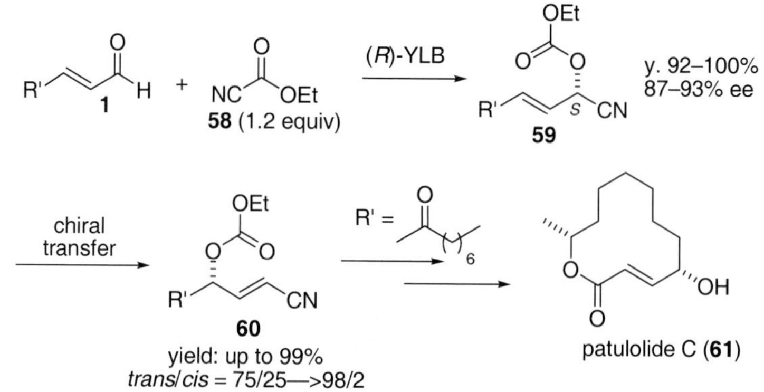

**Scheme 5.14** Catalytic asymmetric total synthesis of patulolide C.

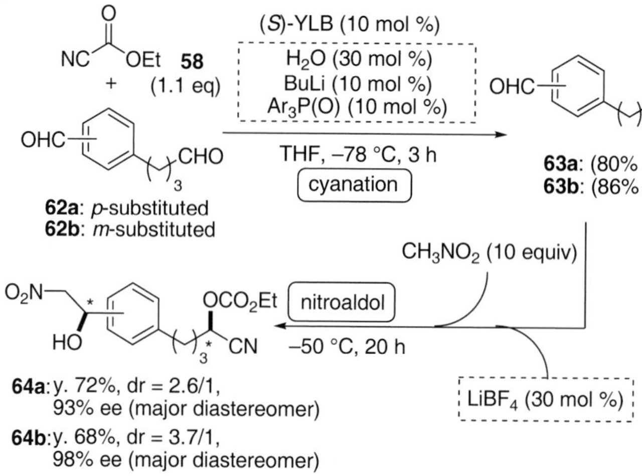

**Scheme 5.15** Sequential catalytic asymmetric cyanation-nitroaldol reaction with suitable catalyst tuning.

both the catalyst structure and reactivity were finely tuned by the addition of achiral additives. In particular, the addition of LiBF$_4$ was found to play a crucial role in achieving sequential asymmetric catalysis.

## 5.8
## Summary

Rare earth–alkali metal heterobimetallic complexes have been found to promote various catalytic asymmetric reactions, as overviewed in this chapter. Synergistic functions between the Lewis acidic metal and Brønsted basic moieties play a key role in these reactions. In most cases, the reaction proceeds through proton transfer, thus affording enantiomerically enriched products without the generation of any waste such as salts. It is noteworthy that the scope of reactions that can be performed with heterobimetallic complexes is still increasing, even though it is more than ten years since the first report in 1992. Modification of the catalysts through appropriate tuning with achiral additives continues to offer the opportunity to further refine the functions of heterobimetallic asymmetric catalysis.

## References and Notes

1   Reviews: (a) E. N. JACOBSEN, A. PFALTZ, H. YAMAMOTO (Eds.), *Comprehensive Asymmetric Catalysis*, Springer, New York, 1999; (b) I. OJIMA (Ed.), *Catalytic Asymmetric Synthesis*, 2nd ed., Wiley, New York, 2000.

2   Reviews: (a) M. SHIBASAKI, N. YOSHIKAWA, *Chem. Rev.* **2002**, *102*, 2187; (b) G. J. ROWLANDS, *Tetrahedron* **2001**, *57*, 1865; (c) M. SHIBASAKI, H. SASAI, T. ARAI, *Angew. Chem. Int. Ed. Engl.* **1997**, *36*, 1236.

3   (a) H. SASAI, T. SUZUKI, S. ARAI, T. ARAI, M. SHIBASAKI, *J. Am. Chem. Soc.* **1992**, *114*, 4418; (b) H. SASAI, T. SUZUKI, N. ITOH, M. SHIBASAKI, *Tetrahedron Lett.* **1993**, *34*, 851.

4   H. SASAI, T. SUZUKI, N. ITOH, K. TANAKA, T. DATE, K. OKAMURA, M. SHIBASAKI, *J. Am. Chem. Soc.* **1993**, *115*, 10372.

5   Ln(O$^i$Pr)$_3$ purchased from Kojundo Chemical Laboratory afforded the best results. The quality of the Ln(O$^i$Pr)$_3$ has an effect on the reactivity and enantio-selectivity of the reactions, probably due to contamination by small amounts of impurities such as halide ion. Kojundo Chemical Laboratory Co., Ltd.; e-mail: sales@kojundo.co.jp.

6   H. SASAI, S. WATANABE, M. SHIBASAKI, *Enantiomer* **1996**, *2*, 267.

7   (a) H. C. ASPINALL, *Chem. Rev.* **2002**, *102*, 1807; (b) H. C. ASPINALL, J. L. M. DWYER, N. GREEVES, A. STEINER, *Organometallics* **1999**, *18*, 1366; (c) H. C. ASPINALL, J. F. BICKLEY, J. L. M. DWYER, N. GREEVES, R. V. KELLY, A. STEINER, *Organometallics* **2000**, *19*, 5416.

8   L. DI BARI, M. LELLI, G. PINTACUDA, G. PESCITELLI, F. MARCHETTI, P. SALVADORI, *J. Am. Chem. Soc.* **2003**, *125*, 5549.

9   (a) H. SASAI, N. ITOH, T. SUZUKI, M. SHIBASAKI, *Tetrahedron Lett.* **1993**, *34*, 855; (b) H. SASAI, Y. M. A. YAMADA, T. SUZUKI, M. SHIBASAKI, *Tetrahedron* **1994**, *50*, 12313; (c) H. SASAI, T. SUZUKI, N. ITOH, M. SHIBASAKI, *Appl. Organomet. Chem.* **1995**, *9*, 421; (d) H. SASAI, W.-S. KIM, T. SUZUKI, M. SHIBASAKI, M. MITSUDA, J. HASEGAWA, T. OHASHI, *Tetrahedron Lett.* **1994**, *35*, 6123; (e) E. TAKAOKA, N. YOSHIKAWA, Y. M. A. YAMADA, H. SASAI, M. SHIBA-SAKI, *Heterocycles* **1997**, *46*, 157.

10  H. SASAI, M. HIROI, Y. M. A. YAMADA, M. SHIBASAKI, *Tetrahedron Lett.* **1997**, *38*, 6031.

11  H. SASAI, T. TOKUNAGA, S. WATANABE, T. SUZUKI, N. ITOH, M. SHIBASAKI, *J. Org. Chem.* **1995**, *60*, 7388.

12  (a) W. A. HERRMANN (Ed.), *Organo-lanthanoid Chemistry: Synthesis, Structure, Catalysis. Topics in Current Chemistry 179*, Springer, Berlin, 1996; (b) S. P. SINHA, *Structure and Bonding*, 25, Springer-Verlag, New York, 1976, p. 69.

13  H. SASAI, T. SUZUKI, N. ITOH, S. ARAI, M. SHIBASAKI, *Tetrahedron Lett.* **1993**, *34*, 2657.

14  T. ARAI, Y. M. A. YAMADA, N. YAMAMOTO, H. SASAI, M. SHIBASAKI, *Chem. Eur. J.* **1996**, *2*, 1368.

15  For a review of catalytic asymmetric aldol reactions, see: (a) C. PALOMO, M. OIARBIDE, J. M. GARCÍA, *Chem. Eur. J.* **2002**, *8*, 37; (b) J. S. JOHNSON, D. A. EVANS, *Acc. Chem. Res.* **2000**, *33*, 325; (c) T. D. MACHAJEWSKI, C. H. WONG, *Angew. Chem. Int. Ed.* **2000**, *39*, 1352; (d) H. GRÖGER, E. M. VOGL, M. SHIBA-SAKI, *Chem. Eur. J.* **1998**, *4*, 1137.

16  B. M. TROST, *Science* **1991**, *254*, 1471.

17 Reviews on direct catalytic asymmetric aldol reaction: (a) B. ALCAIDE, P. ALMENDROS, *Eur. J. Org. Chem.* **2002**, 1595; (b) B. LIST, *Tetrahedron* **2002**, *58*, 5573. See also ref. 2a and references therein.

18 Y. M. A. YAMADA, N. YOSHIKAWA, H. SASAI, M. SHIBASAKI, *Angew. Chem. Int. Ed. Engl.* **1997**, *36*, 1871.

19 N. YOSHIKAWA, Y. M. A. YAMADA, J. DAS, H. SASAI, M. SHIBASAKI, *J. Am. Chem. Soc.* **1999**, *121*, 4168.

20 (a) D. SAWADA, M. SHIBASAKI, *Angew. Chem. Int. Ed.* **2000**, *39*, 209; (b) D. SAWADA, M. KANAI, M. SHIBASAKI, *J. Am. Chem. Soc.* **2000**, *122*, 10521.

21 K. FUJII, K. MAKI, M. KANAI, M. SHIBASAKI, *Org. Lett.* **2003**, *5*, 733.

22 (a) N. YOSHIKAWA, N. KUMAGAI, S. MATSUNAGA, G. MOLL, T. OHSHIMA, T. SUZUKI, M. SHIBASAKI, *J. Am. Chem. Soc.* **2001**, *123*, 2466; (b) N. YOSHIKAWA, T. SUZUKI, M. SHIBASAKI, *J. Org. Chem.* **2002**, *67*, 2556.

23 H. SASAI, T. ARAI, Y. SATOW, K. N. HOUK, M. SHIBASAKI, *J. Am. Chem. Soc.* **1995**, *117*, 6194.

24 H. SASAI, E. EMORI, T. ARAI, M. SHIBASAKI, *Tetrahedron Lett.* **1996**, *37*, 5561.

25 E. EMORI, T. ARAI, H. SASAI, M. SHIBASAKI, *J. Am. Chem. Soc.* **1998**, *120*, 4043.

26 K. FUNABASHI, Y. SAIDA, M. KANAI, T. ARAI, H. SASAI, M. SHIBASAKI, *Tetrahedron Lett.* **1998**, *39*, 7557.

27 N. YAMAGIWA, S. MATSUNAGA, M. SHIBASAKI, *J. Am. Chem. Soc.* **2003**, *125*, 16178.

28 H. SASAI, S. ARAI, Y. TAHARA, M. SHIBASAKI, *J. Org. Chem.* **1995**, *60*, 6656.

29 (a) H. GRÖGER, Y. SAIDA, S. ARAI, J. MARTENS, H. SASAI, M. SHIBASAKI, *Tetrahedron Lett.* **1996**, *37*, 9291; (b) H. GRÖGER, Y. SAIDA, H. SASAI, K. YAMAGUCHI, J. MARTENS, M. SHIBASAKI, *J. Am. Chem. Soc.* **1998**, *120*, 3089; (c) I. SCHLEMMINGER, Y. SAIDA, H. GRÖGER, W. MAISON, N. DUROT, H. SASAI, M. SHIBASAKI, J. MARTENS, *J. Org. Chem.* **2000**, *65*, 4818.

30 K. YAMAKOSHI, S. J. HARWOOD, M. KANAI, M. SHIBASAKI, *Tetrahedron Lett.* **1999**, *40*, 2565.

31 (a) H. SASAI, M. BOUGAUCHI, T. ARAI, M. SHIBASAKI, *Tetrahedron Lett.* **1997**, *38*, 2717; (b) T. YOKOMATSU, T. YAMAGISHI, S. SHIBUYA, *Tetrahedron: Asymmetry* **1993**, *4*, 1783; (c) N. P. RATH, C. D. SPILLING, *Tetrahedron Lett.* **1994**, *35*, 227.

32 (a) K.-I. YAMADA, S. J. HARWOOD, H. GRÖGER, M. SHIBASAKI, *Angew. Chem. Int. Ed.* **1999**, *38*, 3504; (b) N. TURITANI, K.-I. YAMADA, N. YOSHIKAWA, M. SHIBASAKI, *Chem. Lett.* **2002**, 276.

33 M. CHAVAROT, J. J. BYRNE, P. Y. CHAVANT, Y. VALLÉE, *Tetrahedron: Asymmetry* **2001**, *12*, 1147.

34 J. TIAN, N. YAMAGIWA, S. MATSUNAGA, M. SHIBASAKI, *Angew. Chem. Int. Ed.* **2002**, *41*, 3636.

35 J. TIAN, N. YAMAGIWA, S. MATSUNAGA, M. SHIBASAKI, *Org. Lett.* **2003**, *5*, 3021.

# 6
# Catalytic and Stoichiometric Transformations by Multimetallic Rare Earth Metal Complexes

*Zhaomin Hou*

## 6.1
## Introduction

Polynuclear metal complexes which possess multiple metal atoms that are able to participate in a chemical process can, in principle, show unique reactivities that differ from those of the mononuclear analogues, as a result of cooperative effects between the two or more metal atoms. For rare earth metal complexes, however, it was only very recently that chemical transformations on well-defined multimetallic centers became known. This is mainly because organo rare earth metal complexes, and in particular the hydrido and alkyl complexes, which are the true active species in many processes, have long been dominated by the so-called metallocene complexes that bear two sterically demanding cyclopentadienyl ligands [1, 2]. Although such complexes could adopt a bi- or polynuclear structure in the solid state, they tend to fall apart in solution, and as a result reaction most likely occurs at a mononuclear metal center. Very recently, it was found that some mono(cyclo-pentadienyl)-ligated rare earth metal hydrido and hydrocarbyl complexes, such as $[Me_2Si(C_5Me_4)(NR^1)Ln(\mu\text{-}C\equiv CR^2)]_2$ [3], $[Me_2Si(C_5Me_4)(\mu\text{-}PR^1)LnCH_2SiMe_3]_2$ [4], and $[(C_5Me_4SiMe_3)Ln(\mu\text{-}H)_2]_4$ [5], can maintain their polynuclear structures in solution (Scheme 6.1). These new types of complexes can induce various novel transformations which are unique to polynuclear rare earth metal complexes. This chapter focuses on recent progress in this young area.

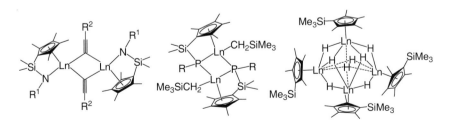

**Scheme 6.1** Examples of multimetallic rare earth metal complexes that are stable in solution.

*Multimetallic Catalysts in Organic Synthesis.* Edited by M. Shibasaki and Y. Yamamoto
Copyright © 2004 WILEY-VCH Verlag GmbH & Co. KGaA, Weinheim
ISBN: 3-527-30828-8

## 6.2
### Binuclear Alkynide Complexes Bearing Silylene-Linked Cyclopentadienyl–Amido Ligands

### 6.2.1
### Synthesis and Structure

Reactions of cyclopentadienyl–arylamido yttrium or lutetium alkyl complexes such as **1a–d** with 1 equiv. of a terminal alkyne afford in a straightforward manner the corresponding alkynide complexes (**2a–f**) (Scheme 6.2) [3]. These half-metallocene alkynide complexes adopt a dimeric structure through μ-alkynide bridges. The binuclear structures are relatively robust and can be maintained even in the presence of THF at high temperatures (80–150 °C) [3]. This is in striking contrast to what was observed in the case of the analogous metallocene or benzamidinate-ligated lanthanide alkynide complexes, which adopted a monomeric structure in THF, and rapidly decomposed or coupled to give trienediyl derivatives such as $[Cp_2^*Ln]$ $[\mu\text{-}\eta^2\text{:}\eta^2\text{-PhC=C=C=CPh}]$ upon heating [6]. As described below, these novel characteristics of the binuclear half-metallocene alkynide complexes enable them (and their alkyl precursors) to serve as a unique catalyst system for the dimerization of terminal alkynes.

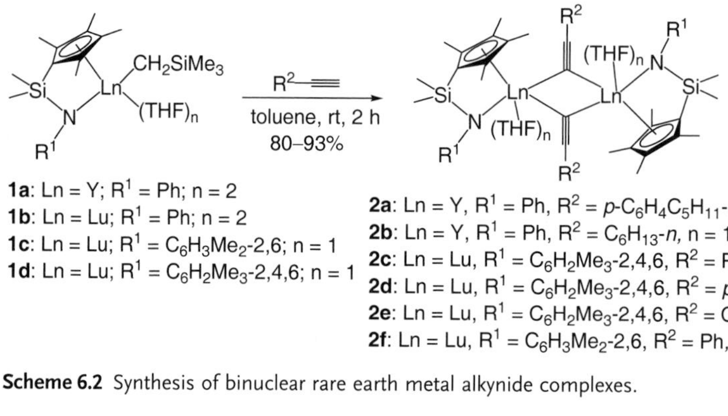

**1a**: Ln = Y; R$^1$ = Ph; n = 2
**1b**: Ln = Lu; R$^1$ = Ph; n = 2
**1c**: Ln = Lu; R$^1$ = C$_6$H$_3$Me$_2$-2,6; n = 1
**1d**: Ln = Lu; R$^1$ = C$_6$H$_2$Me$_3$-2,4,6; n = 1

**2a**: Ln = Y, R$^1$ = Ph, R$^2$ = *p*-C$_6$H$_4$C$_5$H$_{11}$-*n*, n = 1
**2b**: Ln = Y, R$^1$ = Ph, R$^2$ = C$_6$H$_{13}$-*n*, n = 1
**2c**: Ln = Lu, R$^1$ = C$_6$H$_2$Me$_3$-2,4,6, R$^2$ = Ph, n = 0
**2d**: Ln = Lu, R$^1$ = C$_6$H$_2$Me$_3$-2,4,6, R$^2$ = *p*-C$_6$H$_4$Br, n = 0
**2e**: Ln = Lu, R$^1$ = C$_6$H$_2$Me$_3$-2,4,6, R$^2$ = C$_6$H$_{13}$-*n*, n = 0
**2f**: Ln = Lu, R$^1$ = C$_6$H$_3$Me$_2$-2,6, R$^2$ = Ph, n = 0

**Scheme 6.2** Synthesis of binuclear rare earth metal alkynide complexes.

### 6.2.2
### Catalytic Dimerization of Terminal Alkynes

The catalytic dimerization of terminal alkynes is an atom-economic and straightforward method for the synthesis of conjugated enynes, which are important building blocks in organic synthesis and significant components in various biologically active compounds. Although various transition metal and f-block element complexes are known to catalyze the dimerization of terminal alkynes, in most cases a mixture of regioisomers (head-to-head *vs.* head-to-tail) and stereoisomers (*E/Z*-head-to-head) is obtained. In contrast with most previously reported catalysts, the binuclear lutetium alkynide complexes (e.g., **2c–f**) and their alkyl precursors such as **1b–d** showed excellent selectivity in inducing the head-to-head

**Table 6.1** Catalytic dimerization of terminal alkynes by silylene-linked cyclopentadienyl-arylamido lutetium alkyl complex **1d**[a].

| Substrate | Product | Solvent[b] | Temp. (°C) | Time (h) | Conversion (%) | Selectivity (%) |
|---|---|---|---|---|---|---|
| | | $C_6D_6$ | 80 | 5 | > 99 | > 99 |
| | | Toluene-$d_8$ | 110 | 2 | > 99 | > 99 |
| | | Toluene-$d_8$ + THF[c] | 110 | 2 | > 99 | > 99 |
| | | Toluene-$d_8$ | 110 | 2 | 97 | 95[d] |
| | | $C_6D_6$ | 80 | 2 | > 99 | > 99 |
| | | $C_6D_6$ | 80 | 2 | > 99 | > 99 |
| | | $C_6D_6$ | 80 | 1 | > 99 | > 99 |
| | | $C_6D_6$ | 80 | 3 | > 99 | > 99 |
| | | $C_6D_6$ | 80 | 3 | > 99 | > 99 |
| $n\text{-}C_6H_{13}$—≡—H | | THF-$d_8$ | 100 | 14 | > 99 | 95[e] |

[a] Conditions: substrate (1 mmol), **1d** (0.02–0.05 mmol).
[b] 0.45 mL.
[c] THF: 5 equiv. per **1d**.
[d] 1,4,6-Tris[4-trifluoromethyl)phenyl]hexa-3,5-dien-1-yne (5%) was also formed.
[e] 2-Hexyldec-1-en-3-yne (5%, head-to-tail dimer) was also formed.

(Z)-dimerization of terminal alkynes. As shown in Table 6.1, various terminal alkynes can be cleanly dimerized to give the corresponding (Z)-1,3-enynes by using a catalytic amount of **1d** [3]. Aromatic C–Cl, C–Br, and C–I bonds, which are known to be extremely susceptible to reductive cleavage by transition metals, survived in these reactions.

The binuclear alkynide species such as **2c–f** were confirmed as being the true catalyst species in these reactions. A possible reaction mechanism is shown in Scheme 6.3 [3]. An acid-base reaction between an alkyl complex **1** and a terminal alkyne readily gives a dimeric alkynide species such as **2**. Coordination of a further alkyne molecule to a metal center of the dimeric alkynide species could then afford **3** by breaking one of the two alkynide bridges. Attack of the terminal alkynide on the coordinated alkyne in **3** should give **4**, which upon deprotonation by another molecule of alkyne would release the (Z)-1,3-enyne product and regenerate the alkynide catalyst species **2**. Apparently, a dimeric intermediate such as **3**, which leads to the "intermolecular" addition of an alkynide to an alkyne, plays a critically important role in the present (Z)-selective dimerization. This is in sharp contrast to the analogous reactions catalyzed by lanthanide metallocene or benzamidate-ligated catalysts, in which the addition of an alkynide to an alkyne takes place in an "intramolecular" fashion at a monomeric alkynide/alkyne intermediate, and thus always yields the (E)-enyne product whenever head-to-head reaction occurs [6].

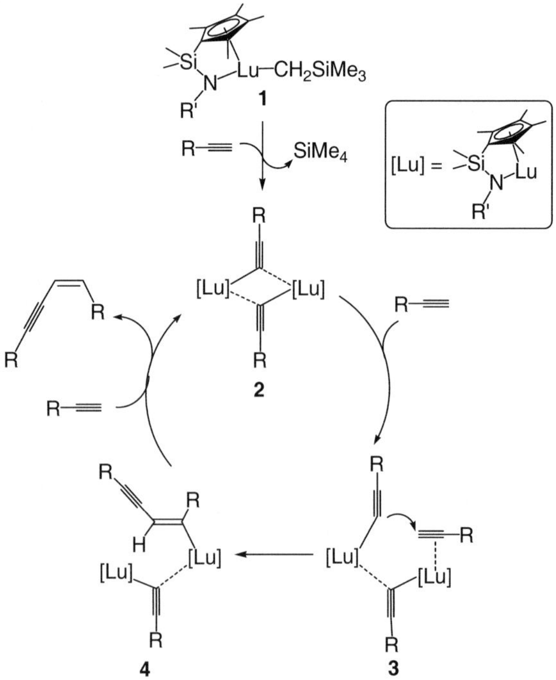

**Scheme 6.3** A possible catalytic cycle for the head-to-head (Z)-dimerization of 1-alkynes.

It is also noteworthy that the dimeric alkynide catalyst species such as **2e,f** are thermally stable and soluble at the reaction temperatures (80–110 °C), but are easily precipitated upon cooling to room temperature after completion of the reaction. Therefore, this catalyst system works homogeneously but can be recovered simply by decantation, and as such it constitutes the first example of a recyclable catalyst system for the dimerization of terminal alkynes (Scheme 6.4) [3].

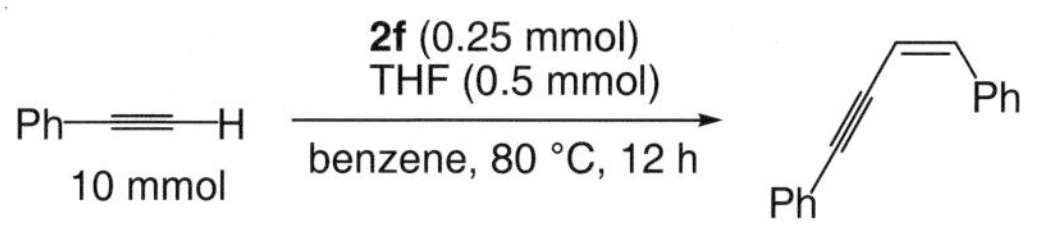

| run | isolated yield |
| --- | --- |
| 1st | 99% |
| 2nd | 99% |
| 3rd | 99% |

**Scheme 6.4** A recyclable catalyst for the dimerization of phenylacetylene.

## 6.2.3
## Polymerization of Aromatic Diynes and Block Copolymerization of Aromatic Diynes with Caprolactone

Half-metallocene lutetium alkyl complexes such as **1c,d**, which were found to serve as novel catalyst precursors for the ($Z$)-selective head-to-head dimerization of terminal alkynes, also showed excellent selectivity in the ($Z$)-head-to-head polymerization of aromatic diynes such as 1,4-diethynyl-2,5-dioctyloxybenzene and 2,7-diethynyl-9,9-dioctylfluorene to selectively afford the corresponding conjugated ($Z$)-polyenynes (Scheme 6.5) [7]. In these polymerization reactions, the propagation might proceed via a bimetallic μ-alkynide species, analogous to the intermediate in

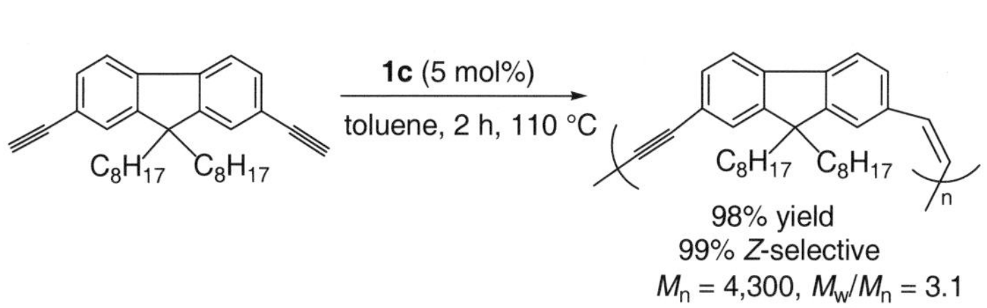

**Scheme 6.5** Polymerization of terminal diynes.

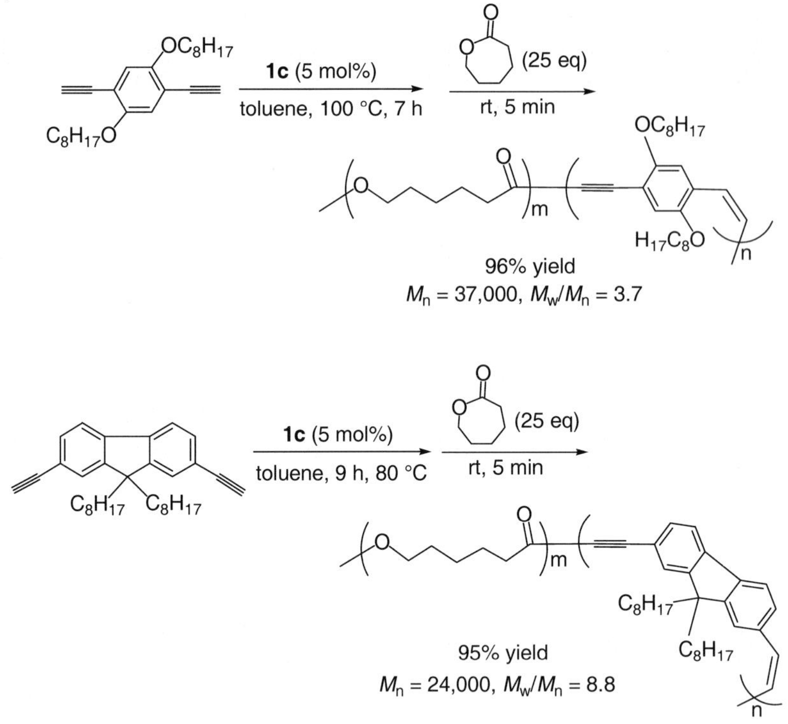

**Scheme 6.6** Copolymerization of diynes with ε-caprolactone.

the dimerization of terminal monoalkynes. When ε-caprolactone or MMA was added to the reaction mixture before quenching with MeOH, selective block copolymerization of the aromatic diynes with these polar momers was achieved through nucleophilic attack of the alkynide species on the ε-caprolactone or MMA (Scheme 6.6) [8]. The resulting block copolymers could be easily cast into films, which are potentially useful as new luminescent materials. When a lanthanide metallocene complex was used in these polymerization reactions, the analogous polymers with (E)-enyne units were formed [7, 8].

## 6.3
## Binuclear Alkyl and Hydrido Complexes Bearing Silylene-Linked Cyclopentadienyl–Phosphido Ligands

### 6.3.1
### Synthesis and Structure

Acid-base reactions between the rare earth metal tris(alkyl) complexes [Ln(CH$_2$SiMe$_3$)$_3$(THF)$_2$] (Ln = Y, Yb, Lu) and 1 equiv. of the silylene-linked

cyclopentadiene–phosphine ligands Me$_2$Si(C$_5$Me$_4$H)(PHR) (R = Ph, Cy) afforded
in a straightforward manner the corresponding cyclopentadienyl–phosphido/alkyl
complexes **5a–f** (Scheme 6.7) [4]. In contrast to the analogous cyclopentadienyl–
amido complexes such as **1a–d**, which adopt monomeric structures, the phosphido
complexes **5a–f** form dimeric structures through phosphido-bridging, in which
the alkyl ligand is placed in a terminal position. The formation of a monomeric
species was not observed, not even in THF solution.

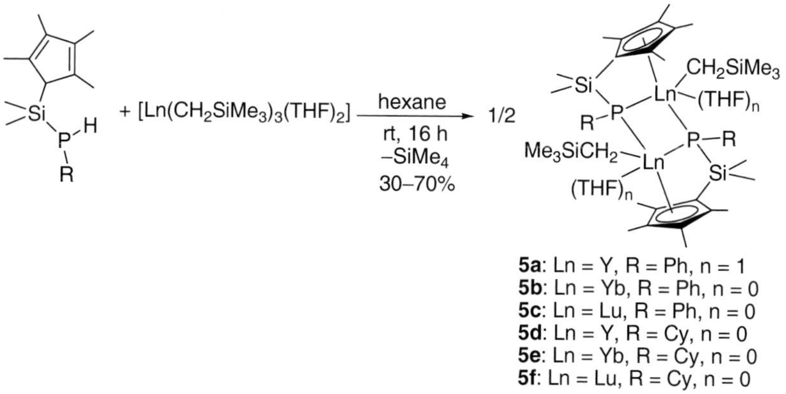

**5a**: Ln = Y, R = Ph, n = 1
**5b**: Ln = Yb, R = Ph, n = 0
**5c**: Ln = Lu, R = Ph, n = 0
**5d**: Ln = Y, R = Cy, n = 0
**5e**: Ln = Yb, R = Cy, n = 0
**5f**: Ln = Lu, R = Cy, n = 0

**Scheme 6.7** Synthesis of binuclear rare earth alkyl complexes
bearing silylene-linked cyclopentadienyl-phosphido ligands.

Reactions of the alkyl complexes **5a–c** with PhSiH$_3$ in THF readily yielded the
corresponding hydrido compounds **6a–c**, which also adopt dimeric structures,
although here the two metal centers are bridged by two hydrido ligands and one
phosphido ligand (Scheme 6.8) [4]. When the alkyl complex **5d** was treated with
2 equiv. of PhSiH$_3$ in benzene, the tetranuclear hydrido complex **7** was isolated
(Scheme 6.9). Complex **7** proved to be insoluble in almost all common organic
solvents, such as benzene, THF, pyridine, or HMPA (hexamethylphosphoric
triamide), demonstrating that the tetranuclear core structure is unusually robust.

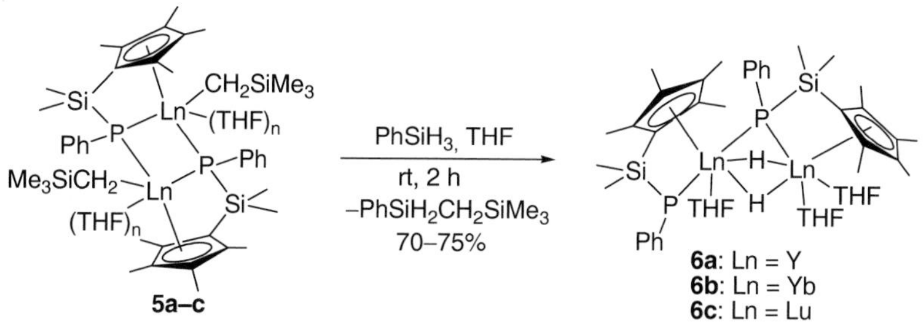

**6a**: Ln = Y
**6b**: Ln = Yb
**6c**: Ln = Lu

**Scheme 6.8** Synthesis of binuclear rare earth hydrido complexes
bearing silylene-linked cyclopentadienyl-phosphido ligands.

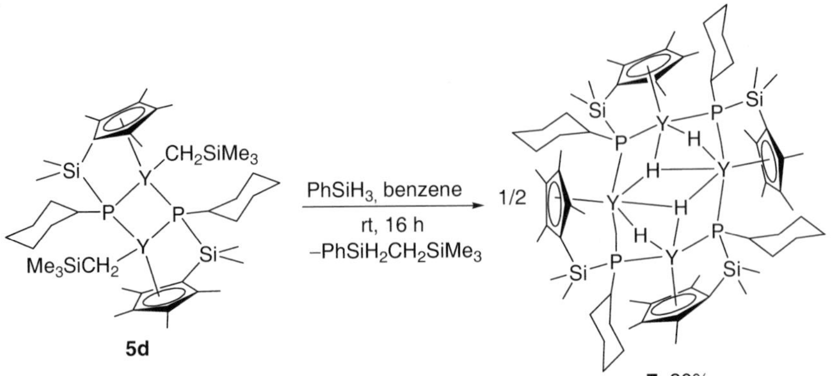
Si
CH2SiMe3
Y
P
P
Y
Si
Me3SiCH2
PhSiH3, benzene
rt, 16 h
–PhSiH2CH2SiMe3
1/2
H
Y
P
Si

**5d** **7**, 80%

**Scheme 6.9** Synthesis of a tetranuclear yttrium hydrido complex bearing silylene-linked cyclopentadienyl-phosphido ligands.

## 6.3.2
## Catalytic Hydrosilylation of Alkenes

The silylene-linked cyclopentadienyl–phosphido rare earth metal alkyl complexes, in particular the cyclohexylphosphido lutetium complex **5f**, were found to act as excellent catalyst precursors for the hydrosilylation of alkenes (Table 6.2) [4]. Complex **5f** is perhaps the most active and regioselective rare earth catalyst (or catalyst precursor) ever reported for the hydrosilylation of 1-alkenes [2c, 9].

**Table 6.2** Catalytic hydrosilylation of alkenes by silylene-linked cyclopentadienyl-phosphido rare earth alkyl complexes[a].

| *Cat.* | *Substrate* | *Time (min)* | *Product* | *Yield (%)* |
|---|---|---|---|---|
| 5a | 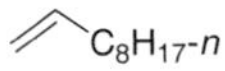 C8H17-n | 60 | 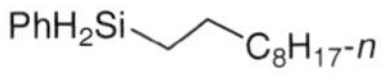 PhH2Si C8H17-n | 95 |
| 5c | 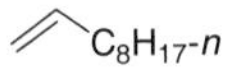 C8H17-n | 180 | 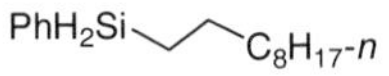 PhH2Si C8H17-n | 100 |
| 5d | 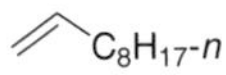 C8H17-n | 45 | 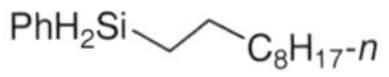 PhH2Si C8H17-n | 100 |
| 5e | 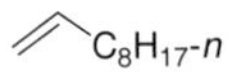 C8H17-n | 45 | 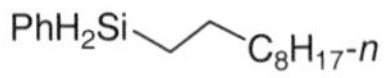 PhH2Si C8H17-n | 100 |
| 5f | 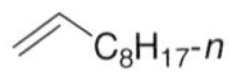 C8H17-n | 10 | 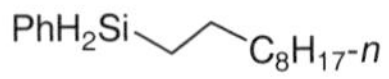 PhH2Si C8H17-n | 100 |
| 5f | 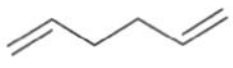 | 45 | 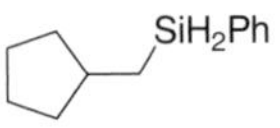 SiH2Ph | 100 |
| 5f | 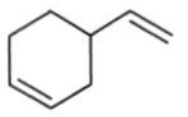 | 30 | 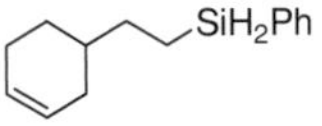 SiH2Ph | 100 |
| 5f | 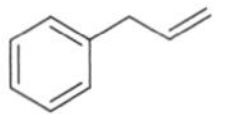 | 15 | 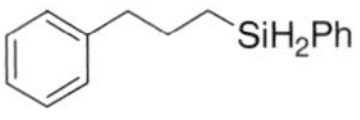 SiH2Ph | 100 |

a) Conditions: substrate (0.4 mmol), $H_3SiPh$ (0.5 mmol), cat. (0.01 mmol), in $C_6D_6$ at rt.

At room temperature, 1-decene was quantitatively hydrosilylated within 10 min by PhSiH$_3$ in the presence of 2.5 mol % of **5f**, selectively affording the linear silylation product 1-(phenylsilyl)decane (Table 6.2). 1,5-Hexadiene was quantitatively converted into the cyclization/silylation product (phenylsilylmethyl)cyclopentane within 45 min. The hydrosilylation of 4-vinyl-1-cyclohexene occurred selectively at the terminal C=C double bond, quantitatively affording the terminal silane product 4-[2-(phenylsilyl)ethyl]cyclohex-1-ene in less than 30 min. The reaction of 3-phenyl-propene was completed in 15 min to cleanly afford the linear silane product 1-phenyl-3-(phenylsilyl)propane. These reactions may proceed via an extremely active terminal hydride species that is generated in situ by σ-bond metathesis between the Lu–CH$_2$SiMe$_3$ bond and PhSiH$_3$. The excellent regioselectivity may result from the steric effects of the sterically demanding binuclear active species. The isolated hydrido complex **7** was found to be inactive under the same conditions, probably owing to its poor solubility and the strong hydrido bridges.

### 6.3.3
**Stereospecific 3,4-Polymerization of Isoprene**

In the presence of 1 equiv. of [Ph$_3$C][B(C$_6$F$_5$)$_4$] in C$_6$H$_5$Cl or toluene, the binuclear yttrium bis(alkyl) complex **5d** showed excellent regio- and stereoselectivity in the 3,4-polymerization of isoprene. At –20 °C, isotactic 3,4-polyisoprene was obtained with both isotacticity (*mmmm*) and 3,4-selectivity greater than 99% (Scheme 6.10) [10]. This regio- and stereoselective polymerization is believed to proceed via a binuclear monoalkyl monocation species such as [Me$_2$Si(C$_5$Me$_4$)(μ-PCy)Y(CH$_2$SiMe$_3$)Y(μ-PCy)(C$_5$Me$_4$)SiMe$_2$][B(C$_6$F$_5$)$_4$], although further studies are required to clarify the mechanistic details. In the absence of [Ph$_3$C][B(C$_6$F$_5$)$_4$], no reaction was observed. When 2 equiv. of [Ph$_3$C][B(C$_6$F$_5$)$_4$] was used, a mixture of 1,4- and 3,4-polyisoprenes was obtained.

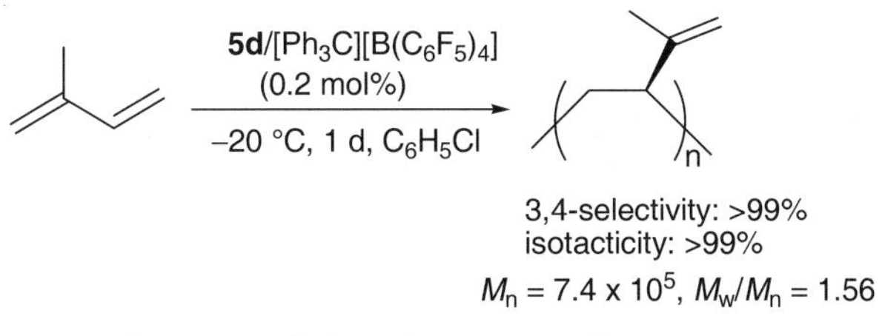

**5d**/[Ph$_3$C][B(C$_6$F$_5$)$_4$]
(0.2 mol%)
—————————→
–20 °C, 1 d, C$_6$H$_5$Cl

3,4-selectivity: >99%
isotacticity: >99%
$M_n = 7.4 \times 10^5$, $M_w/M_n = 1.56$

**Scheme 6.10** Regio- and stereospecific 3,4-polymerization of isoprene.

Although the polymerization of isoprene has been extensively studied for a long time, stereospecific 3,4-polymerization has never been achieved previously. Isotactic 3,4-polyisoprene could serve as a new polymer material with useful properties.

## 6.4
## Polynuclear Hydrido Complexes Bearing the C$_5$Me$_4$SiMe$_3$ Ligand

### 6.4.1
### Synthesis and Structure

Hydrogenolysis of the mono(cyclopentadienyl)/bis(alkyl) complexes [{C$_5$Me$_4$ (SiMe$_3$)}Ln(CH$_2$SiMe$_3$)$_2$(THF)] (8a–d) in THF yielded the tetranuclear polyhydrido complexes 9a–d, which formally consist of two "(C$_5$Me$_4$SiMe$_3$)LnH$_2$" units and two "(C$_5$Me$_4$SiMe$_3$)LnH$_2$(THF)" units (Scheme 6.11) [5, 11]. When the reactions were carried out in toluene, the mono(THF)-coordinated tetranuclear complexes 10a–d were obtained. Complexes 9a–d and 10a–d could be easily interconverted by recrystallization from the appropriate solvents, as shown in Scheme 6.11. For the synthesis of the analogous lutetium polyhydride (10e), treatment of the mixed alkyl/ hydrido complex 11 with H$_2$ in THF proved to be a cleaner route (Scheme 6.12) [5b].

The polyhydrido clusters 9 and 10 are soluble and thermally stable in common organic solvents such as hexane, toluene, and THF. No decomposition or ligand redistribution was observed in [D$_8$]THF or [D$_8$]toluene, as monitored by $^1$H NMR spectroscopy in the case of the diamagnetic Y and Lu complexes.

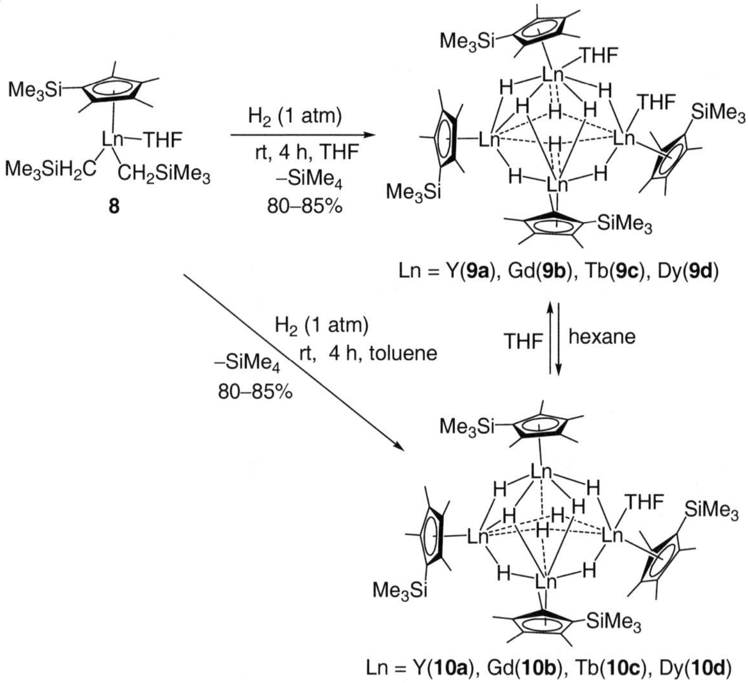

Ln = Y(**9a**), Gd(**9b**), Tb(**9c**), Dy(**9d**)

Ln = Y(**10a**), Gd(**10b**), Tb(**10c**), Dy(**10d**)

**Scheme 6.11** Synthesis of polynuclear rare earth metal polyhydrido complexes.

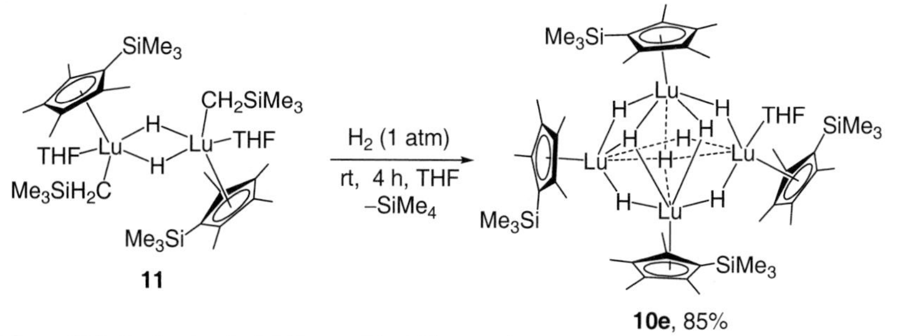

**Scheme 6.12** Synthesis of a lutetium polyhydrido complex.

## 6.4.2
### Hydrogenation of Unsaturated C–C Bonds

Reaction of the yttrium hydrido cluster **10a** with styrene yielded the benzylic allyl complex **12**, in which the allyl part is bonded to one Y atom in a $\eta^3$-fashion, while the phenyl part is bonded to another Y atom in $\eta^2$-form (Scheme 6.13) [5a]. Only one of the eight hydrides in **10a** was able to add to styrene; no further reaction was observed even when an excess of styrene was present. Complex **12** in $C_6D_6$ was seen to be fluxional by NMR. In THF, it decomposed to give a complex mixture. Addition of ca. 10 equiv. of THF to a solution of **12** in $C_6D_6$, however, resulted in clean formation of [(C₅Me₄SiMe₃)Y(μ-H)₂]₄(THF)₂ (**9a**) and styrene. Hydrogenolysis of **12** with H₂ (1 atm.) in $C_6D_6$ afforded ethylbenzene and the THF-free yttrium polyhydrido complex [(C₅Me₄SiMe₃)Y(μ-H)₂]₄. Under 1 atm. of H₂, styrene could be catalytically hydrogenated to give ethylbenzene by [(C₅Me₄SiMe₃)Y(μ-H)₂]₄(THF)₂ or [(C₅Me₄SiMe₃)Y(μ-H)₂]₄.

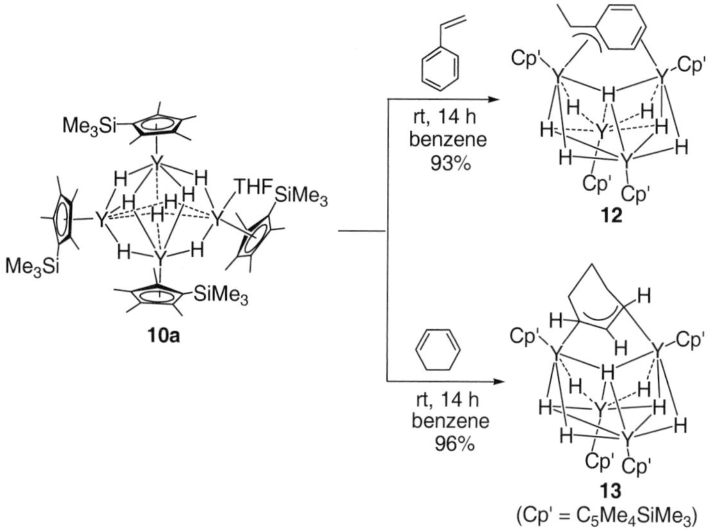

**Scheme 6.13** Reactions of a yttrium polyhydrido complex with C=C double bonds.

Similar reaction of **10a** with 1,3-cyclohexadiene afforded **13**, in which the allyl unit interacts with two Y atoms through the two terminal carbon atoms, each bonding to one metal center in a $\eta^1$-fashion (Scheme 6.13). The structure of **13** in $C_6D_6$ was rigid, as shown by the $^1$H and $^{89}$Y NMR spectra.

When 1,4-bis(trimethylsilyl)-1,3-butadiyne was reacted with **10a**, a novel butene-tetraanion complex **14** was obtained (Scheme 6.14) [5a]. Complex **14** formally consists of a $[(C_5Me_4SiMe_3)YH]_4^{4+}$ unit and a delocalized 2-butene-1,1',4,4'-tetrayl or butene-tetraanion species. The butene-tetraanion moiety is bonded to two Y atoms in a $\pi{:}\eta^4$-"inverse-sandwich" fashion and to the other two Y atoms in a $\sigma_1{:}\mu_1$-terminal form at the two terminal carbon atoms. A possible mechanism for the formation of **14** is shown in Scheme 6.14. Addition of two Y–H units to the two C≡C units in a "2,1-fashion" would yield the 1,3-butadien-1,4-diyl (or butadiene-dianion)/hexahydride species **15**. Subsequent 1,4-Y–H addition to the 1,3-butadiene unit in **15** could then give the 2-butene-1,1',4-triyl/pentahydride species **16**. Deprotonation at the C4 position of the butene species **16** by Y–H should afford the 2-butene-1,1'4,4'-tetrayl/tetrahydride species **17**. This species can be regarded as a resonance structure (a localized form) of **14** (*cf.* also **18** and **19**). The formation of a butene-tetraanion species from a butadiyne derivative has not been reported previously. This reaction is evidently unique to polynuclear polyhydrido rare earth metal complexes.

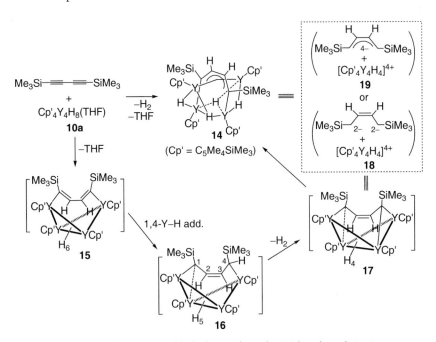

**Scheme 6.14** Reaction of a yttrium polyhydrido complex with a 1,3-butadiyne derivative.

## 6.4.3
## Reduction of Nitriles to Imido Species

Reactions of the polyhydrido clusters [(C$_5$Me$_4$SiMe$_3$)Ln($\mu$-H)$_2$]$_4$(THF) (**10a,e**) with 4 equiv. of nitriles such as acetonitrile or benzonitrile readily afford the corresponding tetranuclear cubane-like imido complexes [(C$_5$Me$_4$SiMe$_3$)Ln($\mu_3$-NCH$_2$Ph)]$_4$ (**20a–d**) (Scheme 6.15) [5a]. In these reactions, the C$\equiv$N triple bond of the nitrile is completely reduced to a C–N single bond by double Ln–H addition. This is in sharp contrast to the reactions of rare earth metallocene hydrido complexes [12] or zirconocene dihydrido complexes [13], which yield only the mono-insertion products, such as [(C$_5$H$_5$)$_2$Y($\mu$-N=CH$^t$Bu)]$_2$ or (C$_5$Me$_5$)$_2$Zr(H)(N=CHR) (R = C$_6$H$_4$Me-$p$). The facile formation of imido species by reduction of nitriles as described here may offer a convenient route to rare earth metal imido complexes, a class of compounds that are of considerable current interest but remain difficult to access by other methods.

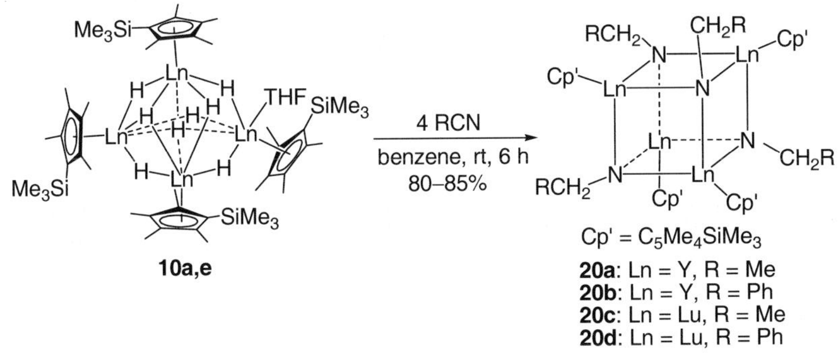

**Scheme 6.15** Reaction of rare earth metal polyhydrido complexes with nitriles.

## 6.4.4
## Reactions with Lactones, Carbon Dioxide, and Isocyanates

The reaction of **10e** with $\gamma$-butyrolactone gave the tetranuclear mixed alkoxo/hydrido complex **21** (Scheme 6.16) [5a]. $\gamma$-Butyrolactone is known as a stable, non-polymerizable cyclic ester. In this reaction, three molecules of $\gamma$-butyrolactone are completely reductively ring-opened by one molecule of **10e** to give the linear diolate species "$^-$O(CH$_2$)$_4$O$^-$", which again demonstrates the unusually high reactivity of the rare earth metal polyhydrido complexes. No further reaction between **21** and $\gamma$-butyrolactone was observed, even when an excess of the lactone was present. Both **21** and **10e**, however, showed high activity in the ring-opening polymerization of $\varepsilon$-caprolactone.

The butene-tetraanion/tetrahydrido complex **14** reacted rapidly with CO$_2$ to give the structurally characterizable bis(methylene diolate) complex **22** (Scheme 6.17) [14]. The reaction occurred only at the hydrido sites, while the butene-tetraanion unit remained unchanged. Further insertion of CO$_2$ into the metal–oxygen bonds

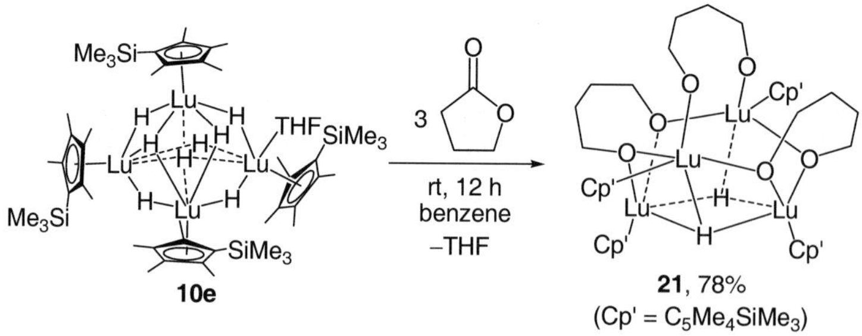

**Scheme 6.16** Reaction of a lutetium polyhydrido complex with γ-butyrolactone.

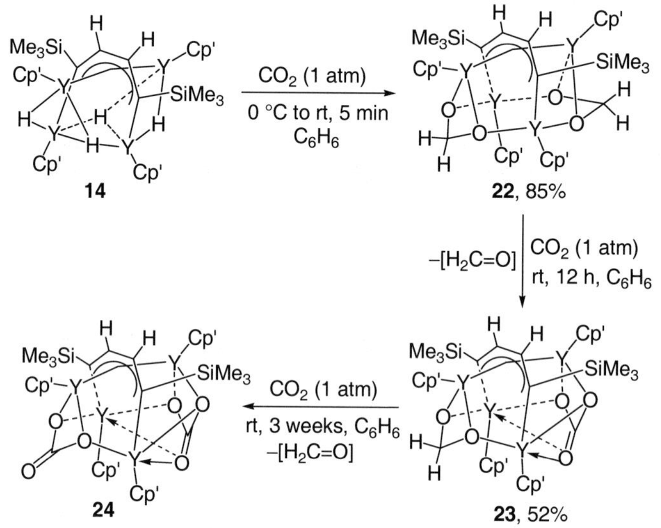

**Scheme 6.17** Reactions of polynuclear yttrium complexes with carbon dioxide.

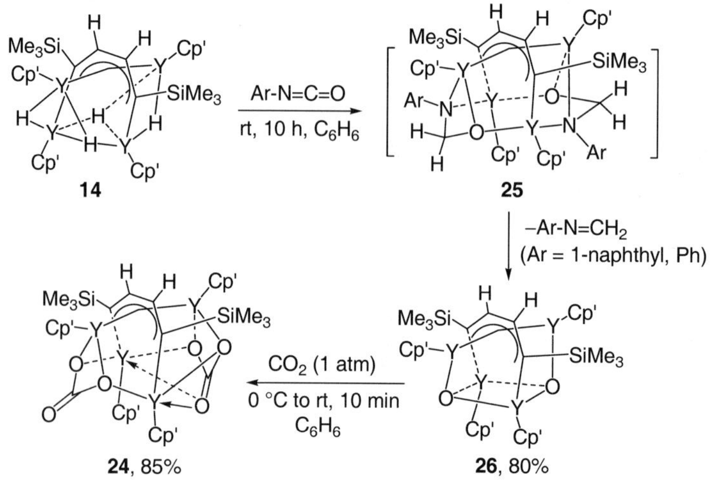

**Scheme 6.18** Formation of a dioxo yttrium complex and its reaction with carbon dioxide.

in **22** was possible to give the mono- and dicarbonate complexes **23** and **24** in a stepwise manner, with possible release of formaldehyde [HCHO]. Alternating copolymerization of $CO_2$ with cyclohexene oxide could be achieved by the use of **14** or **22–24** [15].

When **14** was allowed to react with 2 equiv. of phenyl isocyanate or 1-naphthyl isocyanate, the dioxo complexes **26** were isolated (Scheme 6.18) [14]. The dicarbonate complex **24** could be easily obtained by the reaction of **26** with $CO_2$.

## 6.5
## Polynuclear Imido Complexes Bearing the C$_5$Me$_4$SiMe$_3$ Ligand

### 6.5.1
### Nitrile Insertion and Hydrogen Transfer

The reaction of the benzylimido complex **20d** with 4 equiv. of benzonitrile yielded the benzimidinate-dianion complex **27**, through nucleophilic addition of the imido unit to the CN group of benzonitrile (Scheme 6.19) [16]. Similar treatment of the ethylimido complex **20a** with benzonitrile, however, did not afford an insertion product, but instead gave the mixed benzaldimido/ethylimido complex **28** (Scheme 6.20). In this reaction, four molecules of benzonitrile were hydrogenated to give the benzaldimido species by hydrogen transfer from two ethylimido units in **20a** with the release of two molecules of acetonitrile.

When 10 equiv. of benzonitrile was used to react with **20b**, the benzonitrile tetramerization product **32** was obtained selectively (Scheme 6.21) [16]. Complex **32** could also be prepared by reaction of the polyhydrido complex **10a** with 14 equiv. of benzonitrile. Apparently, both benzonitrile-insertion and hydrogen-transfer reactions must take place in order to form **32**. A possible rationale for the formation of **32** from the imido complex **20b** is shown in Scheme 6.22. Insertion of benzonitrile into the Y–N imido bonds in **20b** would first give **33**, in analogy to the formation of **27** described above. Coordination of one molecule of benzonitrile to each Y atom in **33** should then yield **34**. Intramolecular hydrogen transfer from a methylene group to a coordinated PhCN in **34** (path *a*) should afford **35**. Nucleophilic addition

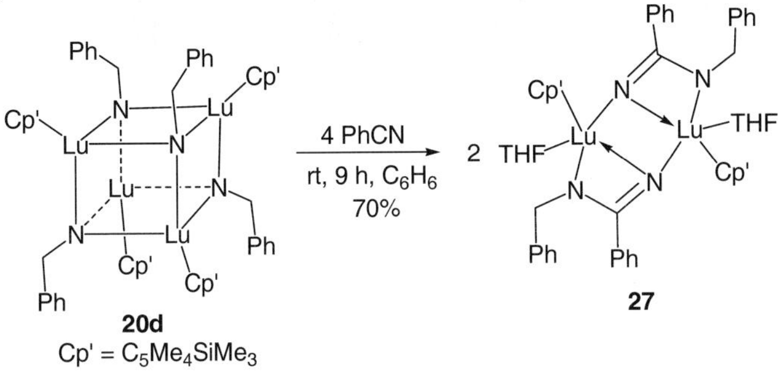

**Scheme 6.19** Reaction of a tetranuclear lutetium benzyl imido complex with benzonitrile.

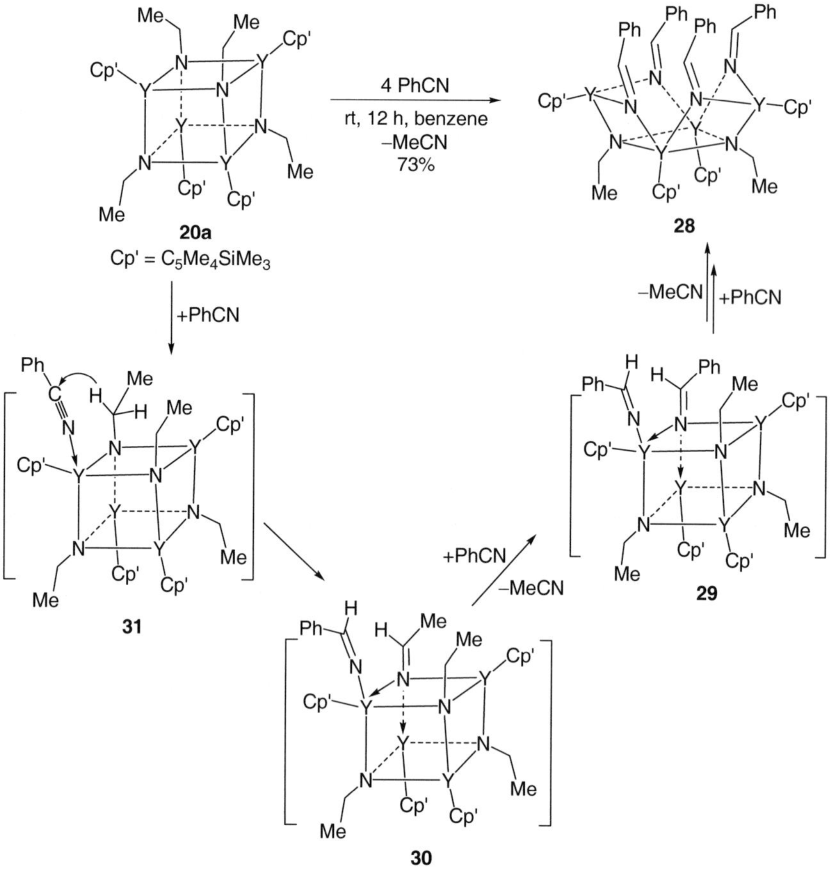

**Scheme 6.20** Reaction of a yttrium ethyl imido complex with benzonitrile.

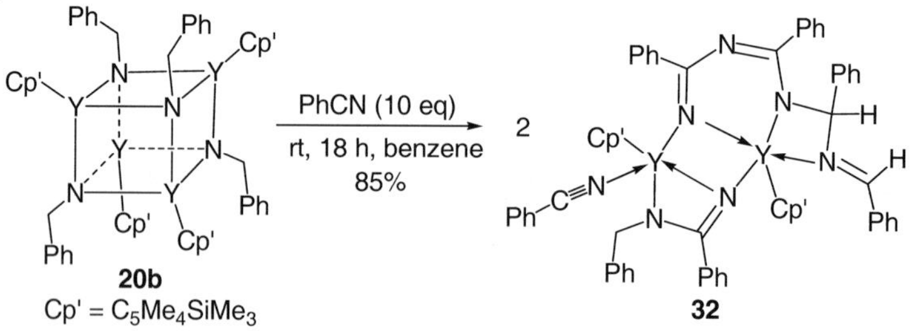

**Scheme 6.21** Reaction of a yttrium benzyl imido complex with excess benzonitrile.

of the newly formed benzaldimido unit to the aldimine group (path *b*) could yield **36**, which on coordination of another molecule of PhCN gives **37**. Migration of the methine hydrogen to the PhCN unit in **37** (path *c*), followed by addition of the resulting aldimido unit to the aldimine group in **38** (path *d*), would give **32**.

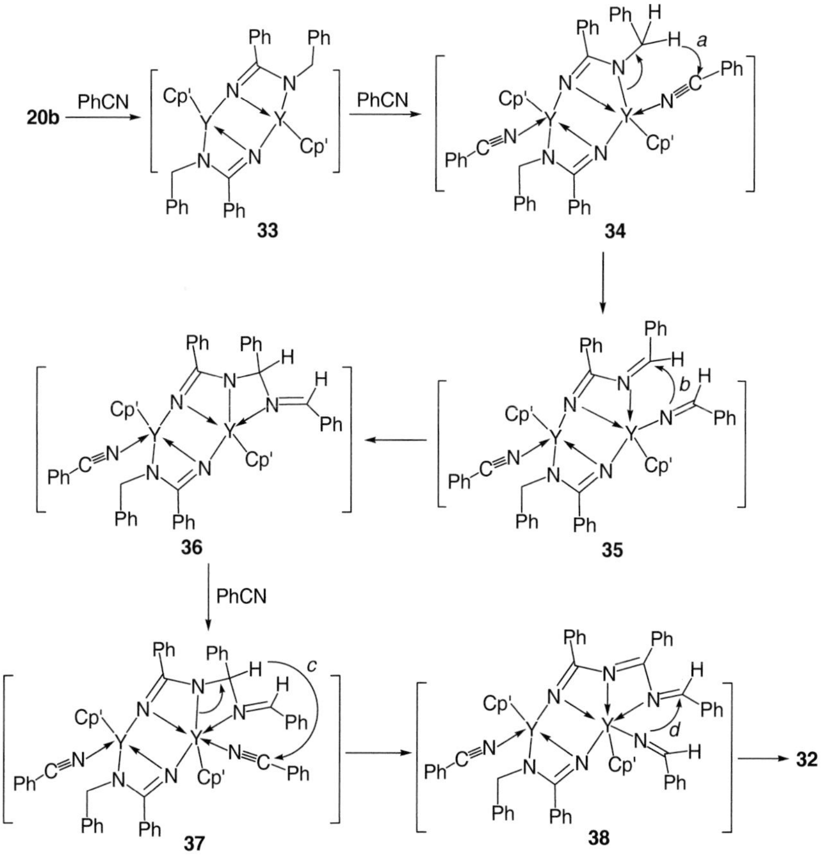

**Scheme 6.22** A possible mechanism for the formation of **32**.

## 6.5.2
## Catalytic Cyclotrimerization of Benzonitrile

In the presence of a catalytic amount of **32**, **20b** or **10a**, the cyclotrimerization of benzonitrile took place (Scheme 6.23) [16]. In all these reactions, the recovery of **32** was confirmed. A possible mechanism is shown in Scheme 6.24; as most of the reaction steps are similar to those described above, they are not discussed further here.

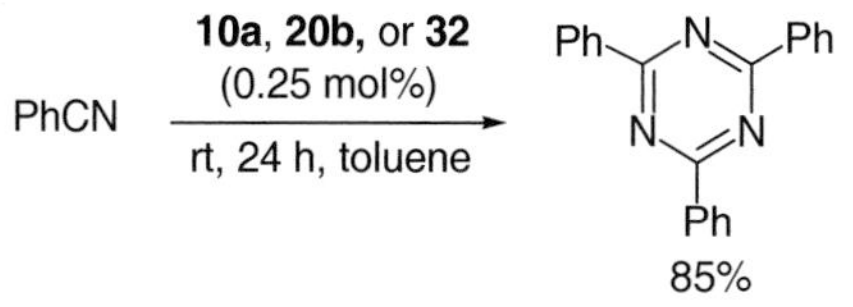

**Scheme 6.23** Catalytic cyclotrimerization of benzonitrile by a polynuclear yttrium complex.

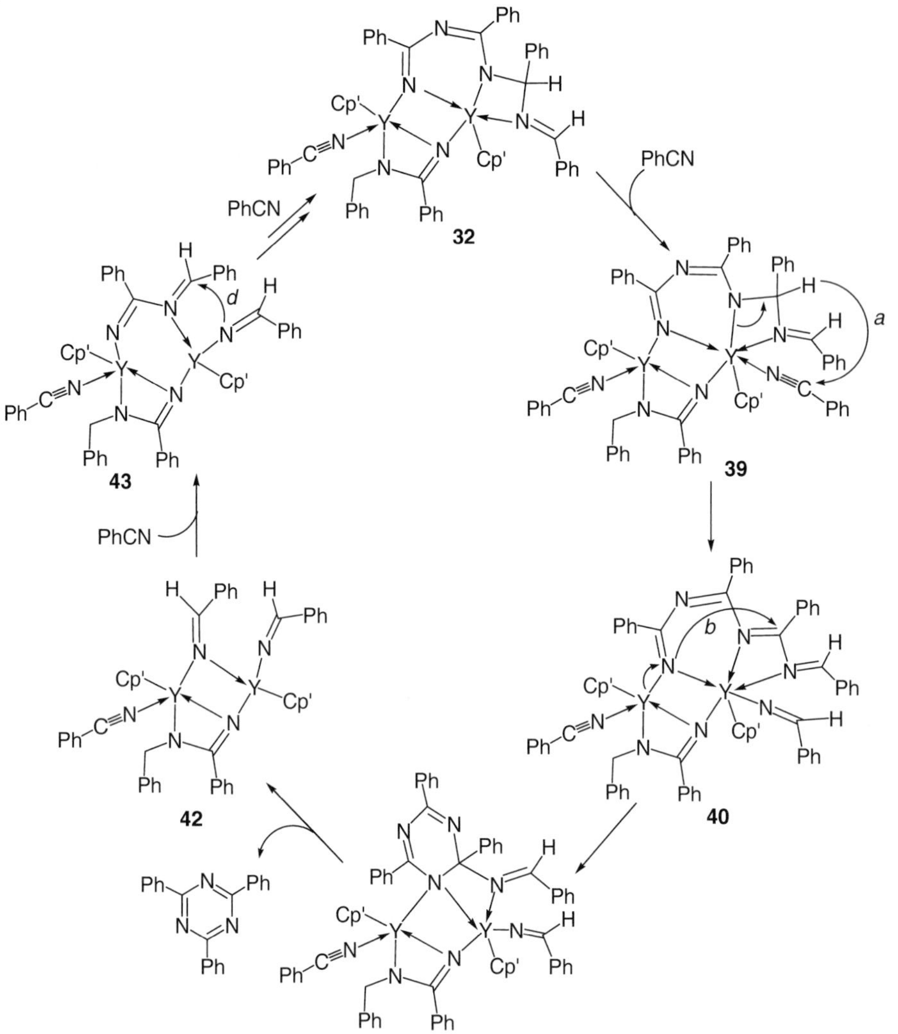

**Scheme 6.24** A possible mechanism for catalytic cyclotrimerization of benzonitrile.

## 6.6
## Outlook

Although synthetic and reactivity studies on well-defined multimetallic rare earth complexes are still in their infancy, it is now evident that such complexes can indeed show unprecedented behavior in catalytic and stoichiometric transformations. As the expertise in this field increases and more complexes are developed, the diverse chemistry available from the rare earth elements will certainly be further extended. An exciting and prosperous future in this area can confidently be predicted.

## References

1 Reviews on lanthanide complexes bearing mono(cyclopentadienyl) ligands: (a) Z. Hou, *Bull. Chem. Soc. Jpn.* **2003**, *76*, 2253; (b) Z. Hou, Y. Wakatsuki, *J. Organomet. Chem.* **2002**, *647*, 61; (c) J. Okuda, *J. Chem. Soc., Dalton Trans.* **2003**, 2367; (d) S. Arndt, J. Okuda, *Chem. Rev.* **2002**, *102*, 1953.

2 Selected general reviews on organolanthanide complexes: (a) Z. Hou, Y. Wakatsuki, in *"Science of Synthesis"* (Eds.: T. Imamoto, R. Noyori), Thieme, Stuttgart, 2002, Vol. 2, p. 849; (b) Z. Hou, Y. Wakatsuki, *Coord. Chem. Rev.* **2002**, *231*, 1; (c) G. A. Molander, J. A. C. Romero, *Chem. Rev.* **2002**, *102*, 2161; (d) M. Ephritikhine, *Chem. Rev.* **1997**, *97*, 2193; (e) H. Yasuda, E. Ihara, *Bull. Chem. Soc. Jpn.* **1997**, *70*, 1745; (f) R. Anwander, in *"Applied Homogeneous Catalysis with Organometallic Compounds"* (Eds.: B. Cornils, W. A. Hermann), VCH, Weinheim, 1996, Vol. 2, p. 866; (g) H. Schumann, J. A. Meese-Marktscheffel, L. Esser, *Chem. Rev.* **1995**, *95*, 865; (h) F. T. Edelmann, in *"Comprehensive Organometallic Chemistry II"* (Eds.: E. W. Abel, F. G. A. Stone, G. Wilkinson, M. F. Lappert), Pergamon, Oxford, 1995, Vol. 4, p. 11; (i) C. J. Schaverien, *Adv. Organomet. Chem.* **1994**, *36*, 283.

3 M. Nishiura, Z. Hou, Y. Wakatsuki, T. Yamaki, T. Miyamoto, *J. Am. Chem. Soc.* **2003**, *125*, 1184.

4 O. Tardif, M. Nishiura, Z. Hou, *Tetrahedron* **2003**, *59*, 10525.

5 (a) D. Cui, O. Tardif, Z. Hou, *J. Am. Chem. Soc.* **2004**, *126*, 1312; (b) O. Tardif, M. Nishiura, Z. Hou, *Organometallics* **2003**, *22*, 1171; (c) K. C. Hultzsch, P. Voth, T. S. Spaniol, J. Okuda, *Z. Anorg. Allg. Chem.* **2003**, *629*, 1272.

6 (a) K. H. den Haan, Y. Wielstra, J. H. Teuben, *Organometallics* **1987**, *6*, 2053; (b) H. J. Heeres, J. H. Teuben, *Organometallics* **1991**, *10*, 1980; (c) W. J. Evans, R. A. Keyer, J. W. Ziller, *Organometallics* **1993**, *12*, 2618; (d) H. J. Heeres, J. Nijhoff, J. H. Teuben, *Organometallics* **1993**, *12*, 2609; (e) C. M. Forsyth, S. P. Nolan, C. L. Stern, T. J. Marks, *Organometallics* **1993**, *12*, 3618; (f) C. J. Schaverien, *Organometallics* **1994**, *13*, 69; (g) R. Duchateau, C. T. van Wee, J. H. Teuben, *Organometallics* **1996**, *15*, 2291; (h) A. Haskel, T. Straub, A. K. Dash, M. S. Eisen, *J. Am. Chem. Soc.* **1999**, *121*, 3014; (i) A. K. Dash, I. Gourevich, J. Q. Wang, J. Wang, M. Kapon, M. S. Eisen, *Organometallics* **2001**, *20*, 5084.

7 For an overview, see: M. Nishiura, Z. Hou, *J. Mol. Catal. A: Chem.* **2004**, *213*, 101.

8 M. Nishiura, Z. Hou, unpublished result.

9 For examples of hydrosilylations of alkenes catalyzed by other types of rare earth complexes, see: (a) Z. Hou, Y. Zhang, O. Tardif, Y. Wakatsuki, *J. Am. Chem. Soc.* **2001**, *123*, 9216; (b) A. A. Trifonov, T. P. Spaniol, J. Okuda, *Organometallics* **2001**, *20*, 4869; (c) A. Z. Voskoboyonikov, A. K. Shestakova, I. P. Beletskaya, *Organometallics* **2001**, *20*, 2794; (d) K. Takaki, K. Sonoda, T. Kousaka, G. Koshoji, T. Shishido, K. Takehira, *Tetrahedron Lett.* **2001**, *42*, 9211; (e) G. A. Molander, E. D. Dowdy, B. C. Noll, *Organometallics* **1998**, *17*, 3754; (f) G. A. Molander, E. D. Dowdy, H. Schumann, *Organometallics* **1998**, *17*, 3386; (g) H. Schumann, M. R. Keitsch, J. Winterfeld, S. Muhle, G. A. Molander, *J. Organomet. Chem.* **1998**, *559*, 181; (h) G. A. Molander, P. J. Nichols, *J. Am. Chem. Soc.* **1995**, *117*, 4415; (i) P.-F. Fu, L. Brard, Y. Li, T. J. Marks, *J. Am. Chem. Soc.* **1995**, *117*, 7157; (j) S. Onozawa, T. Sakakura, M. Tanaka, *Tetrahedron Lett.* **1994**, *35*, 8177; (k) G. A. Molander, M. Julius, *J. Org. Chem.* **1992**, *57*, 6347; (l) T. Sakakura, H.-J. Lautenschlager, M. Tanaka, *J. Chem. Soc., Chem. Commun.* **1991**, 40. See also ref. 2c.

10 L. Zhang, Z. Hou, unpublished result.

11 J. Baldamus, D. Cui, Z. Hou, unpublished result.

**12** W. J. Evans, J. H. Meadows, W. E. Hunter, J. L. Atwood, *J. Am. Chem. Soc.* **1984**, *106*, 1291.

**13** J. E. Bercaw, D. L. Davies, P. T. Wolczanski, *Organometallics* **1986**, *5*, 443.

**14** O. Tardif, D. Hashizume, Z. Hou, *J. Am. Chem. Soc.* **2004**, *126*, in press.

**15** D. Cui, O. Tardif, Z. Hou, unpublished result.

**16** D. Cui, Z. Hou, unpublished result.

# 7
# Bimetallic Transition Metal Catalysts for Organic Oxidation

*Patrick M. Henry*

## 7.1
## Introduction

Of the many catalysts discussed in this book, only a few are capable of acting as oxidation catalysts. This is because this type of catalyst has two requirements. First, it must be capable of performing a particular type of organic transformation (oxidation in the present case) and, secondly, it must be capable of being re-oxidized to its original oxidation state. The overall process is shown in Eq. (7.1).

$$\text{organic} \ + \ M^{m+}\text{—}M^{m+} \ \longrightarrow \ \begin{matrix}\text{oxidized}\\\text{organic}\end{matrix} \ + \ M^{n+}\text{—}M^{n+}$$

$$\xrightarrow[\text{redox system?}]{O_2} \ M^{m+}\text{—}M^{m+}$$

(7.1)

In the case of metal ions that are readily oxidized by air (e.g. $Cu^{I}$, $Fe^{II}$, $Co^{II}$), the second re-oxidation step occurs quite naturally and presents no particular complexity. In fact, in the case of bimetallic complexes, this step often occurs more readily than for monometallic complexes since the two metals can interact simultaneously with the two oxygen atoms of $O_2$. This is the basis for the four-electron reduction of dioxygen in fuel cells [1].

In other cases, the reduced form of the bimetallic may not be directly air-oxidizable. In this case, another redox system, that is itself air-oxidizable, must be present. An example is the Wacker process for the oxidation of $\alpha$-alkenes to ketones [2]. The industrial process employs $Pd^{II}$ salts to oxidize the alkene to a ketone and $Cu^{II}$ to reoxidize the $Pd^0$ back to the $Pd^{II}$ state for further reaction. The $Cu^{I}$ salts formed are converted back to the $Cu^{II}$ form by dioxygen. The complete process is shown in Eq. (7.2).

$$RCH{=}CH_2 + PdCl_4{}^{2-} + H_2O \ \longrightarrow \ \overset{O}{\overset{\|}{RCCH_3}} + Pd(0) + 2\,HCl + 2\,Cl^-$$

$$\xrightarrow{2\,CuCl_2} \ PdCl_4{}^{2-} + 2\,CuCl \ \xrightarrow[2\,HCl]{1/2\,O_2} \ 2\,CuCl_2 + H_2O$$

(7.2)

*Multimetallic Catalysts in Organic Synthesis.* Edited by M. Shibasaki and Y. Yamamoto
Copyright © 2004 WILEY-VCH Verlag GmbH & Co. KGaA, Weinheim
ISBN: 3-527-30828-8

Sometimes, a more elaborate redox system than that shown in Eq. (7.2) is required. Thus, in non-halide containing systems, $Cu^{II}$ does not serve as a re-oxidant for the $Pd^0$ formed in the reaction and often a multi-step electron-transfer procedure involving a two- or three-component system is required. For the monometallic catalysts, a commonly used system consists of $Pd^{II}/Pd^0$-benzoquinone/hydroquinone-$ML_{ox}^m/ML^m$, where $ML^m$ is an oxygen-activating macrocyclic transition metal complex such as Co(salen) [3a,b]. A second redox system uses the heteropoly acid $H_5PMo_{10}V_2O_{40} \cdot 34\,H_2O$ as the re-oxidant [3c,d]. This latter system has the advantage of high stability, easy access to an active catalyst, and simpler work-up procedures.

The justification for using a bimetallic system must be a discernible improvement in either step of the catalytic cycle. Thus, the oxidation products obtained must be different and more valuable than those obtained with the monometallic system, or the rates of product formation must be faster. Alternatively, the improvement could be in the re-oxidation step of the process. Thus, the reduced form of the catalyst may undergo direct air-oxidation. On the other hand, the effects may be subtle. The bimetallic catalyst may produce higher enantioselectivities in an asymmetric synthesis or it may be more soluble in the reaction medium.

## 7.2
## Homobinuclear Systems

### 7.2.1
### $Cu^{II}$ and $Fe^{III}$ Catalysts

These two metal ions are treated together since their chemistries are interrelated. Their complexes have been extensively studied because their reduced forms ($Fe^{II}$ and $Cu^I$) are both air-oxidizable and they are the main metals found in biological dioxygen metabolism [4]. They differ significantly in their redox potentials; $E°$ for the $Fe^{III}/Fe^{II}$ couple is 0.79 V, while that for the $Cu^{II}/Cu^I$ couple is 0.134 V.

As expected from its higher redox potential, $Fe^{III}$ is a much better oxidant than $Cu^{II}$. Its potential is even higher when it is bound by ligands such as 2,2′-bipyridine and 1,10-phenanthroline [5]. Many studies have centered on the oxidation of catechol because of the interest in the high-spin ferric enzymes pyrocatechase and meta-pyrocatechase, which are responsible for two of the most widely studied biological oxygenating processes [6]. Three model systems have shown comparable activity to the natural enzymes, which includes cleavage of catechol to *cis,cis*-muconic acid [7–9]. Most of these studies have employed 3,5-di-*tert*-butylcatechol (DTBC, **1**), since side reactions are avoided. All of these systems use monometallic species. The reaction sequence is shown in Eq. (7.3).

$$(7.3)$$

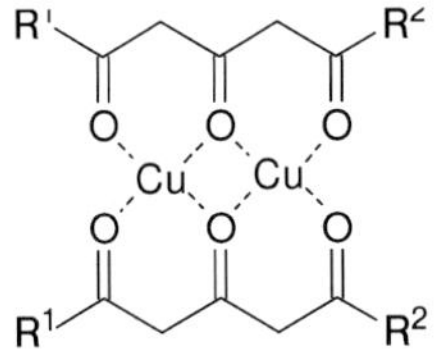

**Figure 7.1** Structures of some copper complexes.

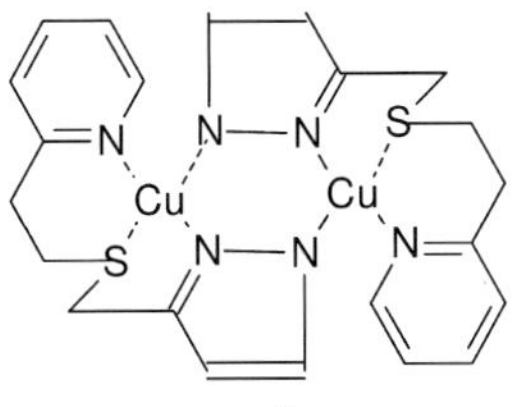

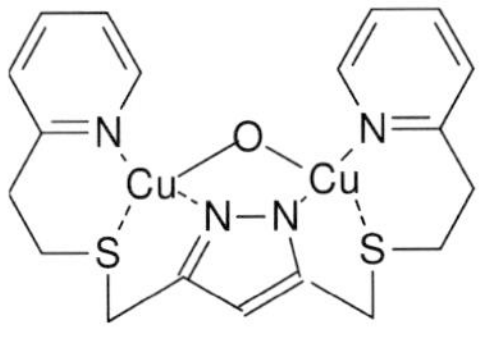

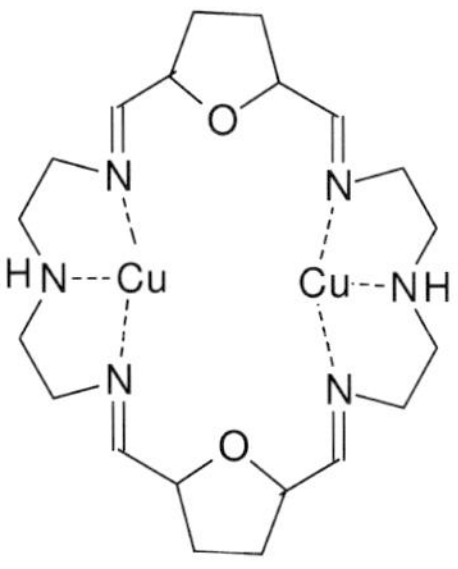

Considerable effort has been expended on studies of model compounds for dioxygen bonding and of dioxygen-activating copper proteins. Natural compounds that bond dioxygen include the hemocyanins, which are found in molluscs and arthropods. Examples of dioxygen-activating enzymes are the tyrosinase mono-oxygenases, which catalyze the *ortho*-hydroxylation of phenols [10]. As these proteins contain bimetallic centers, model systems for their function are also bimetallic and fall within the scope of this chapter. Again, a commonly used substrate for these studies is 3,5-di-*tert*-butylcatechol (DTBC, 1) and the product is 3,5-di-*tert*-butylquinone (DTBQ) (Eq. 7.4) [11–14]. With 2,6-disubstituted phenols, the product is diphenoquinone (Eq. 7.5) [11]. In two cases, the complexes used have been of the triketone type (2) [12, 13]. In one case, the ligands were of the pyrazolate type (3 and 4) [14], while in another study they were of the tetra-Schiff type (5) [11]. These structures are shown in Figure 7.1. The bimetallic catalysts proved to be more reactive than their corresponding monometallic counterparts. In the case of complexes 3 and 4, oxidations were studied in a medium of 1 : 1 $CH_3OH/H_2O$ under dioxygen atmosphere. The reaction was found to be first order in the DTBC and first order in $O_2$. This result is consistent with the rate-determining step being the reaction of the $Cu_2^I$ form of the complex with $O_2$ to give the $Cu_2^{II}$ form [14].

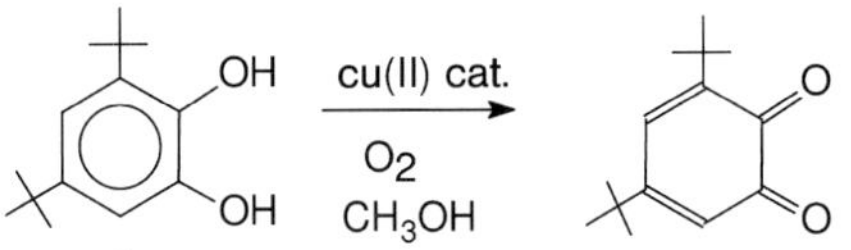

$$(7.4)$$

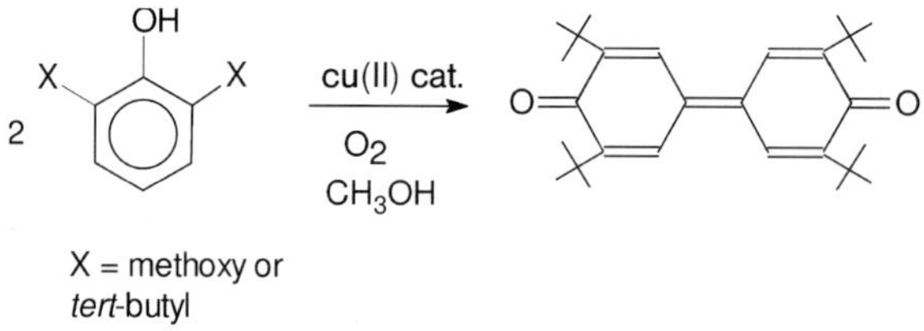

$$(7.5)$$

X = methoxy or
tert-butyl

In the oxidation with the complex **5**, the reaction starts with the $Cu^I$ form of the macrocycle. This reacts with $O_2$ to form the $Cu^{II}$ peroxy-bridged complex, which oxidizes the substrate and produces the $Cu^{II}$ complex, which then oxidizes another molecule of substrate to regenerate the original $Cu^I$ form of the catalyst [11].

Rogiæ and Demming [15], as well as Tsuji and Takayanagi [16], reported a very interesting ring cleavage reaction of catechol. This reaction very likely occurs by way of a $Cu^{II}$ dimer in the presence of dioxygen in pyridine. The catalyst is formed by the addition of water to methoxy(pyridine)copper(II) chloride (Eq. 7.6). This reagent catalyzes the "intradiol" cleavage of catechol itself under both aerobic and anaerobic conditions to produce the monomethyl ester of *cis,cis*-muconic acid (Eq. 7.7).

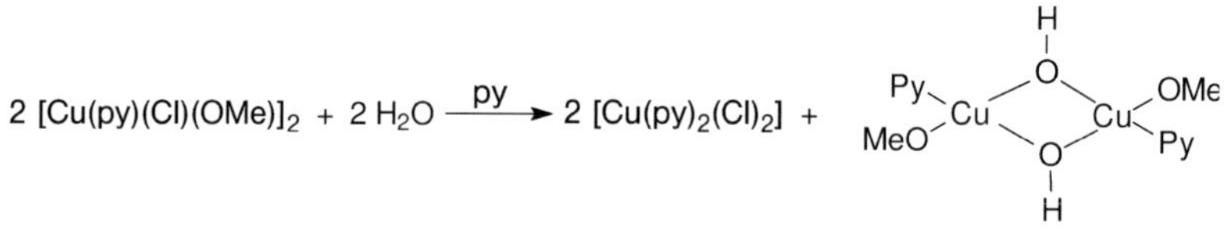

$$2 \ [Cu(py)(Cl)(OMe)]_2 \ + \ 2 \ H_2O \ \xrightarrow{py} \ 2 \ [Cu(py)_2(Cl)_2] \ + \qquad (7.6)$$

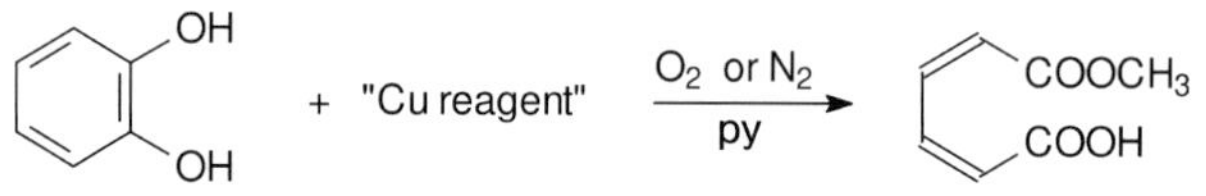

$$(7.7)$$

Another exciting result is that the reaction under strictly *anhydrous* and *anaerobic* conditions leads to a product of "extradiol cleavage". The reaction with 4-*tert*-butylcatechol is shown in Eq. (7.8) [17]. As shown, the 4-*tert*-butyl-1,2-benzoquinone gives the 2,2-dimethoxy-6-carbomethoxy)-4-*tert*-butyloxacyclohexa-3,5-diene (**6**), but the corresponding dimethylmuconic acid does not react to give **6**.

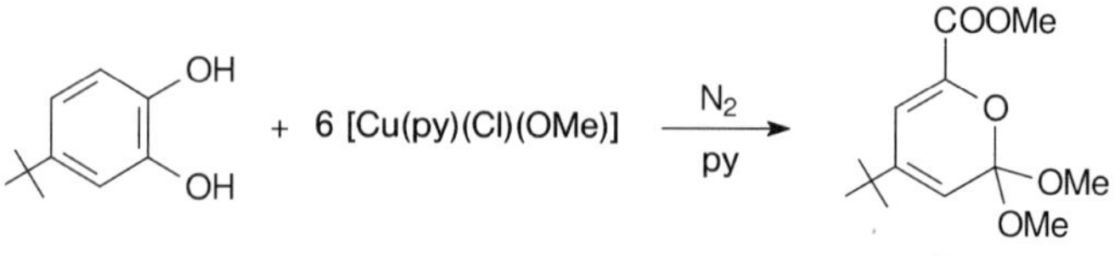

$$(7.8)$$

**6**

The author's introduction to bimetallic metal oxidations occurred while studying the treatment of lignin by-products from sulfite paper mills. Lignins are polyphenols that are non-biodegradable, so the goal of the research was to destroy the phenolic structure and make them biodegradable. A series of monometallic macrocyclic $Fe^{III}$ and $Cu^{II}$ catalysts induced oxidation, but not of the desired type that destroyed the phenolic structure. Instead, the oxidation was of the one-electron type, which gave coupled phenolics.

The design of a successful catalyst required a metal oxidant that would transfer two electrons rather than one and have labile coordination sites so as to allow close approach of the substrate for electron transfer. For easy recovery and re-use of the catalyst, an insoluble system was preferable. The requirements would be met by a polymeric system in which a ligand coordinates two metal ions but still leaves some sites on the metal uncoordinated by strong ligands. The ligand selected was the triketone group, as shown in **2** in Figure 7.1. Although the completely co-ordinated structure **2** would be an oxidant, a metal coordinated by only one triketone ligand would be much more reactive. This half-coordinated structure could only be realized on a surface.

As shown in Eq. (7.9), a suitable support was prepared by polymerizing acetyl-benzenes with diacetylbenzenes to form a polyphenyl polymer [18]. The assembly of this polymer involves the trimerization of three acetyl groups to form a new benzene ring. The resulting polymer is insoluble in all solvents and is in fact used as a stable high-temperature material. The important feature of the material for the present purposes is the fact that it bears acetyl end groups, which can be further modified to produce diketone and triketone coordinating groups on the polymer backbone [19]. Thus, treatment of the acetyl end groups with an acetate ester and a base gave diketones, **7**, and further reaction of these diketone groups with more ester and base gave the triketones, **8** [20]. The resulting polymers were then treated with solutions of metal ions to give either monometallic or bimetallic species on the surfaces. In this way, we were able to compare the activity of a polymer bearing only monometallic catalytic species with that of a polymer bearing only bimetallic catalytic species on its surface. Both species were expected to be more reactive than the fully coordinated homogeneous catalytic species. The overall reaction is shown in Scheme 7.1.

$$(7.9)$$

In actual practice, the pre-polymer was coated onto high surface area Celite, chemically modified, and then exposed to a solution of the metal ions. The modified polymer was then treated with acid and heated in an oven at 150 °C to complete the polymerization. The amount of $Fe^{III}$ or $Cu^{II}$ absorbed was measured by treating the solid with standard solutions of $FeCl_3$ and $Cu(OAc)_2$ and subsequently analyzing the solutions. The amount of metal ions taken up by the triketone-bearing surfaces was about double that taken up by the diketone-bearing surfaces.

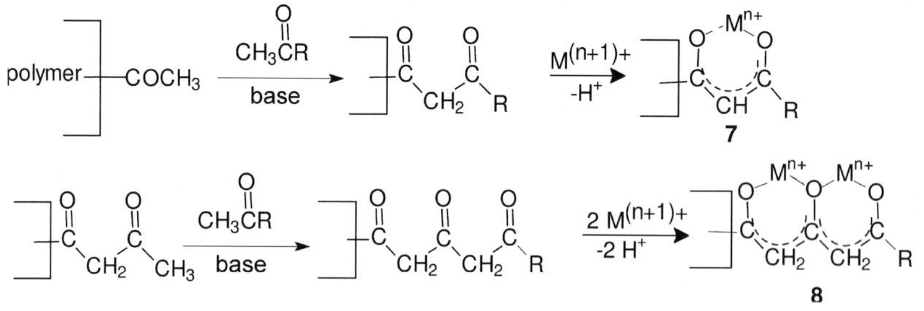

**Scheme 7.1** Procedures for the synthesis of surface diketone and triketone complexes.

The differences between the monometallic and bimetallic catalytic surfaces were dramatic. As shown in Eq. (7.4), the only product found from the oxidation of DTBC was the quinone, DTBQ. On the other hand, the bimetallic surfaces gave, in addition to DTBQ, ring-cleaved products. For example, the product distribution obtained with the $Fe^{III}$ system at the point at which $O_2/DTBC = 1$ is shown in Eq. (7.10). At the point at which $O_2/DTBC = 2$, almost all of the DTBQ had reacted to give **9** and **10**.

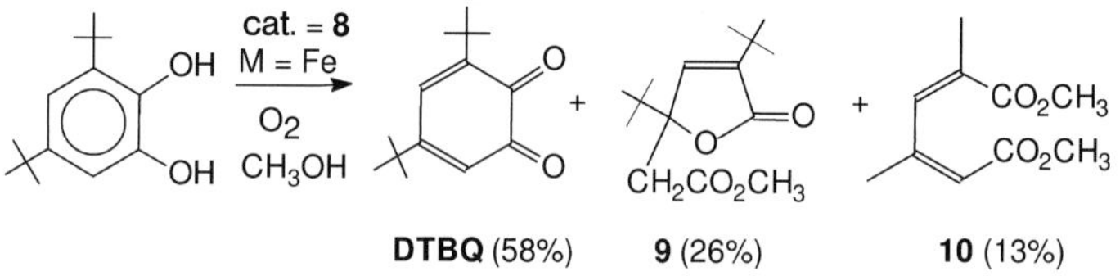

$$\text{DTBQ (58\%)} \qquad \text{9 (26\%)} \qquad \text{10 (13\%)} \tag{7.10}$$

Unfortunately, no kinetic information could be obtained from the dioxygen uptake plots. The reactions showed an induction period followed by a rapid uptake of dioxygen. The total uptake in all cases was double that required for the products formed, indicating that $H_2O_2$ was one product of the dioxygen reduction. In the case of the monometallic surface bearing $Cu^{II}$, the total dioxygen taken up decreased after reaching a maximum. This was attributed to disproportionation of the $H_2O_2$ formed to give $O_2$ and $H_2O$.

In later work, an improved and better-behaved catalytic system was developed [21]. The group used for the chemical modification was the methyl benzoate functionality. The polymer was prepared by the polymerization of methyl *m*- or *p*-acetylbenzoates with *m*- or *p*-diacetylbenzene (Eq. 7.11).

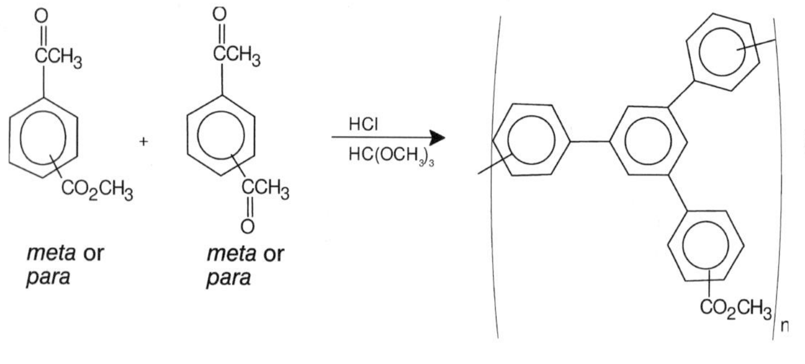

$$\tag{7.11}$$

meta or para        meta or para

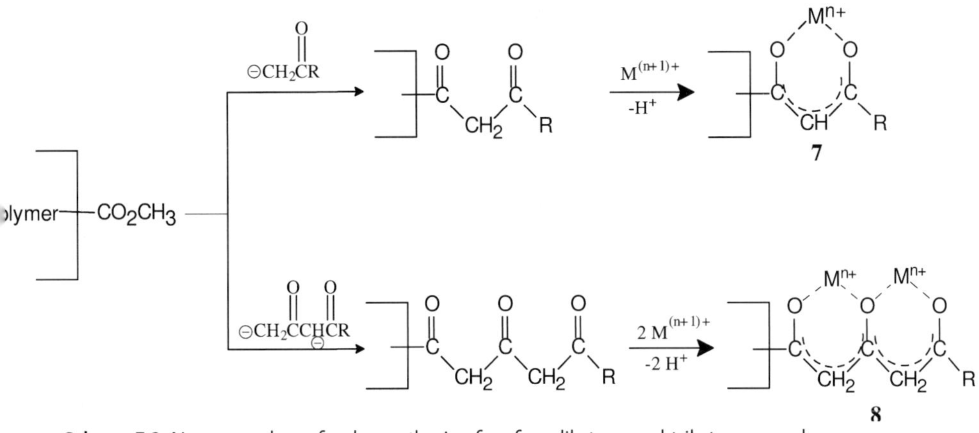

**Scheme 7.2** New procedures for the synthesis of surface diketone and triketone complexes.

Since the methyl benzoate function is not involved in the polymerization, its concentration should be much higher than that of the acetyl function used for chemical modification in the previous studies.

As shown in Scheme 7.2, the chemical modifications now involve the reaction of the methyl benzoate surface species with the anion of a methyl ketone to give a β-diketone and with the dianion of a β-diketone to give a β-triketone. This reaction sequence places the two types of ligand groups on the polymer surface in a selective manner. The first reaction introduces only the diketone species, while the second reaction introduces only the triketone without any contamination from the diketone surface species. Thus, when treated with metal ions, pure **7** and **8** will be formed.

The metal species used for the surface-coating procedure were also more reactive, $[Fe(CH_3CN)_6]^{3+}$ in the case of iron and $[Cu(CH_3CN)_4]^{2+}$ in the case of copper. For comparison purposes, $FeCl_3$ was also studied. The uptake of $Fe^{III}$ from an $FeCl_3$ solution was about 75 times greater than in the previous study, and again the triketone surface adsorbed twice as much as the diketone surface. However, the uptake of $[Fe(CH_3CN)_6]^{3+}$ was only about 10% greater than that of $FeCl_3$. The uptake of $[Cu(CH_3CN)_4]^{2+}$ was about 50 times greater than the uptake of $Cu(OAc)_2$ achieved previously.

As expected from the catalyst-coating results, the behavior of these new surfaces was in sharp contrast to that of those obtained in the original work. There was no induction period, and hence the kinetics of DTBQ formation (Eq. 7.4) could be measured from the initial portion of the reaction curve, assuming the reaction to be first-order in DTBC. When the rate data were plotted as first-order reactions with respect to [DTBC], using the stoichiometry $1 O_2 : 2$ DTBC, linear plots were obtained for the initial portions of the reactions with the triketone catalysts. For the diketone catalysts with $[Fe(CH_3CN)_6]^{3+}$, $[Cu(CH_3CN)_4]^{2+}$, and $FeCl_3$, in which cases $H_2O_2$ was detected, a stoichiometry of $1 O_2 : 1$ oxidized DTBC was assumed. The measured first-order rate constants were converted to second-order rate constants (first order in [DTBC] and first order in $[M^{n+}]$) by dividing by the $[M^{n+}]$ that would be present if the metal ion were in a homogeneous solution of the same volume as the reaction mixture (100 mL). The fact that the reaction is first order in DTBC requires that the slow step is the reaction of the metal with DTBC rather than reaction of the reduced form of the catalyst with dioxygen.

The products of the reaction were, as expected, the same as those reported in the earlier work, but the oxidation was more complete with the improved catalysts. The monometallic systems produced only DTBQ as the product (Eq. 7.4) and half of the $O_2$ absorbed was recovered as $H_2O_2$. At the midpoint of the dioxygen uptake curve ($O_2$/DTBC = 0.5), a product distribution similar to that given in Eq. (7.10) was found, but at the end of the reaction ($O_2$/DTBC = 1) all of the DTBQ had reacted. No $H_2O_2$ was formed in the reactions with the bimetallic system. In the course of this work, the ring-opening reaction of DTBQ to give **9** and **10** was also studied. As expected, only the bimetallic catalyst promoted this reaction. The monometallic system was completely unreactive.

The second-order rate constants displayed by the different catalyst systems proved to be of considerable interest. We first consider the rate constants measured for the initial reaction, that is, the conversion of DTBC to DTBQ. The rates achieved with catalysts formed by treating the surfaces with $[Fe(CH_3CN)_6]^{2+}$ were 20 times faster than those achieved with systems obtained by treatment with $FeCl_3$. This reflects the effect of the presence of the weakly coordinating $CH_3CN$ ligands rather than strongly coordinating chloride ligands. The rate of oxidation with the $[Fe(CH_3CN)_6]^{3+}$ system was 25 times faster than that with the $[Cu(CH_3CN)_4]^{2+}$ system, reflecting the more potent oxidizing power of iron(III).

A real surprise was the differences between the rates achieved with the bimetallic and monometallic systems. These varied from a 30% increase with the $FeCl_3$-bearing catalyst to an approximately fourfold increase for the $[Cu(CH_3CN)_4]^{2+}$ system. If the oxidation of DTBC to DTBQ had occurred by one-electron transfer with the monometallic catalyst and by a two-electron process with the bimetallic catalyst, one would have expected these differences to be much greater. Thus, the monometallic catalysis must occur by a series of two one-electron steps (Eq. 7.12).

The route with the surface bearing the bimetallic catalyst may also involve two-electron steps. However, in this case the reduced bimetallic catalyst can react with $H_2O_2$ to form $H_2O$ and the oxidized form of the complex.

The rates of ring-opening of DTBQ by the bimetallic system were slower than that of the initial reaction to form DTBQ from DTBC, but the surfaces bearing the monometallic catalysts were completely unreactive. From this, it can be concluded that a concerted two-electron process is required for ring cleavage (Eq. 7.13).

$$+ 2\ ROH \longrightarrow \quad + \quad M^{(n-1)+}\!-\!\!-\!M^{(n-1)+} \qquad (7.13)$$

## 7.2.2
## $Pd^{II}$ Catalysis

Catalysis by $Pd^{II}$ compounds is presently one of the most active areas of research. A review of some of the recent advances required a two-volume treatise [22]. Thus, it is impractical to give any more than a brief introduction to the chemistry here. The reaction that initiated interest in this type of catalysis was the Wacker process for the oxidation of $\alpha$-alkenes to aldehydes and ketones, as shown in Eq. (7.14) [23]. Extension of this oxidation to the use of acetic acid as solvent leads to the production of vinyl and allylic esters (Eq. 7.15). Primary and secondary alcohols are converted to aldehydes and ketones. In fact, as shown in Eq. (7.16), the oxidation of allyl alcohol in aqueous solution leads to a combination of two types of transformations. Thus, while hydroxyacetone and 2-hydroxypropanal result from Wacker-type oxidation, acrolein is formed by direct alcohol oxidation [24].

$$CH_2\!=\!CHR + H_2O \xrightarrow[\substack{CuCl_2 \\ O_2}]{PdCl_4{}^{2-}} CH_3\overset{O}{\overset{\|}{C}}R + H\overset{O}{\overset{\|}{C}}CH_2R \qquad (7.14)$$

$$RCH\!=\!CHCH_2R' + HOAc \xrightarrow[\substack{CuCl_2 \\ O_2}]{PdCl_4{}^{2-}} \overset{OAc}{\underset{|}{R}}C\!=\!CHCH_2R' + RCH\!=\!CH\overset{OAc}{\underset{|}{C}}HR \qquad (7.15)$$

$$CH_2\!=\!CHCH_2OH + H_2O \xrightarrow[\substack{CuCl_2 \\ O_2}]{PdCl_4{}^{2-}} \left\{ \begin{array}{l} CH_3\overset{O}{\overset{\|}{C}}CH_2OH + CH_2\!=\!CHCHO \\ HOCH_2CH_2CHO \end{array} \right. \qquad (7.16)$$

## 7.3
## Heterogeneous Catalysts

Extension of the studies discussed above for the $Cu^{II}$ and $Fe^{III}$ heterogeneous systems to include $Pd^{II}$ led the author into a series of studies involving homogeneous as well as heterogeneous catalysis.

Thus, polymeric surfaces, prepared by the procedures described above (Eq. 7.11 followed by Scheme 7.2), were treated with $Pd(CH_3CN)_4(BF_4)_2$ [21]. The metal ion uptake was similar to that found for $Fe^{III}$ and $Cu^{II}$. The surface bearing the diketone catalyst (i.e. the monometallic system) oxidized DTBC to DTBQ at a rate only slightly faster than the surface-bound $FeCl_3$- and $[Cu(CH_3CN)_4]^{2+}$-diketone systems, which was still about a factor of five less than the rate with the surface-bound $[Fe(CH_3CN)_6]^{3+}$ monometallic system. More notable was the rate obtained with the triketone (bimetallic)-bearing surfaces incorporating $Pd^{II}$. The rate was 45 times greater than that with the monometallic catalyst and even three times faster than that with the $[Fe(CH_3CN)_6]^{3+}$ bimetallic catalyst. *However, even more significant was the fact that both the monometallic and bimetallic systems were air-oxidation catalytic systems.* Previously, another redox reagent such as $CuCl_2$ (see Eq. 7.14) had been necessary to make the system an air-oxidation one.

The products obtained with the $Pd^{II}$-loaded surface were similar to those observed with the corresponding $Cu^{II}$- and $Fe^{III}$-loaded surfaces. The monometallic catalyst gave only DTBQ (Eq. 7.4), while the bimetallic surface gave only **10** rather than a mixture of **9** and **10** as found with the bimetallic $Fe^{III}$ and $Cu^{II}$ systems. As expected, the oxidation of DTBQ gave only **10**.

What, if any, are the mechanistic differences between oxidation with the $Pd^{II}$ system and that with the other two metal ion systems? Differences may arise simply due to the fact that $Pd^{II}$ is a very different oxidant compared to the other two metal species. Thus, $Pd^{II}$ is only a two-electron oxidant as a monomeric species; it is incapable of the one-electron transfer shown in Eq. (7.12) since $Pd^I$ monomer is an unstable species. However, $Pd^{II}$ is capable of oxidizing DTBC by the two-electron route shown in Eq. (7.17). In the case of bimetallic $Pd^{II}$, the $Pd^{II}$ reduction product would be a $Pd^I$ dimer, which is a well-known oxidation state in palladium chemistry [25]. Thus, although a $Pd^{II}$ monometallic catalyst can oxidize DTBC by the non-radical two-electron transfer shown in Eq. (7.17), the concerted two-electron transfer shown in Eq. (7.18) must be the more facile process. The formation of a $Pd^I$ dimer as reduction product in Eq. (7.18) has at least one practical advantage. In other studies in this laboratory, it was found that the bimetallic $Pd^{II}$ catalytic species was much more stable to eventual decomposition to $Pd^0$, resulting in many more turnovers than the monometallic catalysts [26]. This reflects the fact that the $Pd^I$ dimer in Eq. (7.18) is much more stable to decomposition to $Pd^0$ than a monomeric $Pd^{II}$ complex.

$$(7.17)$$

$$(7.18)$$

**Table 7.1** Rate constants for the oxidation of several monofunctional alcohols.[a]

| Entry | Substrate[b] | Catalyst type | $k$ (M$^{-1}$s$^{-1}$)[c] |
|-------|-------------|---------------|---------------------------|
| 1 | methanol | monometallic | 0.0013 |
| 2 | methanol[d] | bimetallic | 0.0018 |
| 3 | ethanol | monometallic | 0.0048 |
| 4 | ethanol | bimetallic | 0.025 |
| 5 | 2-propanol | monometallic | 0.0083 |
| 6 | 2-propanol | bimetallic | 0.035 |
| 7 | benzyl alcohol | monometallic | 0.067 |
| 8 | benzyl alcohol | bimetallic | 0.14 |

[a] All reactions are oxidations in aqueous, phosphate-buffered solutions (pH 8.0, $\mu = 0.10$ M) at 25 °C containing 0.5 g of polymer-coated Celite catalyst.
[b] Usually $2.5 \cdot 10^{-2}$ M.
[c] Bimolecular rate constants were calculated by assuming a homogeneous system and dividing the $k_{obs}$ values, obtained by pseudo-first-order treatment, by the [Pd$^{II}$] calculated assuming all the Pd$^{II}$ to be present in homogeneous solution.
[d] Methanol experiments were also run at $12.5 \cdot 10^{-2}$ M and $25.0 \cdot 10^{-2}$ M. The values of $k$ were the same within experimental error.

Next, attention was turned to another Pd$^{II}$-catalyzed reaction, that of alcohol oxidation [27]. Of particular concern was the reaction of molecules having more than one site for coordination to the metal. These include glycols and unsaturated alcohols. For comparison purposes, simple monofunctional alcohols were also studied.

For the simple alcohols listed in Table 7.1, the bimetallic surfaces induced faster reactions than the monometallic ones but, as might be expected, the differences were not large. The largest acceleration, a fivefold increase, was found for ethanol. As shown in Eq. (7.19), coordination of the hydroxyl group to the first metal center may place the hydroxyl hydrogen in a suitable position for easy extraction by the second Pd$^{II}$ center.

$$\text{(7.19)}$$

The fact that methanol oxidation by the bimetallic catalyst was found to be strictly first order with respect to the methanol concentration is a significant result. The methanol oxidation must be the rate-determining step rather than diffusion to the catalyst site or the re-oxidation of Pd$^0$ to Pd$^{II}$. Moreover, the fact that the second-order rate constants remained consistent at several CH$_3$OH concentrations indicates that the results are reproducible.

In contrast to the simple monofunctional alcohols, differences between the mono- and bimetallic catalysts were expected to be larger for polyols, which can coordinate to two sites on the bimetallic catalyst (Eq. 7.20). Kinetic data for the oxidation of a series of diols by $Pd^{II}$ collected in Table 7.2 show the expected trends. The rates achieved with the mono- and bimetallic catalysts in the case of ethylene glycol differ by a factor of 30. This factor decreases to 10 for propane-1,2-diol and to 7 for butane-2,3-diol. The less pronounced rate differences for the latter two substrates are most likely due to steric hindrance by the methyl groups, which makes the chelation less favorable than in the case of ethylene glycol. The factor for butane-1,3-diol is only 2.4. This result must reflect geometric factors in the chelation by the glycol. Thus, energy-minimized estimates (ALCHEMY II) of the oxygen–oxygen distance in ethylene glycol gave a value of 2.8 Å [28]. This is close to the value of 2.72 Å calculated for catechol, a substance that reacts very readily with these bimetallic $Pd^{II}$ catalysts [19, 21]. Values for the oxygen–oxygen distances in the other two 1,2-diols were very similar. On the other hand, the corresponding distance for butane-1,3-diol is 4.4 Å, which is almost certainly too large for effective chelation.

$$(7.20)$$

The most dramatic results were found for allyl alcohol. The first result of note is that allyl alcohol gave the fastest rate of any of the substrates studied with the monometallic catalysts ($0.15 \ M^{-1} \ s^{-1}$). However, in this system, the rate comparisons are complicated by the fact that only 25% of the oxidation is direct alcohol oxidation to give acrolein; the remaining oxidation is alkene oxidation to give 3-hydroxy-1-propanal. Thus, the actual rate of direct alcohol oxidation to acrolein is $0.15 \ M^{-1} \ s^{-1} \cdot 0.25 = 0.0375 \ M^{-1} \ s^{-1}$, a value that is nevertheless still only exceeded by the reactive benzyl alcohol. A similar rate enhancement was found in the

**Table 7.2** Rate constants for oxidations of polyols.[a]

| Entry | Substrate[b] | Catalyst type | k ($M^{-1} \ s^{-1}$)[c] |
|-------|-------------|---------------|--------------------------|
| 1 | ethane-1,2-diol | monometallic | 0.00053 |
| 2 | ethane-1,2-diol | bimetallic | 0.016 |
| 3 | propane-1,2-diol | monometallic | 0.0021 |
| 4 | propane-1,2-diol | bimetallic | 0.020 |
| 5 | butane-2,3-diol | monometallic | 0.0062 |
| 6 | butane-2,3-diol | bimetallic | 0.041 |
| 7 | butane-1,3-diol | monometallic | 0.0032 |
| 8 | butane-1,3-diol | bimetallic | 0.0078 |

[a] All reactions are oxidations in aqueous, phosphate-buffered solutions (pH 8.0, $\mu = 0.20 \ M$) at 25 °C.
[b] $2.5 \cdot 10^{-2} \ M$.
[c] Bimolecular rate constants were calculated by assuming a homogeneous system and dividing the $k_{obs}$ values, obtained by pseudo-first-order treatment, by the $[Pd^{II}]$ calculated assuming all the $Pd^{II}$ to be present in homogeneous solution.

homogeneous oxidation of allyl alcohol by $PdCl_4^{2-}$ in aqueous solution at 25 °C, where approximately the same product distribution was found [24]. Thus, complex formation increases the concentration of the alcohol function in the region around the $Pd^{II}$ so that hydride abstraction becomes a more favorable process.

The rate constant achieved with the bimetallic catalyst was determined to be 1.5 $M^{-1}$ $s^{-1}$. The rate enhancement due to $\pi$-complexation to the monometallic catalyst would be expected to overshadow any real bimetallic rate enhancement obtained through bonding of the alkene to one $Pd^{II}$ followed by hydride extraction by the second $Pd^{II}$. Actually, there is a tenfold increase in rate on going to the bimetallic catalyst. Furthermore, interpretation of the data is again complicated by the fact that the product distribution changes from 25% acrolein with the mono-metallic catalyst to 80% acrolein with the bimetallic catalyst. Thus, the rate of direct alcohol oxidation to acrolein is 1.5 $M^{-1}$ $s^{-1}$ · 0.80 = 1.2 $M^{-1}$ $s^{-1}$ and the actual increase factor for the bimetallic catalyst over the monometallic catalyst is 1.2/0.0375 = 32. This factor is one of the largest observed and suggests that the geometry of the intermediate shown in Eq. (7.21) is ideal for hydride abstraction by $Pd^{II}$. ALCHEMY II calculations indicate the distance between the oxygen of the alcohol and the terminal carbon of the double bond to be 2.83 Å, which is close to the oxygen–oxygen distance in ethylene glycol.

$$CH_2{=}CHCHOH \xrightarrow{-H^+} CH_2{=}CHCHO \xrightarrow{-H^+} \qquad (7.21)$$

$$ + \quad CH_2{=}CHCHO$$

Allyl alcohol was the only allylic alcohol tested to show any appreciable activity. 4-Penten-2-ol and 4-penten-1-ol were completely unreactive with the monometallic catalysts, and 4-penten-2-ol showed only weak reactivity with the bimetallic system. The reason for this must relate to interatomic distances, as discussed above.

As a final observation regarding these homogeneous catalytic systems, the bimetallic systems were efficient air-oxidation catalysts with turnovers of over 1000 in some cases. In contrast, the monometallic systems deposited $Pd^0$ after about 50 cycles.

## 7.4
## Homogeneous Catalysis

### 7.4.1
### In the Absence of Other Redox Agents

The next step was to transfer the insights gained from the heterogeneous system to the development of related homogeneous catalysts. Industrially useful syntheses involving palladium(II) catalysis require that the process be a net air-oxidation one. Thus, the well-known Wacker process (Eq. 7.14) utilizes copper(II) chloride to reoxidize the $Pd^0$ formed back to $Pd^{II}$. In turn, the copper(I) chloride is oxidized to copper(II) chloride by molecular oxygen giving a net air oxidation. However, a serious problem in the use of this regeneration system in the oxidation of alkenes in acetic acid to vinylic and allylic acetates (Eq. 7.15) is the fact that chloroacetates and diacetates become the main products. Thus, a halide-free regeneration system is desirable for oxidations in acetic acid.

There are presently several systems available for the air oxidation of alkenes in chloride-free acetic acid under mild reaction conditions (25 °C and 1 atm. $O_2$). Two of these systems involving multi-step electron transfer were mentioned in the introduction to this chapter. Only one type of system, namely palladium clusters in DMSO, involves direct oxidation of zerovalent palladium [29].

The bimetallic $Pd^{II}$ catalyst in Eqs. (7.20) and (7.21) would seem to be an ideal candidate for the oxidation of alkenes in acetic acid. Unfortunately, the simple heterogeneous catalyst in these equations cannot be prepared in homogeneous media because it would disproportionate to give the bis complex and the free metal. However, the bimetallic homogeneous catalysts 11 shown in Figure 7.2, which contain biphosphine bridging ligands such as 1,4-bis(diphenylphosphino)butane, are able to give allylic and homoallylic esters in acetic acid solution.

The ultimate aim of this research was to prepare reagents that would be useful in natural product synthesis. Thus, the use of a chiral bidentate bridging ligand such as (–)-DIOP (11b) or (S)-METBOX (11c) holds promise for an asymmetric allylic acetate synthesis. The catalyst system was first tested on cyclohexene in acetic acid. As shown in Eq. (7.22), a 95% yield of 2-cyclohexenyl acetate along with a small amount of the homoallylic acetate was obtained [30a]. The enantioselectivity was a respectable 55% with 11c. Previous attempts to achieve asymmetric oxidation of cyclohexene using chiral $Pd^{II}$ complexes produced ee values of less than 5% [30b, c].

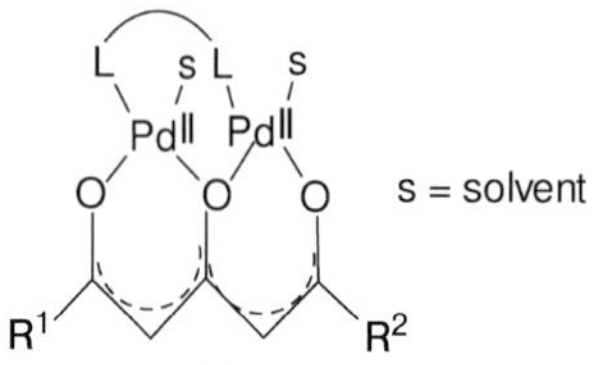

s = solvent

11a  L-L = 1,4-bis(diphenylphosphino)butane
11b  L-L = (–)-2,3-O-isopropylidine-2,3-dihydroxy-
          1,4-bis(diphenylphosphino)butane
11c  L-L = 2,2'-methylbis[4S)-4-tert-butyl-2-oxazoline)]
          ((S)-METBOX)

**Figure 7.2** Homogeneous bimetallic palladium(II) catalysts.

Cyclopentene was also oxidized under the same conditions as cyclohexene using catalyst **11b**. A 92% yield of 2-cyclopenten-1-yl acetate with an ee of 78% was obtained, accompanied by 2% of 3-cyclopenten-1-yl acetate. Using catalyst **11c**, cycloheptene gave a 70% yield of 2-cyclohepten-1-yl acetate with an ee of 70%. No homoallylic acetate was detected. On the other hand, cyclooctene was unreactive.

$$
\begin{array}{c}
\text{11b or 11c} \\
\text{HOAc} \\
\xrightarrow{\hspace{2cm}} \\
\text{O}_2
\end{array}
$$

⬡ + OAc⁻ → ⬡–OAc + ⬡–OAc    (7.22)

95%    2%

**11b**: ee = 52%
**11c**: ee = 55%

The high asymmetric induction obtained with the bimetallic catalysts is probably the most unexpected and interesting result of this work. Formally, the bimetallic catalyst, with one phosphine per $Pd^{II}$, is analogous to the monometallic catalyst $[PdCl_3L]^{3-}$. Two factors almost certainly account for the high enantioselectivity observed. The first is the rigid structure of the bridging diphosphine ligand. Without the free rotation of the monodentate ligand in $[PdCl_3L]^{3-}$, the asymmetric induction is increased. The second factor is the coordination sphere of the $Pd^{II}$, in which there is only one vacant coordination site available to the alkene. This site is adjacent to the chiral phosphine ligand, where the asymmetric induction will be highest.

The use of this direct air-oxidation catalytic system has now been extended to other reactions and to solvents other than acetic acid. Similar bimetallic complexes have been used in the α-hydroxylation of ketones in THF/water mixtures. In a paper from these laboratories, complex **11** incorporating the non-chiral ligand *trans*-1,2-diaminocyclohexane was reported to oxidize cyclohexanone and other ketones to α-hydroxy ketones (Eq. 7.23).

$$
\underset{\text{O}}{\overset{\text{O}}{RCH_2\overset{\|}{C}CR^1}} \xrightarrow[\substack{O_2 \\ THF/H_2O}]{[Pd_2(triketone)(L\text{-}L)]} \underset{\text{HO O}}{RCH\overset{\|}{C}R^1} \quad (7.23)
$$

L-L = 1,2-*trans*-diaminocyclohexane

Addition of acid greatly increases the yields obtained in these substitutions. This means that the $Pd^{II}$ must be reacting with the enol form of the ketone as soon as it is formed since acid will increase the rate of formation of the enol but not its equilibrium position. The next step will be to run this reaction with chiral complexes such as **11b** or **11c**. The asymmetric α-hydroxylation reaction has already been performed in reaction mixtures containing $CuCl_2$, which renders it, like the Wacker reaction itself (Eq. 7.2), a net air-oxidation process [31].

A third reaction that can be performed as a direct air oxidation is the 1,4-di-functionalization of conjugated dienes by $Pd^{II}$ catalysts. Bäckvall and co-workers have reported extensive studies on this reaction [32]. Previous attempts to make this reaction asymmetric employed chiral benzoquinone ligands [33, 34]. The enantiomeric excesses obtained in diacetoxylation and dialkoxylation reactions were modest (about 50%). In this approach, the chiral p-benzoquinone acts as a rate-accelerating ligand through assisting in the removal of $Pd^{II}$ from the π-allyl reaction intermediate (Eq. 7.24) [35]. The factors involved in the asymmetric 1,4-substitution reaction have been studied in detail [36].

$$ (7.24) $$

As described above for the acetoxylation of simple alkenes, various redox reagents have been used to make the reaction a catalytic air-oxidation system in halide-free solvents. We have found that the reaction in acetic acid can be run catalytically in the presence of oxygen without any other redox reagents to give 1,4-diacetoxylation of 1,3-cyclohexadiene [31]. The action of the dimer in an air-oxidation reaction is indicated in Eq. (7.25). The $Pd^I$ dimer is stable in the absence of oxygen, but in air it is readily oxidized to the $Pd^{II}$ form. As a "leaving" group, the $Pd^I$ dimer also functions as a rate-accelerating unit. The roles of $Pd^0$ as a "leaving" group have been discussed in detail [37]. Briefly, $Pd^0$ is a poor "leaving" group because monomeric $Pd^0$ is a very high-energy species. On the other hand, the $Pd^I$ dimer is a stable species in solution and thus is a very good "leaving" group. This function is similar to that of the benzoquinone in Eq. (7.24). The quinone converts the $Pd^0$ into $Pd^{2+}$ as it is leaving the intermediate π-complex, thus avoiding the need for free $Pd^0$ to be the leaving group.

$$ (7.25) $$

L-L = (S)-BINAP

## 7.4.2
### In the Presence of Other Redox Reagents

Asymmetric syntheses using bimetallic catalysts in the presence of copper(II) chloride or bromide will now be considered. In these reactions, the copper salt not only acts as a redox system to make the process an air-oxidation one, but it is also an essential reagent in generating the desired products.

In the Wacker reaction itself (Eq. 7.2), a side product obtained to a minor extent in the reaction with ethylene is 2-chloroethanol. At [Cl⁻] > 3.0 M and [CuCl₂] > 2.5 M, 2-chloroethanol becomes the major product. The addition of a neutral ligand to the coordination sphere of $Pd^{II}$ was also found to increase the yield of 2-chloroethanol. Thus, [PdCl$_3$(pyridine)]$^-$ gave 2-chloroethanol at [Cl$^-$] at low as 0.2 M. At this low [Cl$^-$], [PdCl$_4$]$^{2-}$ gives only acetaldehyde at any [CuCl$_2$] [38].

This finding opens up the possibility of an asymmetric chlorohydrin synthesis from alkene substrates [39, 40]. Pd$^{II}$ catalysts incorporating chiral auxiliaries can be expected to produce optically active products. The logical starting point is the replacement of pyridine with a chiral amine L*. Scheme 7.3 outlines the general reaction scheme with monodentate chiral amines such as (CH$_3$)$_2$C*NH(CH$_3$)Ph. As shown, the two positional isomers, 12 and 13, arise from the two possible modes of hydroxypalladation. The 12/13 ratio of about 4 found for propene and 1-pentene is typical for a number of catalysts.

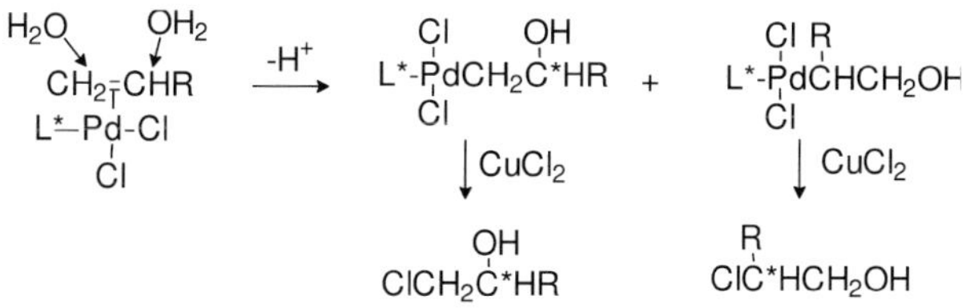

Scheme 7.3 Reaction of α-alkenes with [PdCl$_3$L*] in the presence of CuCl$_2$.

As expected, the optical purity of 12 is low, with ee values of 10–15%. Catalysts with chiral chelating diphosphines can be expected to give much higher optical purities. However, the monometallic Pd$^{II}$ complex incorporating a diphosphine ligand ([PdCl$_2$(L-L)]) is a neutral species and is thus insoluble in the reaction media. Two different approaches were used to solve this problem. One approach relied on sulfonated chiral ligands, while the other involved bimetallic complexes with a bridging diphosphine ligand such as 11 in Figure 7.2.

Table 7.3 lists some representative results. As discussed above, all reactions were catalytic oxidations. The catalytic turnovers, listed in the last column, do not represent the maximum number obtainable. In all cases, the reaction rate had not diminished at the termination point and many more turnovers were possible.

The 12/13 ratios were high in runs 1–3, indicating that the formation of 13 is not a serious problem. Only propene gave a detectable amount of 13. Even in this case it was less than 10% of the total. Steric factors account for this trend in the 12/13 ratios.

The optical purities achieved with the bimetallic catalyst were higher than those achieved with the sulfonated monometallic catalyst, and ranged from modest to good. In all cases, the ee was over 80% and rose to over 90% in two runs (4 and 7). Simple steric factors cannot account for these trends. Thus, the highest value obtained was 94% ee for propene in run 4.

The Cu$^{II}$ reaction was then run in bromide-containing media [41] using 9 with several chiral ligands. Surprisingly, the oxidation of alkenes did not produce the expected bromohydrins. Instead, the main product was the 1,2-dibromo derivative (Eq. 7.26). Furthermore, the dibromo derivative was obtained with high enantioselectivity (ee ~ 95%). The results for several alkenes are listed in Table 7.4.

**Table 7.3** Results for the oxidation of several alkenes by Ar-BINAP catalysts in the presence of CuCl$_2$.[a]

| Entry | Catalyst[b] | [LiCl] (M) | Substrate | 11/12 ratio[c] | % ee | Turnover |
|---|---|---|---|---|---|---|
| 1 | monometallic | 0.20 | propene | 12 | 46 | 60[d] |
| 2 | monometallic | 0.20 | CH$_2$=CHC(O)CH$_3$ | > 95 | 76 | 65[e] |
| 3 | monometallic | 0.20 | CH$_2$=CHCH$_2$OPh | > 95 | 68 | 72[e] |
| 4 | bimetallic | 0.10 | propene | 3.5 | 94 | 195[d] |
| 5 | bimetallic | 0.22 | 1-pentene | 4.0 | 81 | 200[e] |
| 6 | bimetallic | 2.20 | 1-hexene | 18.7 | 87 | 155[e] |
| 7 | bimetallic | 0.30 | CH$_2$=CHCH$_2$OPh | > 95 | 93 | 170[e] |

[a] All runs were carried out with 0.1–0.3 mmol of chiral catalyst in 25–50 mL of solvent and were 3–5 M in CuCl$_2$; temperature 25 °C; the solvent was a H$_2$O/THF mixture containing 30–90% (v/v) THF.
[b] In the case of the monomeric catalyst the Ar groups were sulfonated toluene; for the bimetallic system the Ar were phenyl.
[c] The reaction mixture also contained varying amounts of the Wacker ketone product (5–30%).
[d] Measured by propene uptake using gas burets.
[e] Dioxygen was the oxidant; turnovers measured by O$_2$ uptake using gas burets. In calculating turnovers dioxygen is assumed to be a four-electron oxidant.

**Table 7.4** Results for the oxidation of several substrates.[a]

| Entry | L*-L* | [LiBr] (M) | Substrate | % 3 | %ee of 3 |
|---|---|---|---|---|---|
| 1 | (S)-BINAP | 0.25 | p-CH$_3$O-phenyl allyl ether | 95 | 96 |
| 2 | (S)-Tol-BINAP | 0.13 | p-CN-phenyl allyl ether | 95 | 97 |
| 3 | (S)-BINAP | 0.20 | phenyl allyl ether | 95 | 95 |
| 4 | (S)-METBOX | 0.20 | methyl acrylate | 84 | 94 |
| 5 | (S)-METBOX | 0.30 | methyl crotonate | 80 | 82[b] |
| 6 | (R)-BINDA | 0.20 | cinnamyl alcohol | 77 | 80[b] |
| 7 | (R)-BINDA | 0.0 | cinnamyl alcohol | 75 | 34[b] |
| 8 | (R)-BINDA | 0.20 | methyl trans-cinnamate | 84 | 14[b] |

[a] All runs were carried out with 0.05 – 0.12 mmol of chiral catalyst in 20–30 mL of solvent and were 2.0–2.5 M in CuBr$_2$ and 0.13–0.25 M in LiBr; temperature 25 °C; the solvent was a H$_2$O/THF mixture containing 54–93% (v/v) THF.
[b] These dibromides have the (RS,SR) configuration.

$$RCH{=}CHR^1 + Br^- \text{ or } H_2O \xrightarrow[CuBr_2]{9} \underset{\underset{>95\%}{14}}{RC^*HCHR^1} + \underset{\underset{<5\%}{15}}{RC^*HC^*HR^1} \quad (7.26)$$

As in the chlorohydrin synthesis, the reaction is a net air-oxidation one since the CuBr formed in the dibromo reaction is oxidized back to CuBr$_2$ by oxygen. The catalyst turnovers listed in Table 7.4 range from 10 to 150, but the rate did not decrease during the course of the reaction so many more catalyst turnovers are possible.

The aim of the first three runs with *para*-substituted phenyl allyl ethers was to assess electronic effects on the oxidation. Since the results of all three runs were very similar, electronic effects evidently have little impact on enantioselectivity with this series of substrates. This result is consistent with other palladium(II) catalytic chemistry [42].

The enantioselectivities obtained with the internal alkenes were somewhat poorer than those obtained with the $\alpha$-alkenes. This could not be due to the variation in chiral ligand. Entries 4 and 5 relate to the same ligand, but the ee value decreases from 94% for oxidation of methyl acrylate to 82% for the oxidation of methyl crotonate. Runs 6 and 7 demonstrate the importance of the added bromide ion concentration. When no extra bromide was added, the ee decreased to 34%. The low enantioselectivity observed with methyl *trans*-cinnamate is surprising.

These results can be compared with those for the chlorohydrin reaction, where internal alkenes were completely unreactive. The dibromo product, **14**, obtained from the internal alkenes was found to have the $(RS,SR)$ configuration, which is consistent with anti addition to the double bond. This is the same stereochemistry as found for the chlorohydrin reaction.

The fact that the bromide system produces dibromides rather than bromohydrins can be attributed to the nucleophilicity of bromide. The bromide ion must be a much better nucleophile than chloride or water since, in the examples in Table 7.4, the concentration of water is at least 200 times that of bromide.

## 7.4.3
## Co$^{III}$ Catalysis

Although the reactions of Co$^{II}$ complexes have received considerable attention, the organic oxidation reactions of the resulting dioxygen-bridged bimetallic Co$^{III}$ complexes have been the subject of few studies. However, one report describes investigations of the reactions of these complexes incorporating dipeptides such as glycylglycine and glycyl-L-alanine [43]. The reaction was found to proceed in two steps. The first involves rapid oxidation of one ligand molecule by the two cobalt centers with conversion of bridging dioxygen moiety to water. The second slower step probably involves displacement of the oxidized ligand by the excess dipeptide in the solution.

The oxidation of 3,5-di-*tert*-butylcatechol to the corresponding quinone by bimetallic and monometallic complexes of 2-benzoylacetophenol was studied by optical spectrometry using the DTBC absorption band at 400 nm [13b]. The monometallic complex proved to be unreactive, whereas the bimetallic system catalyzed the oxidation.

### 7.4.4
### Mo$^{VI}$ Catalysis

The bimetallic triketone complex shown in Figure 7.3 was found to catalyze the epoxidation of cyclooctene and 1,5-cyclooctadiene with *tert*-butyl hydroperoxide at 40 and 80 °C, respectively, in dichloroethane [44]. With 1,5-cyclooctadiene, the monoepoxide was formed. The reactivity was compared to that of the monometallic complex MoO$_2$(acac)$_2$.

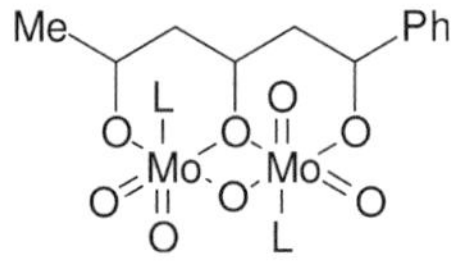

**Figure 7.3** Structure of bimetallic Mo$^{VI}$ catalysts.

### 7.5
### Heterobinuclear Systems

There are a number of heterogeneous catalytic systems that employ mixtures of metals for the oxidation of organics. We shall limit our discussion to cases where evidence exists that a discrete bimetallic complex is the active catalyst.

### 7.5.1
### Pd$^{II}$ Plus Another Metal

There have been two reports of Pd$^{II}$-Cu$^{II}$ complexes being used for Wacker-type oxidation. In one, a PdCuCl(2-PyO)$_3$EtOH complex was prepared (2-PyO$^-$ denotes the 2-hydroxypyridine anion) and its dimeric structure was proven by X-ray analysis [45]. The central Pd and Cu are bridged by three 2-PyO$^-$ ligands. This complex was found to promote a Wacker-type oxidation of 1-octene. The bridging 2-PyO$^-$ ligands, rather than Cl atoms, appear to be essential for the redox process.

A similar but less well-defined PdCl$_2$-CuCl$_2$ bimetallic system was used to oxidize cyclopentene to cyclopentanone in ethanol [46]. The ethanol solvent was also oxidized. The kinetics was consistent with the active catalytic species being a Pd$^{II}$-Cu$^{II}$ bimetallic complex, in which Cu$^I$ serves as a transient O carrier and the Pd$^{II}$OOH species thus formed selectively monooxygenates cyclopentene to give cyclopentanone.

In a somewhat different system from those discussed above, Pd(OAc)$_2$ was reacted with Pb(OAc)$_2$ in HOAc at 60 °C to give PdPb(OAc)$_4$ · AcOH, which was characterized by its UV/visible and IR spectra, as well as by X-ray powder diffractometry [47]. The bimetallic complex was found to catalyze the benzylic acyloxylation of toluene very effectively in the presence of lauric acid.

## 7.5.2
## Fe$^{III}$ Plus Another Metal

The liquid-phase oxidation of Me$_3$CHPh to PhCMe$_2$O$_2$H at 97–98 °C was studied using bimetallic Cu-Fe polyphthalocyanines as catalysts [48]. These catalysts proved to be highly active, giving the hydroperoxide in 98–100% yield.

Fe(acac)$_3$–TiO(acac)$_2$ systems have been used for the oxidation of α-alkenes [49]. Addition of cumene increased the yield by a factor of three.

## 7.5.3
## Ru$^{II}$ Plus Other Metals

In the absence of solvent, first-row transition metal acetylacetonate complexes and RuCl$_2$(PPh$_3$)$_3$ were found to give fairly high turnovers in the allylic oxidation of cyclohexene under 1 atm. of dioxygen [50]. A synergistic effect is observed in the oxidation of cyclohexene in the presence of M(acac)$_m$–RuCl$_2$(PPh$_3$)$_3$.

## 7.5.4
## Rh$^{III}$ and Other Metals

Rh(acac)$_3$–MoO$_5$L (L = hexamethylphosphoric triamide) was found to catalyze the oxidation of α-alkenes in the presence of cumene [49].

## References

1   J. P. COLLMAN, P. DENISEVICH, Y. KONAI, M. MARROCCO, C. KOVAL, F. C. ANSON, *J. Am. Chem. Soc.* **1980**, *102*, 6027–6036.

2   P. M. HENRY, *Palladium-Catalyzed Oxidation of Hydrocarbons*, D. Reidel, Dordrecht, The Netherlands, **1980**, pp. 41–84.

3   (a) J.-E. BÄCKVALL, R. B. HOPKINS, H. GRENNBERG, M. MADER, A. K. AWASTHI, *J. Am. Chem. Soc.* **1990**, *112*, 5160–5166; (b) S. E. BYSTRÖM, E. M. LARSSON, B. ÅKERMARK, *J. Org. Chem.* **1990**, *55*, 5674–5675; (c) K. BERGSTAD, H. GRENNBERG, J.-E. BÄCKVALL, *Organometallics* **1998**, *17*, 45–50; (d) T. YOKOTA, S. FUJIBAYASHI, Y. NISHIYAMA, S. SAKAGUCHI, Y. ISHII, *J. Mol. Catal. A* **1996**, *114*, 113–122.

4   (a) O. HAYAISHI (Ed.), *Molecular Mechanisms of Oxygen Activation*, Academic Press, New York and London, 1974;

(b) T. G. SPIRO (Ed.), *Metal Ion Activation of Dioxygen: Metal Ions in Biology*, Wiley-Interscience, New York, 1980; (c) R. A. SHELDON, J. KOCHI, *Metal-Catalyzed Oxidations of Organic Compounds*, Academic Press, New York, 1981; (d) A. E. MARTELL, D. T. SAWYER (Eds.), *Oxygen Complexes and Oxygen Activation by Transition Metals*, Plenum, New York, 1988.

5   P. M. HENRY, F. T. T. NG, *Can. J. Chem.* **1980**, *58*, 1773–1779.

6   N. NOZAKI, in: *Molecular Mechanisms of Oxygen Activation* (Ed.: O. HAYAISHI), Academic Press, New York and London, 1974, p. 135.

7   M. G. WELLER, U. WESER, *J. Am. Chem. Soc.* **1982**, *104*, 3752.

8   K. OHKUBO, H. ISHIDA, T. SAGAWA, K. MIYATA, K. YOSHINAGA, *J. Mol. Catal.* **1990**, *62*, 107.

9   T. FUNABIKI, A. MIZOGUCHI, T. SUGIMOTO, S. TADA, M. TSUJI,

H. Sakamoto, S. Yoshida, *J. Am. Chem. Soc.* **1986**, *108*, 2921.

10 E. I. Solomon, in *Metal Ions in Biology* (Ed.: T. G. Spiro), Wiley-Interscience, New York, 1981, Vol. 3, p. 44.

11 D. A. Rockcliffe, A. E. Martell, *Inorg. Chem.* **1993**, *32*, 3143.

12 S. Tsuruya, R. L. Lintvedt, Abstracts of Papers, 176th National Meeting of the American Chemical Society, Miami, September 1978; American Chemical Society, Washington, D.C., 1978.

13 (a) N. Oishi, Y. Nishida, K. Ida, S. Kida, *Bull. Chem. Soc. Jpn.* **1980**, *53*, 2847; (b) U. Casellato, S. Tamburini, P. A. Vigato, A. De Stefani, M. Vidali, D. E. Fenton, *Inorg. Chim. Acta* **1983**, *69*, 45–51.

14 J.-P. Chyn, F. L. Urbach, *Inorg. Chim. Acta* **1991**, *189*, 157.

15 M. M. Rogić, M. D. Swerdloff, T. R. Demmin, in: *Copper Coordination Chemistry: Biochemical and Inorganic Perspectives* (Eds.: K. D. Karlin, J. Zubieta), Adenine Press, New York, 1983, pp. 259–279.

16 J. Tsuji, H. Takayanagi, *J. Am. Chem. Soc.* **1974**, *96*, 7349.

17 T. R. Demmin, M. Rogić, *J. Org. Chem.* **1980**, *45*, 4210–4214.

18 (a) V. V. Korshak, M. M. Teplyakov, V. A. Sergeev, *J. Polym. Sci., B. Polym. Lett.* **1973**, *11*, 583; (b) V. V. Korshak, M. M. Teplyakov, V. P. Chebotaryev, *J. Polym. Sci., B. Polym. Lett.* **1973**, *11*, 589.

19 K. Zaw, P. M. Henry, *J. Mol. Catal. A* **1995**, *101*, 187–198.

20 (a) C. R. Hauser, T. M. Harris, *J. Am. Chem. Soc.* **1958**, *80*, 6360; (b) M. L. Miles, T. M. Harris, C. R. Hauser, *J. Org. Chem.* **1965**, *30*, 1007; (c) M. Gall, H. O. House, *Org. Synth.* **1972**, *52*, 39.

21 P. M. Henry, X. Ma, G. Noronha, K. Zaw, *Inorg. Chim. Acta* **1995**, Vol. ICA240/1–2, 205–215.

22 E.-I. Negishi (Ed.), *Handbook of Organopalladium Chemistry for Organic Synthesis*, J. Wiley & Sons, New York, **2002**.

23 P. M. Henry, *Palladium-Catalyzed Oxidation of Hydrocarbons*, D. Reidel, Dordrecht, The Netherlands, **1980**, Chapter 2.

24 K. Zaw, M. Lautens, P. M. Henry, *Organometallics* **1983**, *2*, 197–199.

25 P. M. Henry, *Palladium-Catalyzed Oxidation of Hydrocarbons*, D. Reidel, Dordrecht, The Netherlands, **1980**, p. 5.

26 G. Noronha, unpublished results.

27 G. Noronha, P. M. Henry, *J. Mol. Catal. A: Chem.* **1997**, *120*, 75–87.

28 ALCHEMY II is a registered trademark for a structure modeling program from TRIPOS Associates. The program performs a conjugate-gradient minimization to put the molecule in a minimal energy conformation.

29 (a) R. C. Larock, T. R. Hightower, *J. Org. Chem.* **1993**, *58*, 5298–5300; (b) R. A. T. M. van Benthem, H. Hiemstra, J. J. Michels, W. N. Speckamp, *J. Chem. Soc., Chem. Commun.* **1994**, 357–359; (c) M. Rönn, J.-E. Bäckvall, P. G. Andersson, *Tetrahedron Lett.* **1995**, *36*, 7749–7752.

30 (a) A. El-Qisiari, H. A. Qaseer, P. M. Henry, *Tetrahedron Lett.* **2002**, *43*, 4229–4231; (b) H. Yang, A. K. Khan, K. M. Nicholas, *J. Mol. Catal.* **1994**, *91*, 319–334; (c) N. Yu. Kozitsyna, M. N. Vargaftik, I. I. Moiseev, *J. Organomet. Chem.* **2000**, *593–594*, 274–291.

31 A. K. El-Qisairi, unpublished results.

32 J.-E. Bäckvall, *Palladium-Catalyzed 1,4-Addition to Conjugated Dienes*, in *Metal-Catalyzed Cross-Coupling Reactions* (Eds.: P. Stang, F. Diederich), Wiley-VCH, Weinheim, 1998, pp. 339–385.

33 A. Thorarensen, A. Palmgren, K. Itami, J.-E. Bäckvall, *Tetrahedron Lett.* **1997**, *38*, 8541–8544.

34 K. Itami, A. Palmgren, A. Thorarensen, J.-E. Bäckvall, *J. Org. Chem.* **1998**, *63*, 6466–6471.

35 J.-E. Bäckvall, S. E. Byström, R. E. Nordberg, *J. Org. Chem.* **1984**, *49*, 4619–4631.

36 H. K. Cotton, R. C. Verboom, L. Johansson, B. J. Plietker, J.-E. Bäckvall, *Organometallics* **2002**, *21*, 3367–3375.

37 O. Hamed, P. M. Henry, *Organometallics* **1998**, *17*, 5184–5189.

38 J. W. Francis, P. M. Henry, *J. Mol. Catal. A: Chem.* **1995**, *99*, 77–86.

39 A. El-Qisiari, O. Hamed, P. M. Henry, *J. Org. Chem.* **1998**, *63*, 2790–2791.

**40** A. El-Qisairi, P. M. Henry, *J. Organomet. Chem.* **2000**, *603*, 50–60.

**41** A. K. El-Qisairi, H. A. Qaseer, G. Katsigras, P. Lorenzi, U. Trivedi, S. Tracz, A. Hartman, J. A. Miller, P. M. Henry, *Org. Lett.* **2003**, *5*, 439–441.

**42** P. M. Henry, *Palladium-Catalyzed Oxidation of Hydrocarbons*, D. Reidel, Dordrecht, The Netherlands, **1980**, pp. 73–74.

**43** W. R. Harris, R. C. Bess, A. E. Martell, T. H. Ridgway, *J. Am. Chem. Soc.* **1977**, *99*, 2958–2963.

**44** M. L. D'Amico, K. Rasmussen, D. Sisneros, C. Magnujssen, H. Wade, J. G. Russell, L. L. Borer, *Inorg. Chim. Acta* **1992**, *191*, 167–170.

**45** M. Higashijima, T. Masunaga, Y. Kojima, E. Watanabe, K. Wada, *Research and Development Review – Mitsubishi Kasei Corporation* **1994**, *8*, 14–19.

**46** K. Takehira, T. Hayakawa, S. Orita, M. Shimizu, *J. Mol. Catal.* **1989**, *53*, 15–21.

**47** A. B. Goel, P. E. Throckmorton, R. A. Grimm, *Inorg. Chim. Acta* **1986**, *117*, L15–L17.

**48** T. I. Andrianova, A. I. Sherle, A. A. Berlin, *Izv. Akad. Nauk.* **1973**, *3*, 531–536.

**49** H. Arzoumanian, H. Bitar, J. Metzger, French Patent No. 2310987 (1976).

**50** H. Jiang, Y. Xu, S. Liao, D. Yu, *React. Kinet. Catal. Lett.* **1998**, *63*, 179–183.

# 8
# Bimetallic Oxidation Catalysts: Hydrogen Peroxide Generation and Its Use in Hydrocarbon Oxidation

*Joseph E. Remias and Ayusman Sen*

## 8.1
## Introduction

Bimetallic systems capable of performing two separate chemical reactions in the same vessel are intriguing from a practical standpoint. Of interest in this article is the generation of hydrogen peroxide and its *in situ* use for organic oxidations, by way of another catalyst. However, when developing a reaction using a bimetallic system one must take special care to understand both the independent catalytic reactions occurring on each metal, and the interactions that the two metals can have while present in the same solution. This work attempts to bring to light some of the particulars of *in situ* hydrogen peroxide generation in alkane and aromatic oxidations and outlines some of the aspects that must be considered when a bimetallic system is employed to carry out the two reactions.

Co-reductants serve an important function in enzymatic oxidations. The mono-oxygenases, such as cytochrome P-450 [1, 2] and methane monooxygenase [3], perform difficult oxidations through reductive activation of dioxygen. During monooxygenase-type oxidations, a molecule of water is produced, making the maximum efficiency based on oxygen 50%. The ultimate goal of catalysis would be to use 100% of the oxidant, as in dioxygenases. However, when one considers autoxidation or free radical processes, which use 100% of the available oxygen, they rarely show the necessary selectivity for hydrocarbon oxidations. Consequently, the use of a monooxygenase-type catalyst, one involving a co-reductant, provides an appealing alternative. A wealth of catalysts and oxidants are available which act in this way and provide more selective reactions. Some examples of oxidants include: hydrogen peroxide, peracids, organic hydroperoxides, inorganic and metallorganic peroxides, sodium hypochlorite, iodosobenzene, and nitrous oxide. Many of these oxidants are too expensive for basic chemical production, which is the focus here, and find more use in fine chemical production where expensive and even stoichiometric oxidants still play a large role.

Of the oxidants listed above, hydrogen peroxide shows the most promise for several reasons. First, it delivers the most active oxygen per molecule, 47%, when

*Multimetallic Catalysts in Organic Synthesis.* Edited by M. Shibasaki and Y. Yamamoto
Copyright © 2004 WILEY-VCH Verlag GmbH & Co. KGaA, Weinheim
ISBN: 3-527-30828-8

compared to the others listed above. It is important to note that this is actually ca. 20% when using hydrogen peroxide as an aqueous solution. The atom economy of oxidants is important to consider when trying to minimize the environmental impact of a reaction [4]. Furthermore, hydrogen peroxide produces only water as a by-product, an issue when one considers the need for "green" oxidants (ones that produce minimal and non-toxic waste) [5]. Finally, hydrogen peroxide has proven to be a strong oxidant with good selectivity in many oxidations [6].

The use of metal catalysts for the generation of hydrogen peroxide using dihydrogen (the co-reductant) and dioxygen has generated significant interest recently, as evidenced by the proliferation of publications. This desire is at least partially fueled by an interest in pursuing "greener" routes than the current anthraquinone process to make hydrogen peroxide [7]. Also, generating the oxidant *in situ* maximizes the usable oxygen: since no diluent is needed for the oxidant if it is generated *in situ*, the amount of usable oxygen is ideally increased from ca. 20% to 47%. Furthermore, the cost associated with storing and transporting hydrogen peroxide is eliminated. Another advantage is that a bimetallic catalyst system can be envisaged, in which the hydrogen peroxide is generated in a slow steady-state amount on one catalyst while another performs the oxidation. This should maximize the selectivity and efficiency of the reaction. Finally, the present cost of hydrogen peroxide is about five times greater than that of dihydrogen (the only consumable of cost, assuming 100% catalyst recyclability and air as the oxidant). Thus, a significant cost saving can be realized per mole of oxidant used.

In this short review, we focus on hydrogen peroxide generation *in situ* for hydrocarbon oxidations. Specifically, we are concerned with the selective oxidation of alkanes and aromatics using hydrogen peroxide generated on a noble metal catalyst. In all cases, dihydrogen serves as the co-reductant either directly or indirectly. The use of bimetallic systems in which one catalyst generates the hydrogen peroxide while another performs hydrocarbon oxidation is given particular attention.

Although beyond the scope of this work, it is worthwhile to survey some other examples of bimetallic oxidations with co-reductants. Barton and Doller [8] have developed several 'Gif' systems using iron catalysts in the presence of metallic zinc or iron as a reductant. These methods were developed more to model biological monooxygenases than to serve practical industrial purposes. Furthermore, a substantial body of research exists on the epoxidation of alkenes using *in situ* generated hydrogen peroxide [9–12].

## 8.2
### Metal-Catalyzed Formation of Hydrogen Peroxide

Before discussing the use of bimetallic catalysts for hydrocarbon oxidation with *in situ* generated hydrogen peroxide, the formation of hydrogen peroxide and the reactions associated with it must first be considered. There are numerous reports in the literature detailing the formation of hydrogen peroxide on metal catalysts. During palladium-catalyzed aerobic oxidations, it is known that hydrogen peroxide is formed as a by-product [13]. In some instances, this peroxide remains in solution. For instance, in the aerobic oxidation of alcohols by a combination of palladium acetate and pyridine, it was suggested that the peroxide may actually compete with the substrate for binding to the metal center and, therefore, its removal by decomposition (e.g., by molecular sieves) promoted substrate oxidation [14]. Another notable example is the oxidation of alcohols in the presence of $Pd^{II}$ complexes, which produces a significant amount of peroxide (Eq. 8.1) [15]. These reactions were carried out in a biphasic mixture to prevent oxidation of the catalyst's ligand; however, this methodology may also have served to prevent peroxide decomposition induced by the palladium catalyst. The biphasic conditions may partially explain why in other examples no peroxide was observed in alcohol oxidation (see below). The quick separation of the peroxide from the palladium catalyst via phase separation prevented its subsequent metal-catalyzed decomposition. Surprisingly, the authors claim that equimolar amounts of ketone and peroxide are formed, a result implying no metal-catalyzed decomposition of the latter.

$$R_2CHOH + O_2 \rightarrow R_2CO + H_2O_2 \qquad (8.1)$$

Several patents and papers have detailed the palladium-catalyzed generation of hydrogen peroxide in relatively high concentrations in the absence of an oxidizable substrate (see below). These reactions involve the use of dioxygen and either carbon monoxide or dihydrogen as the co-reductant. Carbon monoxide used under these conditions is an indirect source of dihydrogen through the water-gas shift reaction (Eq. 8.2). Both metallic palladium and discrete palladium compounds have been employed as catalysts for the reaction.

$$H_2O + CO \rightleftharpoons H_2 + CO_2 \qquad (8.2)$$

Pioneering work showed that hydrogen peroxide could be formed in relatively high concentrations (10–20% $w/w$) with a palladium catalyst, dihydrogen, and dioxygen [16]. The need for an acidic solution and the use of halide promoters, making purification difficult and generating a corrosive environment, have led researchers to look at variations on the above. For example, supported heteropoly acids have been used in place of inorganic acids [17]. Furthermore, others have found that halides can be excluded with certain catalysts, albeit only in organic solvents [18]. An interesting recent study sheds light on how palladium catalysts form hydrogen peroxide [19]. The researchers contend that hydrogen peroxide formation occurs only on colloidal Pd particles regardless of their source ($Pd^{2+}$ ions or Pd on a silica support), at least under acidic aqueous conditions.

In contrast to the above report, discrete $Pd^{2+}$ complexes have been found to catalyze the formation of hydrogen peroxide using carbon monoxide as the co-reductant [20]. Here, a biphasic mixture was again employed to isolate the catalyst from the hydrogen peroxide formed. It was necessary to carry out the reaction with a large excess of dioxygen to prevent reduction to palladium metal. By carefully selecting nitrogen-coordinated ligands with appropriate steric and electronic environments, the researchers were able to produce hydrogen peroxide at concentrations of up to 8% ($w/w$).

Though the mechanism of hydrogen peroxide generation on noble metals is still not completely understood, particularly on surfaces in the presence of various promoters and other metals, some conclusions can be made based on results obtained with discrete palladium and platinum complexes. The first step involves the formation of a metal-peroxo complex. The coordinated peroxide can then be displaced from the metal by the addition of a ligand (e.g., Eq. 8.3) [21]. More commonly, in acidic solutions, it is protonated to form hydrogen peroxide [22, 23].

$$(R_3P)_2Pt(O_2) + 2\ R_3P + ROH \rightarrow [(R_3P)_4Pt]^{2+} + HO_2^- + HO_2^- + RO^- \quad (8.3)$$

## 8.3
## Metal-Catalyzed Decomposition of Hydrogen Peroxide

It is pertinent to point out that while hydrogen peroxide can be formed in reasonable yields, once formed, its significant decomposition is also observed. For example, during alkene oxidation with palladium diacetate and hydrogen peroxide, decomposition of the oxidant was observed [24]. This effect was more pronounced in the absence of the substrate. Furthermore, during the aerobic oxidation of alcohols with $Pd(OAc)_2$, it was shown that hydrogen peroxide in the absence of substrate underwent rapid disproportionation to molecular oxygen and water [25]. Further experiments showed that even while alcohol oxidation was occurring, the addition of hydrogen peroxide to the solution resulted only in its decomposition according to Eq. (8.4). No additional oxidation of the alcohol by the peroxide was observed (Eq. 8.5). It is also worthwhile to note that while both palladium and platinum catalysts can effectively form hydrogen peroxide in the presence of a co-reductant, platinum decomposes the oxidant more rapidly [26].

$$H_2O_2 \rightarrow H_2O + 0.5\ O_2 \quad (8.4)$$

$$H_2O_2 + RCH_2OH \rightarrow RCHO + 2\ H_2O \quad (8.5)$$

Figure 8.1 shows our results on the efficiency with which dihydrogen (and hydrogen peroxide formed from it) is used for benzene oxidation with several catalysts [27]. It demonstrates that only about 5.5% of the dihydrogen consumed to make peroxide is used in substrate oxidation. In comparison, under analogous conditions where hydrogen peroxide was added directly, an oxidant efficiency of 20% was observed. The observation that the efficiency with respect to *in situ* generated hydrogen peroxide remains constant regardless of the hydroxylation catalyst used is consistent with the palladium catalyst causing almost all of the observed hydrogen peroxide decomposition. Recent work using palladium, gold, and amalgams of the two has suggested that the decomposition on palladium may limit the amount of usable hydrogen peroxide [28].

A plausible solution to decrease the rate of decomposition may involve simply decreasing the amount of dihydrogen in the medium. This action would limit the availability of dihydrogen for palladium-catalyzed hydrogen peroxide hydrogenation (Eq. 8.6). This has been found to be effective in other cases [7]. However, it seems

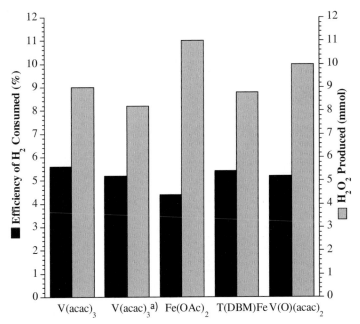

**Figure 8.1** Efficiency in the use of $H_2O_2$ generated for benzene oxidation. Calculated from total $H_2O$ produced vs. amount of phenol and benzoquinone formed. Runs using 3.2 µmol catalyst, 20.0 mg 5% $Pd/Al_2O_3$, 2 mL benzene, and 2.5 mL acetic acid, exposed to 100 psi $H_2$, 1000 psi $N_2$, and 100 psi $O_2$ for 2 h at 65 °C. Each column is the average of four reactions. Abbreviations: acac = acetylacetonate, OAc = acetate, T(DBM) = tris(dibenzoylmethanato). [a] 10.0 mg of 5% $Pd/Al_2O_3$ added.

unlikely that this would completely halt the unwanted reaction and it will not stop the disproportionation reaction. Consequently, one must consider the rate of decomposition when seeking catalysts for the *in situ* generation of hydrogen peroxide.

$$H_2O_2 + H_2 \rightarrow 2\ H_2O \qquad (8.6)$$

## 8.4
## Bimetallic Hydrogen Peroxide Generation and Hydrocarbon Oxidation

While this work focuses on the oxidation of organic compounds using two metals, where one generates peroxide and another uses it for oxidation, it is also feasible to carry out hydrogen peroxide generation and organic oxidation on the same metal [29–37]. A bimetallic example of this reaction, where palladium serves to both generate the oxidant and to use it for alkane oxidation, utilizes copper to enhance selectivity [32]. This system is effective for oxidizing methane and ethane at mild temperatures in the presence of carbon monoxide and dioxygen [31, 32]. In an effort to trap the desired oxidation products at the alcohol stage, trifluoroacetic acid (TFA) was used to generate the corresponding esters. Interestingly, while simply adding TFA did not change the product distribution, the addition of copper salts dramatically increased the selectivity of the reaction, with methanol and its TFA ester being virtually the only products [32]. In the absence of copper, the primary product was formic acid. Based on experimental evidence, it seems that the carbon monoxide present effectively keeps the bulk of the copper in the $Cu^I$ state, and there has been speculation that soluble copper carbonyl complexes may play a role in the reaction. While the exact nature of the copper ions' interactions in the reaction are unknown, the beneficial role of utilizing this bimetallic system to increase selectivity is clear.

Utilizing another catalyst for hydrocarbon oxidations once the hydrogen peroxide has been generated has also been shown to be effective (Scheme 8.1) [31]. Miyake et al. [38] have explored the effect of adding metal oxides to silica-supported palladium catalysts. Numerous metal oxides were found to increase the rate of phenol formation from benzene, with those of vanadium, yttrium, and lanthanum exhibiting the best performance. Though no details were presented, it is hypothesized that the metal oxides may assist in delivering activated oxygen to the palladium catalyst. In a more recent paper, the screening of numerous homogeneous metal salts in conjunction with a heterogeneous palladium catalyst is described; it is proposed that the reaction occurs via the same pathway as above [39]. Of course, a pathway in which hydrogen peroxide forms a catalytically active species from the metal oxide is also a plausible explanation. Palladium and platinum catalysts supported on silica and promoted with vanadium oxides have been compared [40]. The authors noted a significant increase in both yield and selectivity when using the platinum rather than the palladium catalyst, a finding contrary to the body of literature on the generation of hydrogen peroxide, in which palladium tends to be highly favored [16, 41, 42]. This result may suggest a mechanism not involving the simple generation of hydrogen peroxide with its immediate use – an indication of how metals in a bimetallic system may not behave as they do when present alone.

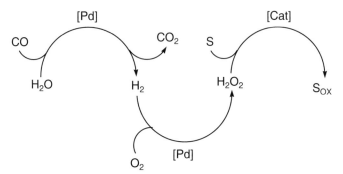

(S = substrate, S<sub>OX</sub> = oxidized substrate, Cat = Pd or second catalyst)

**Scheme 8.1**

The use of palladium/copper supported on silica has been found to be effective in benzene oxidation via the path shown in Scheme 8.2 [43, 44]. It is believed that the actual C–H activation occurs by metal-promoted formation of hydroxyl radicals, which can then attack the benzene nucleus. The pathway shown in Scheme 8.2 has the practical advantage of separating the dihydrogen and dioxygen activation steps, opening up the possibility of avoiding potentially explosive gas mixtures. The reaction has been adapted to the gas phase, conditions which promote higher selectivities as the product phenol is removed from the reaction mixture [45, 46]. Unfortunately, the high temperatures needed for gas-phase oxidation also lead to a reduction in the efficiency of dihydrogen utilization, with much of it forming water. Finally, oxidations of benzene and hexane have also been attempted with palladium supported on titanium silicates with the aim of exploiting the well-known oxidations using TS-1 with generated hydrogen peroxide [47].

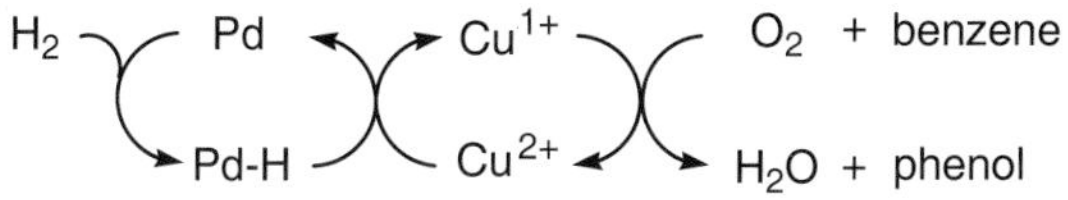

**Scheme 8.2**

$$H_2 \rightarrow Pd \rightarrow Cu^{1+} \rightarrow O_2 + benzene$$
$$\rightarrow Pd\text{-}H \rightarrow Cu^{2+} \rightarrow H_2O + phenol$$

Beyond benzene oxidation, the oxidation of alkanes using *in situ* generated hydrogen peroxide has also received attention. The first report of this nature involved the use of a $Pd^0$- and $Fe^{II}$-doped zeolite for the oxidation of alkanes with dihydrogen and dioxygen [48]. The catalyst exhibited the remarkable shape selectivity achievable with zeolites; an *n*-octane/cyclohexane preference of > 190 was observed. Furthermore, supporting noble metals on TS-1 proved effective in the oxidation of *n*-hexane and *n*-octane in the presence of dihydrogen, with the products consisting exclusively of secondary alcohols and ketones [49]. The efficiency, based on the amount of oxidant used, never exceeded 57%. However, when hydrogen peroxide was added to the reaction mixture from an external source, an efficiency of 86% was obtained, even at a significantly higher temperature. Higher temperatures increase the rate of metal-catalyzed hydrogen peroxide decomposition. Thus, in this case the side reaction of palladium-catalyzed peroxide decomposition was more significant than the reaction producing it. Furthermore, gold supported on titanium mesoporous materials has received some attention in relation to the oxidation of propane and isobutane [50]. While the yield of acetone from propane was meagre when compared to that from propene oxidation, *tert*-butanol was formed in reasonable yield and with reasonable selectivity from isobutane. Otsuka observed a very interesting increase in both the rate of methane oxidation and selectivity in favor of methanol when hydrogen was added to certain iron catalysts [51]. Experiments showed that the hydrogenation of carbon monoxide (formed through methane oxidation) was not responsible for the formation of methanol, and that similar selectivities were observed when hydrogen peroxide was used in place of the dihydrogen/dioxygen gas mixture.

The use of platinum heteropoly catalysts has been investigated for the oxidation of cyclohexane [52]. Catalyst redox studies showed that activity is associated with both $Pt^0$ and $Pt^{II}$ being present along with a redox couple of $Mo^V/Mo^{VI}$ in the phosphomolybdate. The GoAgg$^{II}$ system, originally developed by Barton [8], has been modified and implemented for the oxidation of cyclohexane using hydrogen peroxide generated on a palladium surface from a dihydrogen/dioxygen mixture [53]. Based on the typical use of acidic conditions for palladium-catalyzed hydrogen peroxide generation, acetone and acetone/acetic acid mixtures were employed (instead of the typical basic conditions for the GoAgg$^{II}$ reaction). Under these conditions, an increase in the production of the alcohol of over an order of magnitude was observed compared to the basic conditions.

We have recently completed work showing the persistence of palladium-generated hydrogen peroxide in solution. The oxidant, though showing marked decomposition, is formed sufficiently rapidly to interact with another catalyst (e.g., vanadium or iron) and effectively oxidizes aromatic and aliphatic hydrocarbons [13, 27]. Control experiments showed that very little oxidation occurs in the absence of the second, hydroxylation catalyst (e.g., a vanadium species) and that V(acac)$_3$ (acac = acetylacetonate) catalyzes the hydroxylation of benzene by hydrogen peroxide, even in the absence of metallic palladium, dihydrogen, and dioxygen. Furthermore, it was shown that no oxidation occurs when the vanadium hydroxylation catalyst is used without some source of hydrogen peroxide. Thus, metallic palladium is active in the formation of hydrogen peroxide, which, in turn, generates an active catalytic species from the vanadium complex.

It was of interest to ascertain what, if any, selectivity advantage could be obtained by employing *in situ* generated hydrogen peroxide for hydrocarbon oxidations. One can envisage that by carefully adjusting the ratio of the two catalysts (palladium and vanadium) present, better selectivities might be obtained since no excess of hydrogen peroxide would ever be present during the reaction. We had previously established that the oxidations were limited by the amount of hydrogen peroxide formed [13, 27]. The results of benzene oxidations carried out with hydrogen peroxide generated *in situ* and with directly added hydrogen peroxide are compared in Tables 8.1 and 8.2. It is clear that in all cases a remarkable selectivity advantage

**Table 8.1** Conversion and selectivity to phenol using a low initial benzene concentration and *in situ* H$_2$O$_2$ generation.[1]

| Hydroxylation catalyst | Benzene (μmol) | Phenol (μmol) | Benzo-quinone (μmol) | Conversion to phenol (%) | Selectivity in favor of phenol (%)[2] |
|---|---|---|---|---|---|
| V(acac)$_3$ | 1400 | 200 (10) | trace | 14 | > 99 |
| V(acac)$_3$ | 520 | 130 (10) | trace | 26 | > 99 |
| FeT(DBM) | 1400 | 59 (9) | 1.7 (0.2) | 4.2 | 97 |

[1] Performed using benzene as indicated with 3.2 μmol of V(acac)$_3$ or FeT(DBM) = tris(dibenzoylmethanato)Fe$^{III}$ and 20.0 mg of 5% Pd/Al$_2$O$_3$ in acetic acid (4.5 mL) exposed to 100 psi H$_2$, 1000 psi N$_2$, and 100 psi O$_2$ for 2 h at 65 °C. Data are average values for two runs.

[2] Selectivity calculated as mmol phenol/(mmol phenol + mmol benzoquinone).

**Table 8.2** Oxidation of benzene using $H_2O_2$ added by means of a syringe pump.[1]

| Hydroxylation catalyst | Benzene (μmol) | Phenol (μmol) | Benzo-quinone (μmol) | Conversion to phenol (%) | Selectivity (%)[2] | $H_2O_2$ efficiency (%) |
|---|---|---|---|---|---|---|
| V(acac)₃ | 22000 | 600 (30) | 190 (10) | 2.7 | 76 | 20 |
| V(acac)₃ | 5900 | 220 (60) | 41 (2) | 3.7 | 84 | 6.2 |
| V(acac)₃ | 520 | 47 (3) | 12 (2) | 9.0 | 80 | 1.4 |
| FeT(DBM) | 5900 | 9.7 | tr. | 0.16 | 69 | 0.37 |
| FeT(DBM) | 520 | 1.7 | 1 | 0.33 | 64 | 0.075 |

[1] Performed using 3.2 μmol V(acac)₃ or FeT(DBM) = tris(dibenzoylmethanato)Fe$^{III}$, benzene as indicated, and acetic acid (to make the total volume 4.5 mL) for 2 h at 65 °C. During this time, 50% $H_2O_2$ (0.28 mL, 4.9 mmol; diluted to a total volume of 0.68 mL with acetic acid) was added dropwise by means of a syringe pump. Data are average values for two runs unless indicated.
[2] Selectivity calculated as mmol phenol/(mmol phenol + mmol benzoquinone).

can be obtained by *in situ* hydrogen peroxide generation. For the vanadium-catalyzed oxidations the selectivity difference is around 20%, even when the substrate is in huge excess. It is important to note that due to phase separation problems the syringe-pumped hydrogen peroxide reactions employed about one-half the amount of oxidant as in the *in situ* case (Figure 8.1), a condition that should favor selectivity in the added hydrogen peroxide case. In the case of the iron catalyst, the selectivity advantage is nearly 30%.

Another aspect needing consideration in systems where two catalysts operate simultaneously, one generating the hydrogen peroxide and another performing hydrocarbon oxidation, is whether the hydrogen peroxide generating catalyst participates in the substrate oxidation. As described above, there are numerous examples in which a palladium catalyst generates the oxidant and also uses it for substrate oxidation. In order to truly exploit the available range of catalysts for hydrocarbon oxidations with hydrogen peroxide, it must be ensured that the hydrogen peroxide generator does not interfere with the catalyst responsible for the actual substrate oxidation. We have explored this for the vanadium-catalyzed oxidation of benzene using palladium-generated hydrogen peroxide [27]. The mechanism of benzene oxidation in this system was found to be identical to that when hydrogen peroxide was added directly. Clearly, the palladium catalyst does not play a part in the benzene oxidation step. However, this may not be the case under all conditions. In an effort to increase the oxidant efficiency, we examined the use of soluble halide salts to retard hydrogen peroxide decomposition by palladium [16]. Rather than the desired effect, a trend was observed showing that the amount of phenol produced decreased in the order $Cl^- > Br^- > I^-$, mirroring the coordinative ability of the halide. This result suggests that these species might be showing coordinative inhibition of the vanadium catalyst. Again, this is an example of how in a bimetallic system the two reactions being carried out are not simply the sum of the two separate reactions.

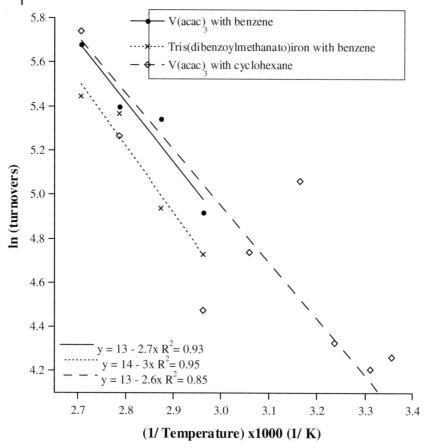

**Figure 8.2** Arrhenius plots for two catalysts and two substrates. For reactions with benzene, benzene (2 mL) was allowed to react with V(acac)$_3$ (3.2 µmol) or tris(dibenzoylmethanato)iron (3.2 µmol) and 5% Pd on Al$_2$O$_3$ (20.0 mg) in glacial acetic acid (2.5 mL). For reactions with cyclohexane, cyclohexane (1.5 mL) was allowed to react with V(acac)$_3$ (3.2 µmol) and 5% Pd on Al$_2$O$_3$ (20.0 mg) in propionic acid (3 mL). Reactions were run in glass liners in 300 mL stainless steel autoclaves under an atmosphere consisting of H$_2$ (100 psi; 82 mmol), N$_2$ (1000 psi), and O$_2$ (100 psi; 82 mmol) for 2 h. Reactions were repeated at least three times and the quoted data are average values.

A key issue in bimetallic hydrocarbon oxidations, such as those described above [27], is the identity of the rate-determining step. As shown in Figure 8.2, the activation energy is similar regardless of the hydroxylation catalyst or the substrate. Thus, the values obtained are 5(1), 6(1), and 5(1) kcal mol$^{-1}$ for the hydroxylation of benzene with V(acac)$_3$, the hydroxylation of benzene with tris(dibenzoylmethanato)Fe$^{III}$, and the hydroxylation of cyclohexane with V(acac)$_3$, respectively. This suggests that the slow step under these conditions is the *in situ* generation of hydrogen peroxide by palladium. The activation energy values do not change significantly when the stirring rate is changed, indicating that gas diffusion is not the rate-limiting step [44]. Furthermore, the values are close to that calculated from a report of gas-phase benzene to phenol oxidation by a palladium/copper composite catalyst, carried out at around 200 °C in the presence of dihydrogen and dioxygen: 6 kcal mol$^{-1}$ [46].

## 8.5
## Conclusion

Performing selective hydrocarbon oxidations with hydrogen peroxide and generating this oxidant *in situ* are currently receiving considerable attention. The feasibility of oxidizing aromatic and aliphatic molecules using *in situ* generated hydrogen peroxide has clearly been demonstrated. We have also shown that a significant increase in selectivity can be achieved by generating this oxidant *in situ*. However, the reaction is markedly less efficient in its use of oxidant. It is also crucial to address the influence of each catalyst and other components present when using bimetallic conditions. The observation that conditions promoting one reaction in the system may inhibit or facilitate the formation of alternative products in the second reaction points to a need for careful scrutiny of how the two systems interact. Ideally, the catalyst involved in substrate oxidation should work at its maximum rate and through a mechanism unhindered by, or even assisted by, the presence of the hydrogen peroxide generating catalyst.

A particular concern in reactions generating hydrogen peroxide in situ is promotion of its concomitant decomposition or hydrogenation by the catalyst used to produce it. Although the use of a large excess of oxidant can minimize hydrogenation and performing the reaction in a biphasic medium can retard disproportionation, it seems unlikely that these methods will completely eliminate these competitive reactions. Moreover, in some situations the use of a biphasic mixture is impossible or impractical, such as in a gas-phase reaction. Consequently, there is currently a significant need for catalysts that are less active in promoting the unwanted reactions. It is important to note that efficient use of the generated oxidant is important due to the fact that the efficiency with which the co-reductant is consumed has a significant impact on the economics of the process.

## Acknowledgement

JER thanks the NCER Star/EPA fellowship for financial support. This work was funded by a grant from the NSF.

## References

1 P. R. Ortiz de Montellano, *Cytochrome P-450: Structure, Mechanism, and Biochemistry*, Plenum Press, New York, 1986.

2 L. Que, R. Y. N. Ho, *Chem. Rev.* **1996**, *96*, 2607.

3 K. E. Liu, S. J. Lippard, *Adv. Inorg. Chem.* **1995**, *42*, 263.

4 R. A. Sheldon, *Chem. Ind. (London)* **1997**, 12.

5 W. R. Sanderson, *Pure Appl. Chem.* **2000**, *72*, 1289.

6 G. Strukul (Ed.), *Catalytic Oxidations with Hydrogen Peroxide as Oxidant*, Kluwer, Dordrecht and Boston, 1992.

7 D. Hancu, J. Green, E. J. Beckman, *Acc. Chem. Res.* **2002**, *35*, 757.

8 D. H. R. Barton, D. Doller, *Acc. Chem. Res.* **1992**, *25*, 504.

9 A. K. Sinha, S. Seelan, S. Tsubota, M. Haruta, *Stud. Surf. Sci. Catal.* **2002**, *143*, 167.

10 J. Q. Lu, M. F. Luo, H. Lei, X. H. Bao, C. Li, *J. Catal.* **2002**, *211*, 552.

11 J. Q. Lu, M. F. Luo, H. Lei, C. Li, *Appl. Catal., A* **2002**, *237*, 11.

12 G. Jenzer, T. Mallat, M. Maciejewski, F. Eigenmann, A. Baiker, *Appl. Catal., A* **2001**, *208*, 125.

13 J. E. Remias, A. Sen, *J. Mol. Catal., A: Chem.* **2002**, *189*, 22.

14 T. Nishimura, T. Onoue, K. Ohe, S. Uemura, *J. Org. Chem.* **1999**, *64*, 6750.

15 R. Bortolo, D. Bianchi, R. D'Aloisio, C. Querci, M. Ricci, *J. Mol. Catal., A: Chem.* **2000**, *153*, 25.

16 L. W. Gosser, U.S. Patent, 4,681,751 (1987).

17 H. Nagashima, Y. Ishiuchi, Y. Hiramatsu, M. Kawakami, U.S. Patent, 5,320,821 (1994).

18 V. V. Krishnan, A. G. Dokoutchaev, M. E. Thompson, *J. Catal.* **2000**, *196*, 366.

19 D. P. Dissanayake, J. H. Lunsford, *J. Catal.* **2002**, *206*, 173.

20 D. Bianchi, R. Bortolo, R. D'Aloisio, M. Ricci, *Angew. Chem. Int. Ed. Engl.* **1999**, *38*, 706.

21 A. Sen, J. Halpern, *J. Am. Chem. Soc.* **1977**, *99*, 8337.

22 S. S. Stahl, J. L. Thorman, R. C. Nelson, M. A. Kozee, *J. Am. Chem. Soc.* **2001**, *123*, 7188.

23 S. Muto, K. Tasaka, Y. Kamiya, *Bull. Chem. Soc. Jpn.* **1977**, *50*, 2493.

24 M. Roussel, H. Mimoun, *J. Org. Chem.* **1980**, *45*, 5387.

25 B. A. Steinhoff, S. R. Fix, S. S. Stahl, *J. Am. Chem. Soc.* **2002**, *124*, 766.

26 M. Lin, A. Sen, *J. Am. Chem. Soc.* **1992**, *114*, 7307.

27 J. E. Remias, T. A. Pavlosky, A. Sen, *J. Mol. Catal., A: Chem.* **2003**, *203*, 179.

28 P. Landon, P. J. Collier, A. J. Papworth, C. J. Kiely, G. J. Hutchings, *Chem. Commun.* **2002**, 2058.

29 W. Laufer, W. F. Hoelderich, *Chem. Commun.* **2002**, 1684.

30 N. I. Kuznetsova, L. G. Detusheva, L. I. Kuznetsova, M. A. Fedotov, V. A. Likholobov, *J. Mol. Catal., A: Chem.* **1996**, *114*, 131.

31 A. Sen, *Acc. Chem. Res.* **1998**, *31*, 550.

32 M. R. Lin, T. Hogan, A. Sen, *J. Am. Chem. Soc.* **1997**, *119*, 6048.

33 A. Pifer, T. Hogan, B. Snedeker, R. Simpson, M. R. Lin, C. Y. Shen, A. Sen, *J. Am. Chem. Soc.* **1999**, *121*, 7485.

34 E. G. Chepaikin, G. N. Boyko, A. P. Bezruchenko, A. A. Leshcheva, E. H. Grigoryan, *J. Mol. Catal., A: Chem.* **1998**, *129*, 15.

35 E. G. Chepaikin, A. P. Bezruchenko, A. A. Leshcheva, G. N. Boyko, I. V. Kuzmenkov, E. H. Grigoryan, A. E. Shilov, *J. Mol. Catal., A: Chem.* **2001**, *169*, 89.

36 M. R. Lin, T. E. Hogan, A. Sen, *J. Am. Chem. Soc.* **1996**, *118*, 4574.

37 E. C. Baker, D. E. Hendriksen, R. Eisenberg, *J. Am. Chem. Soc.* **1980**, *102*, 1020.

38 T. Miyake, M. Hamada, Y. Sasaki, M. Oguri, *Appl. Catal., A* **1995**, *131*, 33.

39 T. Miyake, M. Hamada, H. Niwa, M. Nishizuka, M. Oguri, *J. Mol. Catal., A: Chem.* **2002**, *178*, 199.

40 H. Ehrich, H. Berndt, M. M. Pohl, K. Jahnisch, M. Baerns, *Appl. Catal., A* **2002**, *230*, 271.

41 Y. Izumi, M. Hidetaka, S.-I. Kawahara, U.S. Patent, 4,009,252 (1977).

42 Y. Izumi, M. Hidetaka, S.-I. Kawahara, U.S. Patent, 4,279,883 (1981).

**43** A. Kunai, T. Wani, Y. Uehara,
F. Iwasaki, Y. Kuroda, S. Ito, K. Sasaki,
*Bull. Chem. Soc. Jpn.* **1989**, *62*, 2613.

**44** T. Kitano, Y. Kuroda, A. Itoh, J. Lifen,
A. Kunai, K. Sasaki, *J. Chem. Soc., Perkin
Trans. 2* **1990**, 1991.

**45** T. Kitano, Y. Kuroda, M. Mori, S. Ito,
K. Sasaki, M. Nitta, *J. Chem. Soc., Perkin
Trans. 2* **1993**, 981.

**46** T. Kitano, T. Nakai, M. Nitta, M. Mori,
S. Ito, K. Sasaki, *Bull. Chem. Soc. Jpn.*
**1994**, *67*, 2850.

**47** T. Tatsumi, K. Yuasa, H. Tominaga,
*J. Chem. Soc., Chem. Commun.* **1992**, 1446.

**48** N. Herron, C. A. Tolman, *J. Am. Chem.
Soc.* **1987**, *109*, 2837.

**49** M. G. Clerici, P. Ingallina, *Catal.
Today* **1998**, *41*, 351.

**50** Y. A. Kalvachev, T. Hayashi, S. Tsubota,
M. Haruta, *J. Catal.* **1999**, *186*, 228.

**51** Y. Wang, K. Otsuka, *J. Chem. Soc.,
Chem. Commun.* **1994**, 2209.

**52** N. V. Kirillova, N. I. Kuznetsova,
L. I. Kuznetsova, V. I. Zaikovskii,
S. V. Koscheev, V. A. Likholobov,
*Catal. Lett.* **2002**, *84*, 163.

**53** S. B. Kim, K. W. Jun, S. B. Kim,
K. W. Lee, *Chem. Lett.* **1995**, 535.

# 9
# Two Approaches to Multimetallic Catalysis: Combined Use of Metal Complexes and Multinuclear Complex Catalysts

*Youichi Ishii and Masanobu Hidai*

## 9.1
## Introduction

In recent years, homogeneous multimetallic catalysis has been recognized as an effective synthetic tool in organic chemistry [1], because chemical transformations induced by multimetallic catalyst systems often show higher reaction rates and selectivities than those induced by monometallic and mononuclear complex catalysts. In more specialized cases, multimetallic catalysts give rise to new reactions which are inaccessible with monometallic systems. In reactions where the use of multimetallic catalysts enhances the reaction rate or better controls the reaction course, cooperative or successive interaction of two or more metal centers with the substrates is envisaged as taking place.

A simple and straightforward way to make multimetallic catalyst systems is to use two different metal complexes in combination. Catalysts of this type have sometimes been utilized for one-pot multistep (tandem) catalytic reactions, where the overall chemical transformation is composed of a series of independent catalytic processes, and each component complex promotes an individual catalytic step. In such cases, however, each metal species participates in the catalytic process as a monometallic catalyst, and synergism between the metal species is not necessarily involved. We have been more attracted by catalytic systems in which synergistic effects between the two metal complexes are observed, because the associated reactions are expected to involve the formation of a heterobimetallic intermediate, the reactivity of which differs considerably from that of the corresponding monometallic species. From this point of view, a more sophisticated approach would be the use of preorganized multinuclear complex catalysts, in which the metal centers are suitably organized and possess potentially vacant coordination sites. Needless to say, the availability of synthetic methods for the designed multinuclear complexes is a prerequisite for this approach.

We have been engaged in the development of homogeneous multimetallic catalysis based on the two approaches outlined above for over two decades, and our work is summarized in this account. Much attention with regard to the combined

*Multimetallic Catalysts in Organic Synthesis.* Edited by M. Shibasaki and Y. Yamamoto
Copyright © 2004 WILEY-VCH Verlag GmbH & Co. KGaA, Weinheim
ISBN: 3-527-30828-8

use of metal complexes has been directed towards carbonylation reactions. On the other hand, various catalytic transformations of acetylenic substrates have been developed by taking advantage of multinuclear complex catalysts.

## 9.2
## Combined Use of Metal Complexes

### 9.2.1
### Homologation and Hydroformylation by Co–Ru Catalysts

Our earliest studies on multimetallic catalysis were concerned with the homologation of methanol to ethanol (Eq. 9.1), because we were interested in some reports that the selectivity of ethanol formation in the homologation of methanol was improved by the addition of small amounts of ruthenium species to the $[Co_2(CO)_8]$ catalyst [2]. Detailed investigation into the homologation of methanol by the $[Co_2(CO)_8]$–$RuCl_3$ catalyst in the presence of methyl iodide revealed that the simultaneous use of $RuCl_3$ as the second component of the catalyst sharply decreases the yields of acetaldehyde and its dimethyl acetal, while the yields of ethanol and its ethers are improved [3, 4]. Interestingly, the yield of total $C_2$ products is also increased by $RuCl_3$, and the yields of both total $C_2$ and total ethanol reach a maximum at an Ru : Co ratio of 1 : 3. On the basis of the time dependence of yield and selectivity, it was concluded that the formation of acetaldehyde from methanol, which is the first step of the reaction, is accelerated by the synergistic effect of cobalt and ruthenium, and that the subsequent hydrogenation of acetaldehyde to ethanol is catalyzed by a ruthenium species.

$$CH_3OH + CO + 2H_2 \xrightarrow{\text{Co or Co-Ru}} CH_3CH_2OH + H_2O \quad (9.1)$$

| | selectivity of total $C_2$ (%) |
|---|---|
| $[Co_2(CO)_8]$ | 26.2 |
| $(PPh_4)[RuCo_3(CO)_{12}]$ | 54.7 |
| $Ru_3(CO)_{12}$ | 11.3 |

This result prompted us to synthesize an anionic $RuCo_3$ mixed-metal cluster $[RuCo_3(CO)_{12}]^-$ (**1**) (Eq. 9.2), the PPN (PPN = $(Ph_3P)_2N^+$) salt of which was structurally characterized by means of an X-ray diffraction study [3]. Cluster **1** was also found to be an effective catalyst for the homologation of methanol to give ethanol; the selectivity of ethanol formation by the catalyst $(NEt_4)[RuCo_3(CO)_{12}]$ can be as high as 64% at ca. 25% conversion of methanol, while that by $[Co_2(CO)_8]$ alone is below 5%. Unfortunately, cluster **1** is not stable under the harsh catalytic conditions (180 °C, CO 40 atm, $H_2$ 80 atm); the monometallic ruthenium complex $[RuI_3(CO)_3]^-$ was recovered after a catalytic run. Consequently, it is not clear as to whether the synergism originates from the bimetallic cluster catalysis. Nevertheless, these findings stimulated us to investigate other types of carbonylation reactions with the Co–Ru bimetallic catalyst.

$$RuCl_3 + 4 Na[Co(CO)_4] \longrightarrow Na[RuCo_3(CO)_{12}] \quad (1) \qquad (9.2)$$

$$+ 3 NaCl + 1/2 [Co_2(CO)_8]$$

Hydroformylation of cyclohexene in the presence of the $[Co_2(CO)_8]$–$[Ru_3(CO)_{12}]$ catalyst in THF was found to give cyclohexanecarbaldehyde in much higher yield than with the monometallic catalysts $[Co_2(CO)_8]$ or $[Ru_3(CO)_{12}]$ alone (Eq. 9.3) [4–6]. Table 9.1 shows that the initial reaction rate increases with increasing Ru : Co ratio. When the Ru : Co ratio is 9.9 : 1, the initial rate is 27 times greater than that with $[Co_2(CO)_8]$ alone, while $[Ru_3(CO)_{12}]$ is not an effective catalyst for the hydroformylation of cyclohexene. Interestingly, a large solvent effect was observed for the hydroformylation with the Co–Ru bimetallic catalyst, whereas the initial rate with $[Co_2(CO)_8]$ was almost independent of the nature of the solvent. The most notable acceleration was observed in alcohols such as methanol and ethanol, and the major carbonylation products in these alcohols were the corresponding acetals.

$$\text{cyclohexene} + CO + H_2 \xrightarrow[\quad[Co_2(CO)_8]-[Ru_3(CO)_{12}]\quad]{[Co_2(CO)_8] \text{ or}} \text{CyCHO} \qquad (9.3)$$

Similar synergistic effects were also observed for the hydroformylations of 1-hexene and styrene, but the acceleration of the initial rate was not so remarkable compared with that in the case of cyclohexene [6]. Thus, the initial rates for the hydroformylations of 1-hexene and styrene in the presence of the Co–Ru (Co : Ru = 1 : 1) catalyst in benzene are about 2–3 times faster than those with $[Co_2(CO)_8]$, whereas the rate for the hydroformylation of cyclohexene is enhanced by a factor of 9.1. It is interesting to note that cyclohexene, the least reactive substrate in the hydroformylation by $[Co_2(CO)_8]$, exhibited the greatest rate enhancement effect.

**Table 9.1** Hydroformylation of cyclohexene by $[Co_2(CO)_8]$–$[Ru_3(CO)_{12}]$ catalyst.[a]

| Catalyst | Ru : Co | Yield of CyCHO (%) | Relative initial rate[b] |
|---|---|---|---|
| $[Co_2(CO)_8]$ | – | 14 | 1.0 |
| $[Ru_3(CO)_{12}]$[c] | – | 3 | 0.3 |
| $[Co_2(CO)_8]$–$[Ru_3(CO)_{12}]$ | 0.95 | 52 | 5.9 |
| $[Co_2(CO)_8]$–$[Ru_3(CO)_{12}]$ | 3.2 | 62 | 8.4 |
| $[Co_2(CO)_8]$–$[Ru_3(CO)_{12}]$ | 9.9 | 100 | 27 |
| $[Co_2(CO)_8]$–$[Ru_3(CO)_{12}]$[d] | 1.0 | 73 | 9.1 |
| $[Co_2(CO)_8]$–$[Ru_3(CO)_{12}]$[e] | 1.0 | 93[f] | 15 |
| $[Co_2(CO)_8]$–$[Ru_3(CO)_{12}]$[g] | 1.0 | 87[h] | 19 |

[a] Reaction conditions: $[Co_2(CO)_8]$, 0.1 mmol; cyclohexene, 80 mmol; THF, 10 mL; CO : $H_2$ 40 : 40 atm; 110 °C, 4 h. [b] Relative to $[Co_2(CO)_8]$.
[c] $[Ru_3(CO)_{12}]$, 0.067 mmol. [d] Solvent, benzene (10 mL). [e] Solvent, EtOH (10 mL).
[f] Yield of CyCHO + CyCH(OEt)$_2$. [g] Solvent, MeOH (10 mL). [h] Yield of CyCHO + CyCH(OMe)$_2$.

A high-pressure IR spectral study of the catalytic reaction mixture did not show the presence of Co–Ru bimetallic hydrido clusters. We consider that the synergistic effect of the Co–Ru bimetallic catalyst may be explained in terms of dinuclear reductive elimination of the aldehyde between an acylcobalt intermediate and a ruthenium hydride species. Thus, the activation of the alkene and the subsequent CO insertion into the metal–carbon bond are conducted at the cobalt center, while the hydrogenolysis of the resultant acyl complex is promoted by the ruthenium species. To gain additional support for this explanation, the reactions between $[n\text{-}C_5H_{11}COCo(CO)_4]$ and several metal carbonyl hydrides, such as $[HCo(CO)_4]$, $[HRu(CO)_4]^-$, and $[HRu_3(CO)_{11}]^-$, were followed by means of $^1H$ NMR (Eq. 9.4). As expected, $[HRu(CO)_4]^-$ reacted with the acylcobalt complex ca. 4 times faster than $[HCo(CO)_4]$ at room temperature [7]. This observation suggests that, in the catalytic hydroformylation reaction with the Co–Ru bimetallic catalyst, ruthenium is involved in the aldehyde-forming step.

$$[n\text{-}C_5H_{11}COCo(CO)_4] + H\text{-}M \xrightarrow{\text{under CO, 20 °C}} n\text{-}C_5H_{11}CHO \qquad (9.4)$$

| H-M | rate constant / $M^{-1}s^{-1}$ |
|---|---|
| $[HCo(CO)_4]$ | $2.1 \times 10^{-2}$ |
| $PPN[HRu(CO)_4]$ | $7.8 \times 10^{-2}$ |
| $NEt_4[HRu_3(CO)_{11}]$ | $1.6 \times 10^{-4}$ |

Another phenomenon which supports the dinuclear reductive elimination mechanism was observed in the hydroformylation of norbornene [8]. In the reaction with $[Co_2(CO)_8]$ alone, lactone **2** containing two norbornene units was obtained as the major product, with 2-norbornanecarbaldehyde being formed in a low yield (Eq. 9.5). In contrast, the $[Co_2(CO)_8]–[Ru_3(CO)_{12}]$ catalyst affords the aldehyde as the major product. Since norbornene readily inserts into a metal–acyl bond, hydrogenolysis of the acylcobalt intermediate to give the aldehyde cannot be the major process in the $[Co_2(CO)_8]$-catalyzed reaction. The addition of $[Ru_3(CO)_{12}]$ as the second component of the catalyst is believed to promote the hydrogenolysis of the Co–acyl bond and thereby change the product distribution.

$$\text{(9.5)}$$

**2**

However, the synergistic effects observed with the Co–Ru catalyst cannot be fully explained in terms of the dinuclear reductive elimination mechanism alone. In the hydroesterification of cyclohexene, where the hydrogenolysis of an acylmetal intermediate is not involved, the $[Co_2(CO)_8]–[Ru_3(CO)_{12}]$ catalyst exhibited a small but significant rate enhancement in comparison with catalysis by $[Co_2(CO)_8]$ or

[Ru$_3$(CO)$_{12}$] alone (Eq. 9.6) [6]. Moreover, the rate-determining step in the [Co$_2$(CO)$_8$]-catalyzed hydroformylation was proposed to be the initial interaction of an alkene with [HCo(CO)$_4$] [9]. Ruthenium might also promote insertion of the alkene into the Co–H bond.

$$\begin{array}{c} [Co_2(CO)_8] \text{ or} \\ [Co_2(CO)_8] - [Ru_3(CO)_{12}] \end{array}$$

(9.6)

Recently, a detailed study on dinuclear reductive elimination in the [Rh$_4$(CO)$_{12}$]–[Mn$_2$(CO)$_{10}$]-catalyzed hydroformylation of 3,3-dimethylbut-1-ene has been described (Eq. 9.7) [10]. In this case, the addition of [Mn$_2$(CO)$_{10}$], which is ineffective as a hydroformylation catalyst, enhances the rate of the [Rh$_4$(CO)$_{12}$]-catalyzed hydroformylation by up to a factor of five (Rh : Mn = 3.7). Detailed kinetic study and in situ IR measurements indicate that the reaction rate obeys a two-term linear-bilinear equation:

$$\text{rate} = k_1[\text{RCORh(CO)}_4][\text{H}_2][\text{CO}]^{-1} + k_2[\text{RCORh(CO)}_4][\text{HMn(CO)}_5][\text{CO}]^x$$

where $x = -1.5 \pm 0.1$. The second term is fully consistent with operation of the dinuclear reductive elimination process involving [HMn(CO)$_5$], and [HMn(CO)$_5$] has been concluded to be 170 times more efficient for the hydrogenolysis of the Rh–acyl bond to give the aldehyde than H$_2$ under the catalytic conditions. [HMn(CO)$_5$] was also shown to react with alkylmanganese complexes [RMn(CO)$_5$] to form a dinuclear $\eta^1$-aldehyde complex, which may constitute a model of the intermediate species involved in the dinuclear reductive elimination (Eq. 9.8) [11]. Furthermore, it should be mentioned that intramolecular hydride transfer between an Rh–H and an Rh–acyl center was proposed to be responsible for the rate enhancement effect observed in the hydroformylation of 1-alkenes catalyzed by a dinuclear rhodium complex [12].

$$\begin{array}{c} [Rh_4(CO)_{12}] \text{ or} \\ [Rh_4(CO)_{12}] - [Mn_2(CO)_{10}] \end{array}$$

(9.7)

$$[\text{RMn(CO)}_5] + [\text{HMn(CO)}_5] \longrightarrow$$

(9.8)

## 9.2.2
## Carbonylation Reactions of Aryl Iodides by Pd-Ru and Pd-Co Systems

In order to obtain further information about the role of dinuclear reductive elimination in catalytic carbonylation reactions, we turned the focus of our research to palladium-based bimetallic catalyst systems. We first examined the formylation of aryl halides, which is known to be promoted by palladium complexes. Usually, this type of reaction is conducted at temperatures of 80–150 °C, because hydrogenolysis of an acylpalladium intermediate requires relatively high temperature. It was anticipated that an improved bimetallic catalyst could be developed, if dinuclear reductive elimination from the acylpalladium and a metal hydride species could be effectively accomplished under the catalytic conditions.

As expected, when $[PdCl_2(PPh_3)_2]$ and $[Ru_3(CO)_{12}]$ were used simultaneously (Pd : Ru = 1 : 2) as the catalyst for the formylation of iodobenzene at 70 °C in the presence of a base ($NEt_3$), the yield of benzaldehyde was increased fourfold compared to that obtained with $[PdCl_2(PPh_3)_2]$ alone (Eq. 9.9) [13]. Since $[Ru_3(CO)_{12}]$ itself failed to show catalytic activity for this formylation, the improvement in the catalytic activity implies a synergistic effect between palladium and ruthenium. Other metal carbonyls such as $[Cr(CO)_6]$, $[Mo(CO)_6]$, $[W(CO)_6]$, $[Mn_2(CO)_{10}]$, $[Fe(CO)_5]$, and $[Co_2(CO)_8]$ exhibited essentially no effect as the second component of the catalyst.

$$PhI + CO + H_2 \xrightarrow[\text{NEt}_3]{[PdCl_2(PPh_3)_2]\text{ -metal carbonyl}} PhCHO \quad (9.9)$$

The origin of the synergism was investigated through stoichiometric reactions. When a benzene solution containing $[Ru_3(CO)_{12}]$ and $NEt_3$ was kept under catalytic formylation conditions, the formation of $[HRu_3(CO)_{11}]^-$ as the major ruthenium species was observed by means of IR spectroscopy [14]. Taking this fact into consideration, we examined the reactions between carbonyl hydride complexes and the acylpalladium complex $[PdI(CO\text{-}p\text{-}Tol)(PPh_3)_2]$ (3), the latter of which is readily derived from a $Pd^0$ carbonyl complex and iodobenzene [15] and may be regarded as a model intermediate for the catalytic formylation. Interestingly, ruthenium hydrides, especially $[HRu_3(CO)_{11}]^-$, were found to react effectively with complex 3 to give $p$-tolualdehyde under very mild conditions (0–40 °C) (Eq. 9.10). In contrast, 3 failed to react with $H_2$ at 40 °C even at a pressure of 40 atm. This result strongly suggests that in the catalytic formylation by the Pd–Ru catalyst the dinuclear reductive elimination between the acylpalladium intermediate and a ruthenium hydride such as $[HRu_3(CO)_{11}]^-$ competes favorably with the monometallic hydrogenolysis involving $H_2$. It is interesting to note that complex 3 reacts readily with other anionic carbonyl hydride complexes such as $[HCr_2(CO)_{10}]^-$ and $[HMo_2(CO)_{10}]^-$ to give $p$-tolualdehyde. However, these anionic complexes cannot be produced from $[Cr(CO)_6]$ and $[Mo(CO)_6]$ with $H_2$, and it is for this reason that only ruthenium exhibits a notable synergistic effect.

$$[Pd(CO\text{-}p\text{-}Tol)I(PPh_3)_2] + [HRu_3(CO)_{11}]^- \xrightarrow[\text{under CO, 40 °C}]{} p\text{-TolCHO (6)} \quad (9.10)$$
$$\mathbf{3} \qquad\qquad\qquad\qquad\qquad\qquad\qquad\qquad 68\%$$

To expand the scope of the Pd–Ru bimetallic catalyst system, we next investigated silylative carbonylation of iodides in consideration of the formal similarity between $H_2$ and hydrosilanes in their reactivities toward organometallic compounds. An attempted reaction of $p$-TolI with CO (50 atm)/$HSiEt_3$ in the presence of $[PdCl_2(PPh_3)_2]$ at 80 °C gave no more than 5% total yield of the carbonylation products. In contrast, p-TolI was selectively carbonylated to afford p-TolCH$_2$OSiEt$_3$ (4) in 76% yield (at 85% conversion) by using [Co$_2$(CO)$_8$] in combination with [PdCl$_2$(PPh$_3$)$_2$], although [Co$_2$(CO)$_8$] alone was ineffective (Eq. 9.11) [16, 17]. Interestingly, the addition of NEt$_3$ to the reaction system dramatically changed the distribution of the carbonylation products, such that 1,2-di(p-tolyl)-1,2-bis(triethylsiloxy)ethane (5) was obtained as the major product (57% yield, at 70% conversion), and other products (4 and p-tolualdehyde (6)) were observed in no more than marginal yields (Eq. 9.12). [Ru$_3$(CO)$_{12}$] was also found to be effective as the second component of the catalyst, but the selectivity was lower. The formation of ISiEt$_3$ was confirmed in the Pd–Co-catalyzed reactions, and the stoichiometry of each reaction was deduced to be as shown in Eqs. (9.11) and (9.12).

$$\text{p-TolI} \; + \; \text{CO} \; + \; 2\,\text{HSiEt}_3$$

$$\xrightarrow[\text{[Co}_2\text{(CO)}_8]}{\text{[PdCl}_2\text{(PPh}_3\text{)}_2]} \quad \underset{\mathbf{4}}{\text{p-Tol}\diagdown\text{OSiEt}_3} \quad + \; \text{ISiEt}_3 \qquad (9.11)$$

$$2\,\text{p-TolI} \; + \; 2\,\text{CO} \; + \; 3\,\text{HSiEt}_3 \; + \; \text{NEt}_3$$

$$\xrightarrow[\text{[Co}_2\text{(CO)}_8]}{\text{[PdCl}_2\text{(PPh}_3\text{)}_2]} \quad \underset{\mathbf{5}}{\overset{\text{p-Tol} \quad \text{p-Tol}}{\underset{\text{Et}_3\text{SiO} \quad \text{OSiEt}_3}{\diagup\diagdown}}} \quad + \; \text{ISiEt}_3 + \text{NHEt}_3\text{I} \qquad (9.12)$$

Similar results have also been obtained with several other aryl iodides, although the reaction rate shows some dependence on the aryl groups of the substrates. We set out to investigate the mechanisms of these reactions to elucidate: (1) why the specific combination of two catalytically ineffective complexes (Pd–Co) displays a remarkable synergistic effect, and (2) how the product selectivity is controlled depending upon the presence of NEt$_3$.

**Table 9.2** Silylative carbonylation of p-TolI catalyzed by [PdCl$_2$(PPh$_3$)$_2$] and/or metal carbonyls.[a]

| Catalyst | NEt$_3$ (mmol) | Conv. (%) | GLC yield (%) | | |
|---|---|---|---|---|---|
| | | | 4 | 5 | 6 |
| [PdCl$_2$(PPh$_3$)$_2$] | – | 10 | 0 | 0 | 3 |
| [PdCl$_2$(PPh$_3$)$_2$] | 3 | 4 | 0 | 0 | 2 |
| [Co$_2$(CO)$_8$] | – | 4 | 0 | 0 | 0 |
| [Co$_2$(CO)$_8$] | 3 | 5 | 0 | 0 | 0 |
| [PdCl$_2$(PPh$_3$)$_2$]–[Co$_2$(CO)$_8$] | – | 85 | 76 | 0 | 0 |
| [PdCl$_2$(PPh$_3$)$_2$]–[Co$_2$(CO)$_8$] | 3 | 70 | 6 | 57 | 2 |
| [PdCl$_2$(PPh$_3$)$_2$]–[Ru$_3$(CO)$_{12}$][b] | – | 79 | 40 | 4 | 10 |

[a] Reaction conditions: p-TolI, 2.5 mmol; HSiEt$_3$, 7.5 mmol; [PdCl$_2$(PPh$_3$)$_2$], 0.05 mmol; [Co$_2$(CO)$_8$], 0.025 mmol; benzene, 5 mL; CO, 50 atm; 80 °C, 3 h.
[b] [Ru$_3$(CO)$_{12}$], 0.017 mmol.

Since one possible intermediate of the above reactions is an aldehyde, hydrosilylation of **6** under the conditions adopted for the catalytic carbonylation reactions was examined in detail. Although both [PdCl$_2$(PPh$_3$)$_2$] and [Co$_2$(CO)$_8$] behave as catalysts for the hydrosilylation of **6** to give **4** in the presence of *p*-TolI under N$_2$, high CO pressure (50 atm) seriously suppresses their activities. However, the [PdCl$_2$(PPh$_3$)$_2$]–[Co$_2$(CO)$_8$] bimetallic catalyst was found to be quite effective for the formation of **4** from **6** even under CO pressure. Surprisingly, the hydrosilylation of **6** was strongly inhibited by the addition of NEt$_3$. These results indicate that aldehyde **6** is a highly plausible intermediate for the reaction to give **4** (Eq. 9.11) but cannot be the intermediate for the production of **5** (Eq. 9.12).

Scheme 9.1 depicts a proposed mechanism for the formation of **4**. Oxidative addition of an aryl iodide (ArI) to a Pd$^0$ species followed by CO insertion yields an aroylpalladium complex [PdI(COAr)(PPh$_3$)$_2$]. On the other hand, the reaction of [Co$_2$(CO)$_8$] with HSiEt$_3$ has been proposed to give a hydridosilylcobalt complex [H(Et$_3$Si)Co(CO)$_3$] as the primary product [18]. Since neither the palladium nor the cobalt monometallic catalyst exhibited activity for the silylative carbonylation of ArI, it is reasonable to assume that the bimetallic reaction between the acylpalladium and hydridocobalt complexes to produce the aldehyde is involved in the catalysis as demonstrated in Eq. (9.10).

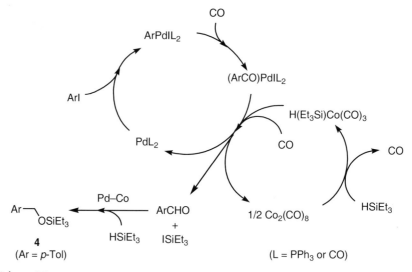

**Scheme 9.1**

On the other hand, the [Co(CO)$_4$]$^-$ anion is considered to be generated from [Co$_2$(CO)$_8$] and HSiEt$_3$ under the basic conditions of the reaction according to Eq. (9.12) [18]. We have examined the model reaction between ArI and K[Co(CO)$_4$] in the presence of PPh$_3$ and observed that [Pd(PPh$_3$)$_4$] indeed catalyzes the formation of the aroylcobalt complex [(ArCO)Co(CO)$_3$(PPh$_3$)] (Eq. 9.13) [19]. This reaction is believed to proceed by initial oxidative addition of the ArI to a Pd$^0$ complex with subsequent migration of the aryl (or aroyl) group from the palladium center to cobalt, where a Pd–Co dinuclear complex is involved as a key intermediate. Several Pt–W, Pt–Mo, and Pt–Fe dinuclear complexes with an alkyl group on the platinum center have been reported to undergo alkyl migration to afford the corresponding W-, Mo-, and Fe–alkyl complexes [20, 21].

$$\text{ArI} + \text{K[Co(CO)}_4\text{]} + \text{PPh}_3 \xrightarrow{\text{Pd(PPh}_3)_4} \text{[(ArCO)Co(CO)}_3(\text{PPh}_3)\text{]} + \text{KI} \quad (9.13)$$

Although our attempts to obtain the Pd–Co bimetallic intermediate bearing $\text{PPh}_3$ ligands were unsuccessful, a series of Pt–Co complexes $[(\text{PPh}_3)(\text{CO})(\text{R})\text{PtCo}(\text{CO})_3(\text{PPh}_3)]$ (R = *p*-Tol (7), Ph, Me) have been obtained from the reaction of $[\text{Pt}(\text{R})(\text{OTf})(\text{PPh}_3)_2]$ (OTf = $\text{OSO}_2\text{CF}_3$) with $\text{K[Co(CO)}_4\text{]}$ (Eq. 9.14) [19, 22]. Thermolysis of 7 under a CO atmosphere at 50 °C afforded $[(p\text{-TolCO})\text{Co(CO)}_3(\text{PPh}_3)]$ (8) in 47% yield. These facts unambiguously indicate the intermediacy of a Pd–Co dinuclear species analogous to 7 in the catalytic reaction according to Eq. (9.13) [19].

$$\begin{array}{c}
\overset{\displaystyle \text{PPh}_3}{\underset{\displaystyle \text{PPh}_3}{p\text{-Tol}-\text{Pt}-\text{OTf}}} \quad + \quad \text{K[Co(CO)}_4\text{]}
\end{array}$$

$$\longrightarrow \quad \underset{\displaystyle p\text{-Tol}}{\overset{\displaystyle \text{CO}}{\text{Ph}_3\text{P}-\text{Pt}-\text{Co(CO)}_3(\text{PPh}_3)}} \quad + \text{KOTf} \quad (9.14)$$

7

Further investigation of complex 8 revealed that its treatment with $\text{HSiEt}_3$ under high CO pressure, especially in the presence of $\text{NEt}_3$, results in the formation of 5 as the major product (Eq. 9.15) [16]. Therefore, it is highly probable that complex 8 is an intermediate for the catalytic formation of 5 in the carbonylation of *p*-TolI with $\text{CO/HSiEt}_3$ in the presence of $\text{NEt}_3$. Catalyzed and uncatalyzed reactions of acyl complexes with hydrosilanes have previously been reported to afford siloxyalkyl complexes [23–26]. In addition, it has been documented that some siloxyalkyl complexes, such as $[\text{Mn}\{\text{CHPh(OSiEt}_3)\}(\text{CO})_5]$, undergo thermolysis to give 1,2-disiloxyethanes via a siloxyalkyl radical [27]. Reaction according to Eq. (9.15) can be explained in terms of a similar process involving formation of the siloxybenzylcobalt complex $[\text{Co}\{\text{CHAr(OSiEt}_3)\}(\text{CO})_3(\text{PPh}_3)]$, homolytic cleavage of the Co–C bond to liberate the siloxybenzyl radical $(\text{Et}_3\text{SiO})\text{ArCH}\cdot$, and coupling of the radicals.

$$[(p\text{-TolCO})\text{Co(CO)}_3(\text{PPh}_3)] \;+\; \text{HSiEt}_3 \xrightarrow[80\,°\text{C}]{\text{CO 50 atm}} \underset{\displaystyle \text{Et}_3\text{SiO}}{\overset{\displaystyle p\text{-Tol}}{\diagdown}}\!\!\!\bowtie\!\!\!\underset{\displaystyle \text{OSiEt}_3}{\overset{\displaystyle p\text{-Tol}}{\diagup}} \quad (9.15)$$

8                          5

The mechanism for the catalytic reaction according to Eq. (9.12) is summarized in Scheme 9.2. In this mechanism, ArI is first activated by a $\text{Pd}^0$ species through oxidative addition, and the aryl or aroyl group is then transferred from the palladium

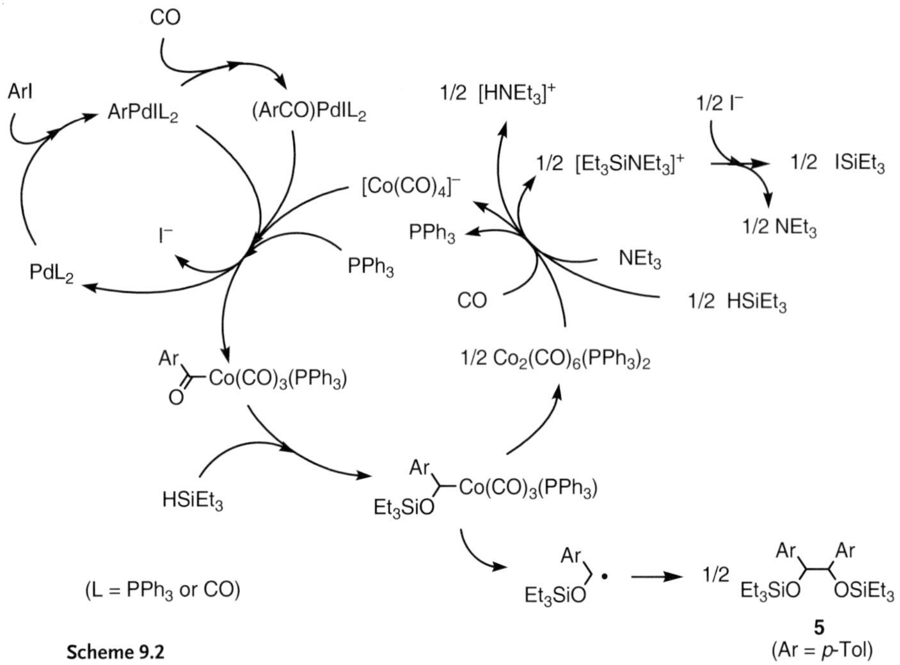

**Scheme 9.2**

(L = PPh$_3$ or CO)

**5**

(Ar = p-Tol)

to the cobalt center by way of the Pd–Co heterobimetallic complex. The aroylcobalt species then undergoes hydrosilylation to yield the siloxybenzylcobalt complex, thermolysis of which affords the final product **5**. The Co$^0$ species generated is transformed into the [Co(CO)$_4$]$^-$ anion by the action of HSiEt$_3$, NEt$_3$, and CO to complete the Co-catalyzed cycle. It should be emphasized that this mechanism involves the transfer of an organic group from one metal center (Pd) to another (Co) during the catalysis, and this is the origin of the synergistic effects between palladium and cobalt.

### 9.2.3
### Selective Hydroformylation of Internal Alkynes by Pd-Co Catalysts

The hydroformylation of alkynes to give α,β-unsaturated aldehydes has been poorly exploited [28], primarily because it is usually difficult to suppress the formation of the corresponding saturated aldehydes and noncarbonylated hydrocarbons. We therefore set out to achieve the hydroformylation of alkynes by adopting the bimetallic catalyst approach [29].

Preliminary examination of the hydroformylation of 4-octyne (Eq. 9.16) with catalyst systems composed of [PdCl$_2$(PPh$_3$)$_2$] and/or metal carbonyls resulted in low selectivities, but the bimetallic [PdCl$_2$(PPh$_3$)$_2$]–[Co$_2$(CO)$_8$] catalyst showed some synergistic effect and gave the desired unsaturated aldehyde **9** in up to 49% yield after a reaction time of 24 h. On the other hand, the monometallic catalyst [PdCl$_2$(PCy$_3$)$_2$] (Cy = cyclohexyl) exhibited very high selectivity in the presence of

**Table 9.3** Hydroformylation of 4-octyne by Pd–Co catalysts.[a]

| Catalyst | Time (h) | Conv. (%) | GLC yield (%) | | |
|---|---|---|---|---|---|
| | | | 9 | 10 | 11 |
| [PdCl$_2$(PPh$_3$)$_2$] | 24 | 85 | 7 | 0 | 4 |
| [Co$_2$(CO)$_8$] | 24 | 74 | 2 | 0 | 39 |
| [PdCl$_2$(PPh$_3$)$_2$]–[Co$_2$(CO)$_8$] | 24 | 100 | 39 | 14 | 19 |
| [PdCl$_2$(PPh$_3$)$_2$]–[Co$_2$(CO)$_8$][b] | 24 | 80 | 49 | 8 | 4 |
| [PdCl$_2$(PCy$_3$)$_2$][c] | 1 | 20 | 16 | < 1 | 0 |
| [PdCl$_2$(PCy$_3$)$_2$][c] | 6 | 84 | 83 | < 1 | < 1 |
| [PdCl$_2$(PCy$_3$)$_2$]–[Co$_2$(CO)$_8$][c] | 1 | 100 | 95 | 2 | 3 |

[a] Reaction conditions: 4-octyne, 5 mmol; [PdCl$_2$(PR$_3$)$_2$], 0.1 mmol; [Co$_2$(CO)$_8$], 0.05 mmol; CO, 25 atm; H$_2$, 25 atm; C$_6$H$_6$ 5 mL, 150 °C.
[b] NEt$_3$, 5 mmol. [c] NEt$_3$, 3 mmol; CO, 35 atm; H$_2$, 35 atm.

NEt$_3$, although its catalytic activity was moderate. The yield of **9** reached 83% after 6 h at 150 °C, and the formation of the saturated aldehyde and the simple hydrogenation products was almost negligible. This is a relatively rare example of palladium-catalyzed hydroformylation. The combined use of [PdCl$_2$(PCy$_3$)$_2$] and [Co$_2$(CO)$_8$] (Pd : Co = 1 : 1) remarkably improved the catalytic activity, albeit with little change in the selectivity; the reaction was complete within 1 h, and aldehyde **9** (95% *E*) was obtained in 95% yield (Table 9.3). It is obvious that the palladium and cobalt metal centers participate cooperatively in the selective production of the unsaturated aldehyde. As the second component of the catalyst coupled with [PdCl$_2$(PCy$_3$)$_2$], [Co$_2$(CO)$_8$] accelerated the reaction most effectively, but quite unexpectedly enhancement of the reaction rate was also observed with [Fe$_3$(CO)$_{12}$] and [W(CO)$_6$].

$$ (9.16) $$

The [PdCl$_2$(PCy$_3$)$_2$] and [PdCl$_2$(PCy$_3$)$_2$]–[Co$_2$(CO)$_8$] catalyst systems proved to be applicable to the hydroformylation of various internal alkynes, although in the reaction of diphenylacetylene a significant amount of *cis*-stilbene was formed through hydrogenation caused mainly by cobalt species. Hydroformylation of 1-cyclohexyl-1-propyne in the presence of the [PdCl$_2$(PCy$_3$)$_2$]–[Co$_2$(CO)$_8$] catalyst produced predominantly (*E*)-3-cyclohexyl-2-methyl-2-propenal, with good regio-selectivity (Eq. 9.17). Obviously, the more sterically demanding cyclohexyl group effectively directs the CO insertion to occur at the remote acetylenic carbon.

$$\text{[PdCl}_2\text{(PCy}_3\text{)}_2\text{] – [Co}_2\text{(CO)}_8\text{]} \atop \text{H}_2\text{, CO, NEt}_3$$

(9.17)

52% isolated yield
(10 : 1)

The chemoselectivity of this reaction is noteworthy. When a 1 : 1 mixture of 5-decyne and cyclooctene was hydroformylated with the $[PdCl_2(PCy_3)_2]–[Co_2(CO)_8]$ catalyst, the unsaturated aldehyde **12** was obtained in 92% yield and cyclooctene was recovered quantitatively (Eq. 9.18). Thus, the $[PdCl_2(PCy_3)_2]–[Co_2(CO)_8]$ catalyst promotes only the hydroformylation of alkynes.

$$n\text{-Bu}\!\!\!\equiv\!\!\!n\text{-Bu} + \bigcirc \quad \xrightarrow[\text{H}_2\text{, CO, NEt}_3]{\text{[PdCl}_2\text{(PCy}_3\text{)}_2\text{] – [Co}_2\text{(CO)}_8\text{]}}$$

(9.18)

| **12** | 3% | 4% | 100% recovery |

92% yield

Although further investigations are required to fully elucidate the cooperative effects of palladium and cobalt, some comments can be made here. Since the product selectivity shown by the bimetallic $[PdCl_2(PCy_3)_2]–[Co_2(CO)_8]$ catalyst is similar to that shown by the monometallic $[PdCl_2(PCy_3)_2]$ catalyst, it is presumed that the chemical transformation of the alkyne occurs mainly at the palladium metal center and that the role of the cobalt species is to accelerate a part of the Pd-catalyzed cycle. Several mechanisms may be responsible for the rate enhancement effect of cobalt. Preliminary kinetic studies showed that both the $[PdCl_2(PCy_3)_2]$ and $[PdCl_2(PCy_3)_2]–[Co_2(CO)_8]$ catalysts exhibited some dependence of the reaction rates on the $H_2$ pressure, which may indicate the participation of a hydridocobalt species in the hydrogenolysis of the Pd–acyl bond through a dinuclear reductive elimination process. On the other hand, we have observed that the reaction of [PdPh(OTf) $(PMe_3)_2$] with the $[Co(CO)_4]^-$ anion induces facile CO insertion into the Pd–C bond to ultimately afford the dinuclear benzoyl complex $[(PMe_3)_2(PhCO)PdCo(CO)_4]$ [19]. More recently, it has been reported that CO insertion into the Pd–C bond in $[(dppe)MePdCo(CO)_4]$ is much faster than that in the case of $[PdMeCl(dppe)]$ [30, 31]. Interestingly, a theoretical study on a model complex $[(PH_2CH_2CH_2PH_2) MePdCo(CO)_4]$ has revealed that the most favorable CO insertion process involves initial methyl migration from the palladium center to the cobalt, CO insertion at the cobalt center, and acyl group migration from the cobalt center to the palladium [31]. These results favor a catalytic mechanism in which cobalt promotes CO insertion into the Pd–C bond of a vinylpalladium intermediate.

**9.3**
**Multimetallic Complex Catalysts**

9.3.1
**Reactions Catalyzed by Thiolato-Bridged Dinuclear Ruthenium Complexes**

In the preceding section, we have described how the combined use of two different metal complexes can provide a better catalyst system than the constituent mono-metallic units. The synergistic phenomena observed originate from reactions between the different metal species, and multimetallic species generated in situ are considered to be involved in such processes as the key intermediates. Thus, the use of well-designed multimetallic complexes can be reasonably considered as providing a rational strategy for the development of new catalysts based on synergistic effects between the metal centers. However, metal–metal bonds are usually more labile than most chemical bonds in organic substrates, and un-supported multimetallic structures often undergo fragmentation into monometallic species under the conditions used for catalytic reactions. To avoid this, it is necessary to introduce bridging ligands that can stabilize the multimetallic structures. On the other hand, it is well-known that unique and highly efficient chemical transformations are achieved on sulfur-bridged multimetallic active sites in metalloenzymes such as nitrogenase and hydrogenase. In these systems, the sulfur ligands play an important role both in tuning the electronic states of the metal centers and in stabilizing the multinuclear structures. These biological multimetallic systems inspired us to investigate the preparation and catalytic reactivities of sulfur-bridged multimetallic complexes [32, 33]. In particular, we focussed our attention on those of noble metal elements of groups 8–10, because these metals are not used in biological systems but are known to serve as active catalysts in many synthetic reactions.

Initially, we synthesized a series of thiolato-bridged dinuclear ruthenium complexes such as **13–15**. An important feature of these diruthenium complexes is that their dinuclear cores possess two potentially vacant coordination sites in close proximity, which should function as a bimetallic reaction site. As expected, our initial investigations into the reactivities of the diruthenium complexes revealed that they undergo fascinating stoichiometric reactions with alkynes to give a variety of ruthenacycles, the structures of which depend upon the substituents of the alkynes [34].

From a catalytic point of view, the cationic Ru$^{III}$–Ru$^{III}$ complex [Cp*RuCl($\mu$-S$^i$Pr)$_2$-Ru(OH$_2$)Cp*] (**15**), which can be easily derived from the dichloro complex [Cp*RuCl-($\mu$-S$^i$Pr)$_2$RuClCp*] (**14**), exhibited the most intriguing reactivities (Scheme 9.3).

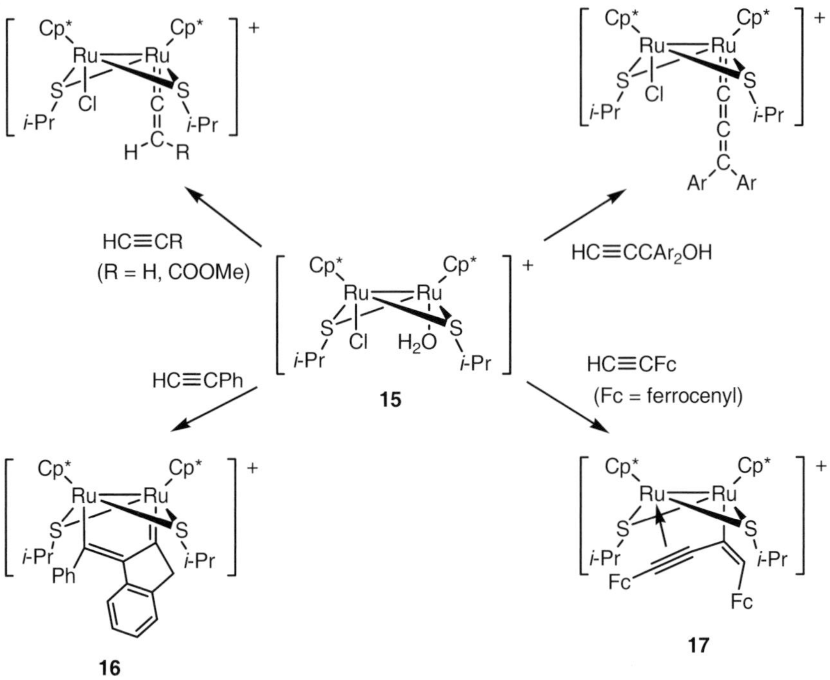

**Scheme 9.3**

Thus, its reaction with acetylene or methyl propiolate affords the corresponding vinylidene complex [Cp*RuCl(μ-S$^i$Pr)$_2$Ru(C=CHR)Cp*]$^+$ (R = H, COOMe) [35], while that with 1,1-diarylpropargyl alcohols yields the allenylidene complexes [Cp*RuCl(μ-S$^i$Pr)$_2$Ru(C=C=CAr$_2$)Cp*]$^+$ [36]. When terminal arylacetylenes are used, two molecules of the alkyne are incorporated at the diruthenium center to form unique diruthenacycle complexes 16 [36]. This reaction is considered to proceed via the alkynyl-vinylidene and butenynyl intermediates derived from 15. In fact, the butenynyl complex 17 has been isolated from the reaction of 15 with ferrocenylacetylene [37].

Interestingly, complex 17 behaves as an efficient catalyst for linear di- and trimerization of ferrocenylacetylene (Eq. 9.19), although it is not effective for the dimerization of other acetylenes [36]. This drawback could be overcome by fine-tuning of the bridging ligands. We found that the Ru$^{III}$–Ru$^{III}$ complex [Cp*RuCl(μ-SMe)$_2$RuClCp*] (18) with bridging MeS ligands catalyzes the head-to-head (Z)-dimerization of various aliphatic terminal alkynes in the presence of NH$_4$BF$_4$ (Eq. 9.20) [38]. The strong dependence of the catalytic activity upon the size of the bridging thiolato ligands is of special interest; primary alkanethiolato ligands are required to attain good catalytic activities.

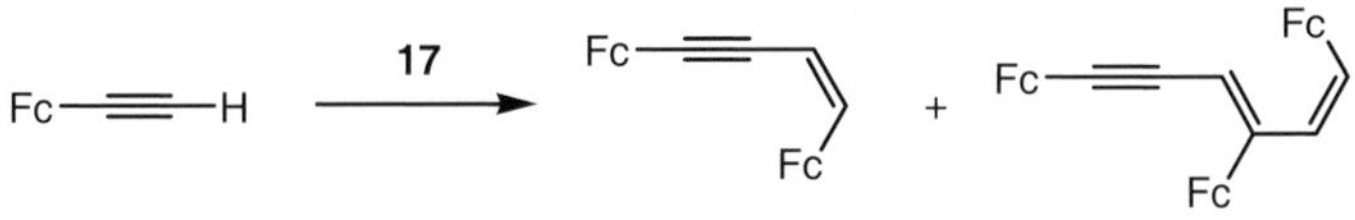

(9.19)

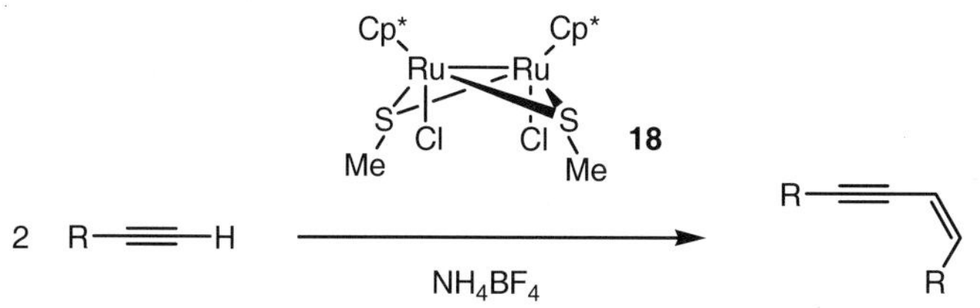

$$2 \quad R\!\!\equiv\!\!H \xrightarrow[\text{NH}_4\text{BF}_4]{\textbf{18}} \qquad (9.20)$$

A plausible mechanism is shown in Scheme 9.4. Two molecules of the terminal alkyne are incorporated at the dinuclear core in the initial step to form a vinylidene-alkynyl intermediate. The vinylidene-alkynyl coupling then takes place at the dinuclear site, and the butenynyl complex is produced. As already mentioned, this type of complex has been isolated and characterized in the reaction with ferro-cenylacetylene. Further protonolysis of the butenynyl complex affords the head-to-head Z-dimer. The bridging coordination mode of the butenynyl ligand together with the steric effect of the Cp* ligands effectively controls the stereochemistry of the vinylidene–alkynyl coupling reaction, which gives rise to the high stereo-selectivity of this catalytic reaction.

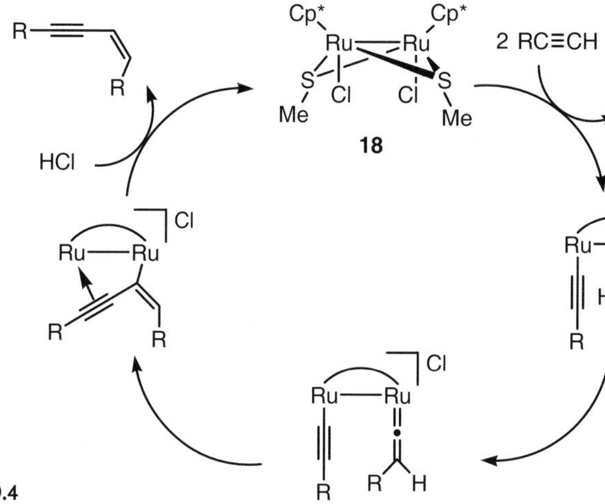

**Scheme 9.4**

This dimerization reaction has been successfully applied to the catalytic cyclization of α,ω-diynes. Treatment of 1,15-hexadecadiyne with a catalytic amount of **18** gives the *endo*-macrocyclic enyne, (Z)-1-cyclohexadecen-3-yne (Eq. 9.21) [39]. This cyclization provides a rare synthetic route to *endo*-cyclic (Z)-1-en-3-ynes. Interestingly, a related palladium-catalyzed cyclization of α,ω-diynes has been reported to produce the corresponding *exo*-cyclic 1-en-3-ynes [40].

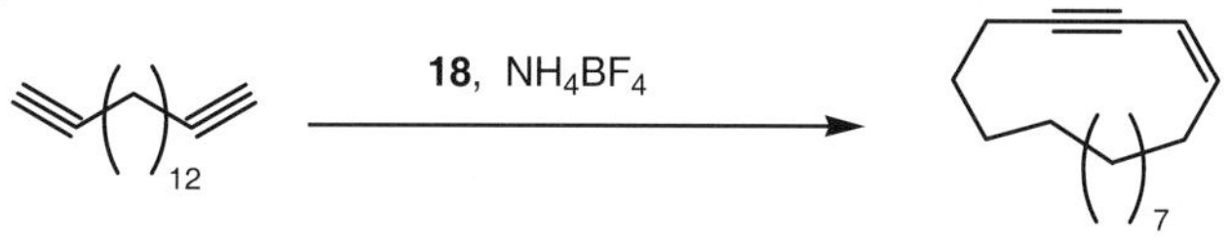

$$ \xrightarrow{\textbf{18, NH}_4\text{BF}_4} \qquad (9.21)$$

The allenylidene complex was also found to act as a highly effective catalytic intermediate, and novel propargylic substitution reactions of propargyl alcohols have been developed from it. Thus, the reactions of propargylic alcohols with nucleophiles in the presence of **18** afford the corresponding propargylic substituted products (Scheme 9.5). Interestingly, this reaction is applicable to a very wide range of nucleophiles; not only alcohols, amines, amides, and thiols [41], but also carbon nucleophiles such as simple ketones [42], alkenes [43], phenols [44], and hetero-aromatic compounds [45] can be used as substrates. Even simple ketones such as acetone and 2-butanone participate in the propargylic substitution, and the γ-keto acetylenes are obtained under mild and essentially neutral conditions [42]. It should be noted that in the reactions of unsymmetrical ketones the sterically more encumbered α-carbon is substituted predominantly, which probably reflects the equilibrium ratio of the enol isomers. The propargyl substitution of ketones has been further applied to the catalytic synthesis of furans and pyrroles, where the propargyl substitution by **18** and the hydration of the resultant alkyne by PtCl$_2$ take place sequentially [46]. It should be emphasized that the diruthenium core is essential for the above catalysis. Conventional mononuclear ruthenium complexes failed to catalyze these reactions, although they have been well-documented to react with propargylic alcohols to form allenylidene species.

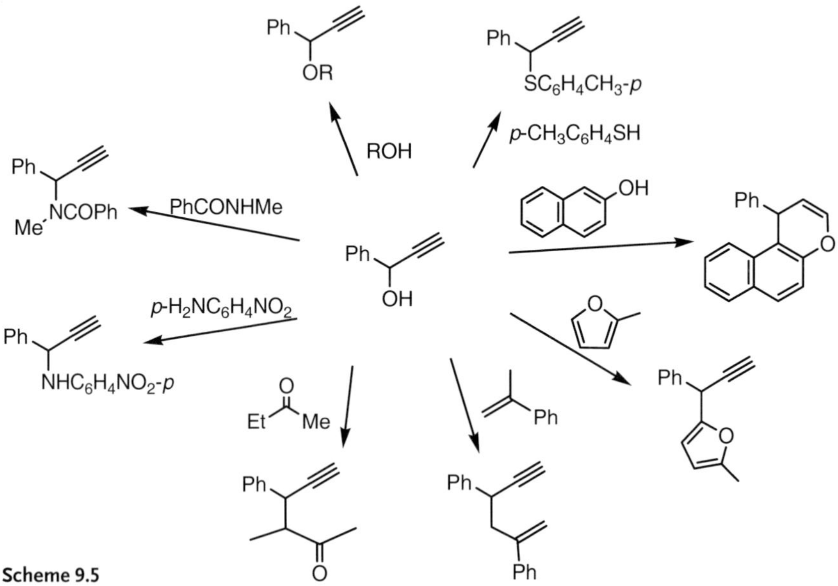

Scheme 9.5

This substitution reaction has many synthetic advantages. The substitution proceeds with very high regioselectivity. Allenic by-products are not observed at all, whereas they are commonly observed in classical propargylic substitution reactions. Readily available propargyl alcohols can be used directly as substrates, and derivatization to halides or esters is not necessary. This point is particularly important with regard to atom economy. The overall course of the reaction is formally related to the Nicholas reaction, which is known to be an efficient method for propargylic substitution [47]. However, the Nicholas reaction requires a stoichiometric amount of $[Co_2(CO)_8]$, and so the synthetic utility of the present catalytic reaction is obvious.

A proposed reaction mechanism is depicted in Scheme 9.6. The propargylic alcohol is incorporated at the ruthenium center as a vinylidene ligand in the initial step, and its dehydration leads to an allenylidene species. The attack of a nucleophile at the γ-carbon of the allenylidene ligand gives the substituted vinylidene complex. Finally, the newly formed vinylidene ligand isomerizes into the corresponding alkyne and is liberated as the product. This mechanism is supported by the observation that the allenylidene complex $[Cp*RuCl(\mu\text{-}SMe)_2Ru\{C=C=C(p\text{-}Tol)_2\}Cp*]^+$ can be isolated from the reaction of **18** with 1,1-di-$p$-tolylpropargyl alcohol, and subsequent reaction with ethanol affords the ethyl propargyl ether (Eq. 9.22). On the other hand, substitution reactions of propargylic alcohols with internal acetylene groups have recently been achieved by switching the catalyst to the cationic diruthenium complex $[Cp*RuCl(\mu\text{-}SMe)_2Ru(OH_2)Cp*][OTf]$ [48]. In this case, a dinuclear η-propargyl complex rather than the allenylidene species is believed to be involved in the catalysis.

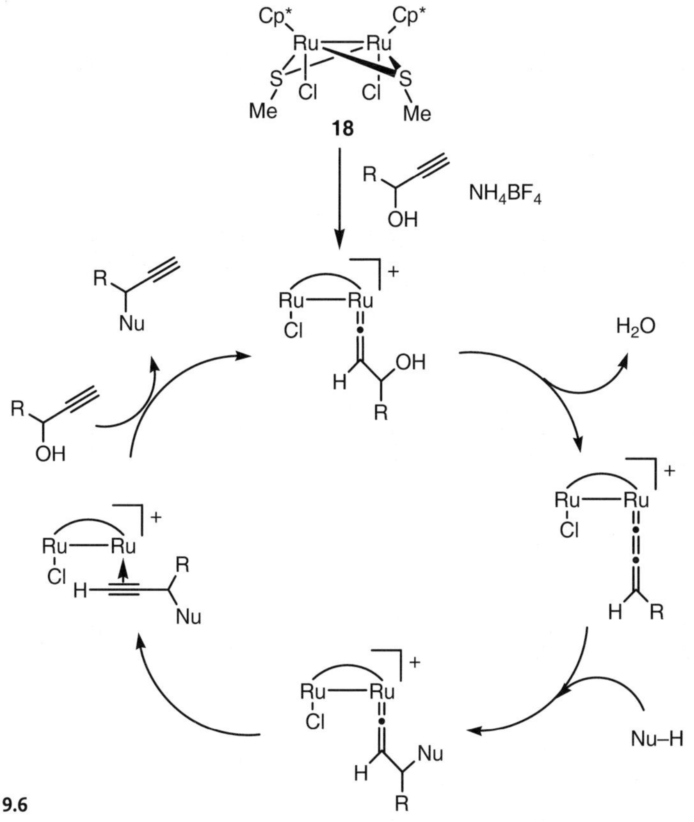

Scheme 9.6

$$\text{(9.22)}$$

The reaction scheme shows complex **18** reacting with $HC\equiv C-C(p\text{-Tol})_2OH$ and $NH_4BF_4$ to give the bracketed cationic product, followed by reaction with EtOH at 60 °C to give $HC\equiv C-C(p\text{-Tol})_2OEt$.

Some other catalytic reactions have also been developed with the thiolato-bridged dinuclear ruthenium complexes. Complex **15** is an active catalyst for the allylation of aromatic hydrocarbons with cinnamyl alcohols [49] and for the silylative dimerization of aromatic aldehydes with hydrosilanes [50]. Complex **13** catalyzes the disproportionation of hydrazine into ammonia and dinitrogen according to the stoichiometry shown in Eq. (9.23), where the diazene complex $[Cp*Ru(\mu\text{-}\eta^1\text{:}\eta^1\text{-}HN=NH)(\mu\text{-}S^iPr)_2RuCp*]$ is considered to be a key intermediate [51]. The latter reaction is of special interest in relation to the mechanism of biological nitrogen fixation [52].

$$3\,N_2H_4 \quad \xrightarrow[\;40\,°C\;]{\textbf{13}} \quad N_2 + 4\,NH_3 \qquad \text{(9.23)}$$

## 9.3.2
### Catalytic Transformations of Alkynes by Tri- and Tetranuclear Sulfido Clusters

Besides the above-mentioned thiolato-bridged diruthenium complexes, we have also engaged in the development of rational synthetic methods for sulfido-bridged noble metal clusters [32], and some of the newly synthesized clusters have been found to show high catalytic activities in transformations of alkynes.

Our synthetic strategy for sulfido-bridged clusters is based on utilization of pre-assembled metal–sulfur aggregates with $M_2S_2$ and $M_3S_4$ cores as building blocks. The readily accessible hydrosulfido-bridged dinuclear complexes $[Cp*MCl(\mu\text{-}SH)_2 MClCp*]$ (M = Ru, Rh, Ir) have been found to provide excellent $M_2S_2$-type starting materials; incorporation of an additional metal fragment leads to a variety of tri-, tetra-, and pentanuclear clusters [32]. For example, the diiridium complex $[Cp*IrCl(\mu\text{-}SH)_2IrClCp*]$ reacts with $[MCl_2(cod)]$ (M = Pd, Pt; cod = cycloocta-1,5-diene) to give the trinuclear sulfido clusters $[(Cp*Ir)_2(\mu\text{-}S)_2MCl_2]$ (Eq. 9.24) [53]. The $Ir_2Pd$ cluster is an active catalyst for the regioselective addition of alcohols to internal 1-aryl-1-alkynes, and the corresponding 1-aryl-2,2-dimethoxyalkane is obtained in good yield (Eq. 9.25) [54]. The $Ir_2Pt$ cluster is also effective but much less regio-selective. Although several transition metal compounds have been reported to catalyze the addition of alcohols to alkynes, regioselective addition to internal alkynes has not been documented. The detailed reaction mechanism has not been clarified, but a plausible mechanism involves regioselective nucleophilic attack of an alcohol molecule on the coordinated alkyne at the palladium center followed by protonolysis of the resultant alkoxyvinyl cluster.

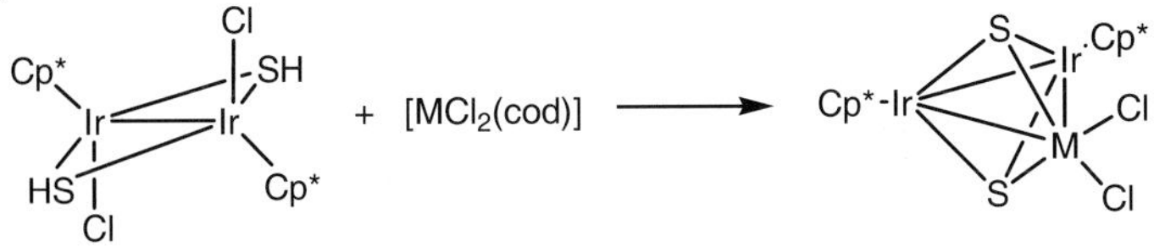

(9.24)

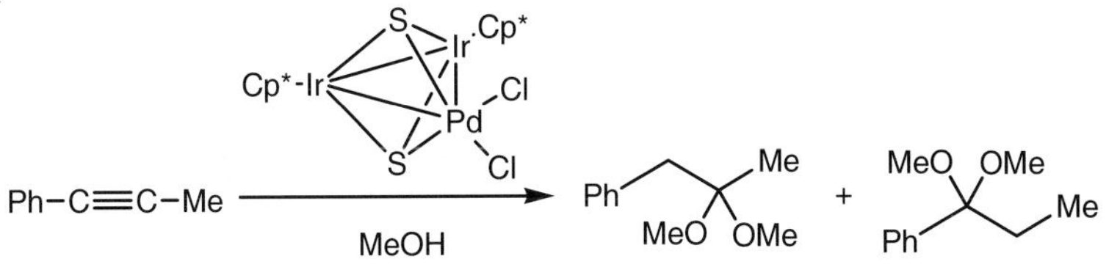

(9.25)

Another catalytically active sulfido cluster has been synthesized by using an $M_3S_4$-type building block. It has been well-documented that the cationic trimolybdenum cluster $[\{Mo(H_2O)_3\}_3(\mu_3\text{-}S)(\mu_2\text{-}S)_3]^{4+}$ incorporates heterometals into the incomplete cubane core to form a series of cubane-type tetranuclear clusters [55]. We have applied this method to the synthesis of clusters containing noble metals [56, 57] and have developed the $Mo_3Pd$ cluster $[\{Mo(tacn)\}_3(PdCl)(\mu_3\text{-}S)_4](PF_6)_3$ (**19**; tacn = 1,4,7-triazacyclononane) (Scheme 9.7) [56]. From a structural point of view, the palladium center embedded in the cubane core is highly interesting; it adopts a tetrahedral geometry in spite of its apparent $Pd^{II}$ nature.

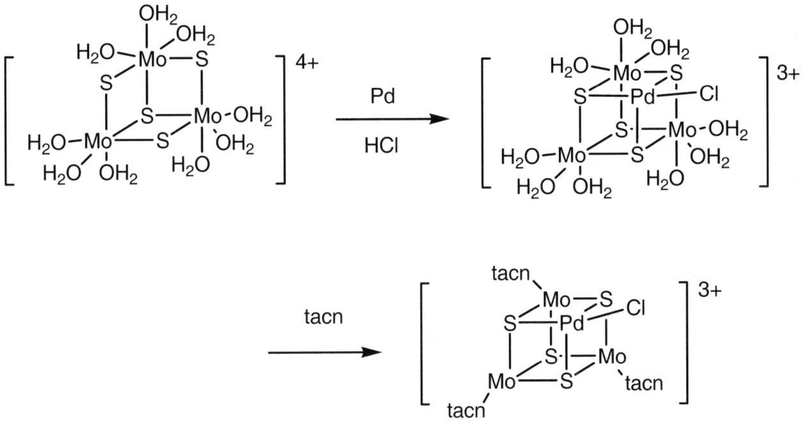

**Scheme 9.7**                                                                 **19**

Cluster **19** was found to exhibit several unique catalytic activities. It catalyzes the stereoselective addition of alcohols to electron-deficient alkynes such as methyl propiolate (Eq. 9.26) [56]. The corresponding (Z)-vinyl ethers are obtained with very high stereoselectivity. It is noteworthy that neither conventional mononuclear palladium complexes nor the $Mo_3$ cluster $[\{Mo(H_2O)_3\}_3(\mu_3\text{-}S)(\mu\text{-}S)_3]^{4+}$ is effective for this reaction. This fact implies that the $Mo_3Pd$ mixed-metal cluster structure is essential for the catalytic activity, although the catalytic site is probably the palladium center.

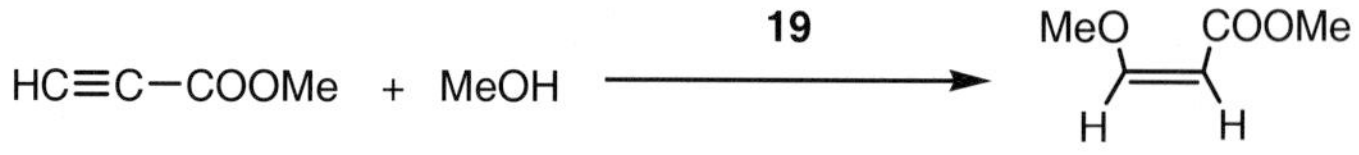
19
HC≡C−COOMe + MeOH
MeO
COOMe
H
H

(9.26)

As depicted in Scheme 9.8, the catalytic cycle is presumed to be initiated by nucleophilic addition of MeOH from the outer coordination sphere to a coordinated alkyne at the palladium center to form the vinylpalladium intermediate, which undergoes protonolysis of the Pd–C bond to liberate the (*Z*)-vinyl ether. The $Mo_3S_4$ moiety is not directly involved in the catalytic cycle, but it obviously plays an important role in controlling the electronic state of the palladium center.

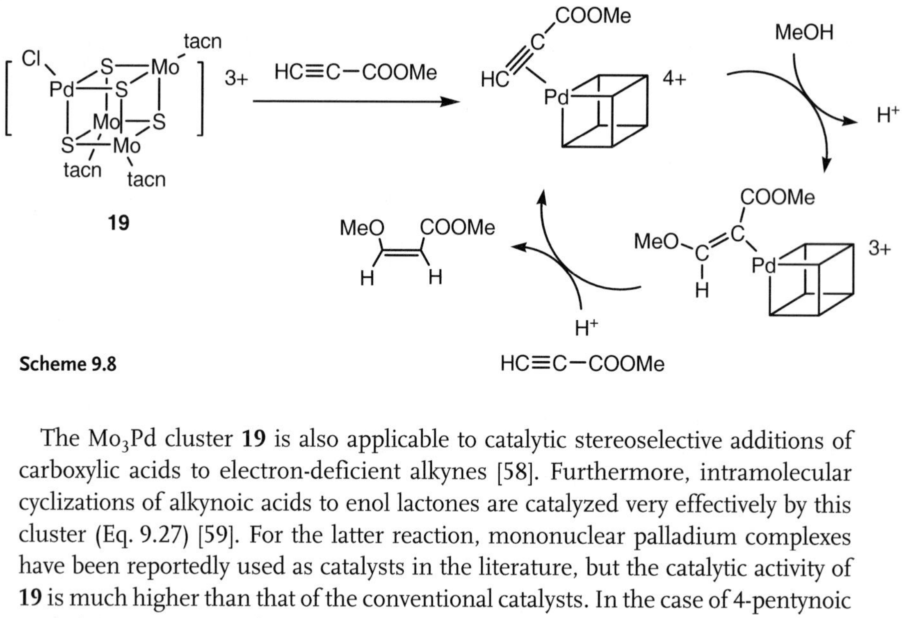
tacn
Cl
Pd
S
Mo
S
Mo
S
S
Mo
tacn
tacn
3+
19
HC≡C−COOMe
COOMe
C
HC
Pd
4+
MeOH
H+
COOMe
MeO
C
C
H
Pd
3+
MeO
COOMe
H
H
H+
HC≡C−COOMe

**Scheme 9.8**

The $Mo_3Pd$ cluster **19** is also applicable to catalytic stereoselective additions of carboxylic acids to electron-deficient alkynes [58]. Furthermore, intramolecular cyclizations of alkynoic acids to enol lactones are catalyzed very effectively by this cluster (Eq. 9.27) [59]. For the latter reaction, mononuclear palladium complexes have been reportedly used as catalysts in the literature, but the catalytic activity of **19** is much higher than that of the conventional catalysts. In the case of 4-pentynoic acid, the turnover number reaches 100000 after 19 h. Another interesting feature of this lactonization is that it can be conducted in water owing to the good solubility and stability of **19** in aqueous media. Very recently, a related $(Mo_3NiS_4)_2$ cubane cluster has been derived from $[(Cp^{*}Mo)_3(\mu_2\text{-}S)_3(\mu_3\text{-}S)](PF_6)$ and $[Ni(cod)_2]$ (Eq. 9.28) and was shown to possess good catalytic activity for similar lactonization reactions [60].

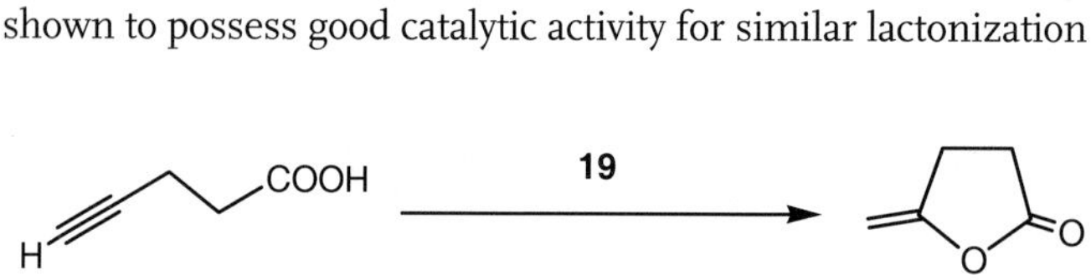
COOH
H
19
O
O

(9.27)

$$(PF_6)_2 \quad (9.28)$$

## 9.4
## Concluding Remarks

We have described the development of two types of multimetallic catalyst systems, namely the combined use of metal complexes and multinuclear complex catalysts, both of which have been demonstrated to promote unique and effective catalyses. Bimetallic cooperativity is an important factor in such reactions, and this is the most characteristic feature of the multimetallic catalysis that cannot be seen in reactions induced by conventional monometallic and mononuclear catalysts. It is expected that continued exploration of rational synthetic methods for multinuclear complexes and the discovery of novel bimetallic reactivities will lead to further expansion of this field.

## References

1   (a) P. Braunstein, J. Rose, in *Metal Clusters in Chemistry* (Eds.: P. Braunstein, L. A. Oro, P. R. Raithby), Vol. 2, Wiley-VCH, Weinheim, **1999**, pp. 616; (b) *Catalysis by Di- and Polynuclear Metal Cluster Complexes* (Eds.: R. D. Adams, F. A. Cotton), Wiley-VCH, New York, **1998**.

2   (a) D. Butter, T. Haute, U.S. Patent 3,285,948, 1966; (b) M. Mizoroki, T. Matsumoto, A. Ozaki, *Bull. Chem. Soc. Jpn.* **1979**, *52*, 479.

3   M. Hidai, M. Orisaku, M. Ue, Y. Koyasu, T. Kodama, Y. Uchida, *Organometallics* **1983**, *2*, 292.

4   M. Hidai, H. Matsuzaka, *Polyhedron* **1988**, *7*, 2369.

5   M. Hidai, A. Fukuoka, Y. Koyasu, Y. Uchida, *J. Chem. Soc., Chem. Commun.* **1984**, 516.

6   M. Hidai, A. Fukuoka, Y. Koyasu, Y. Uchida, *J. Mol. Catal.* **1986**, *35*, 29.

7   Y. Koyasu, A. Fukuoka, Y. Uchida, M. Hidai, *Chem. Lett.* **1985**, 1083.

8   Y. Ishii, M. Sato, H. Matsuzaka, M. Hidai, *J. Mol. Catal.* **1989**, *54*, L13.

9   R. Whyman, *J. Organomet. Chem.* **1974**, *81*, 97.

10  C. Li, E. Widjaja, M. Garland, *J. Am. Chem. Soc.* **2003**, *125*, 5540.

11  R. M. Bullock, B. J. Rappoli, *J. Am. Chem. Soc.* **1991**, *113*, 1659.

12  M. E. Broussard, B. Juma, S. G. Train, W.-J. Peng, S. A. Laneman, G. G. Stanley, *Science* **1993**, *260*, 1784.

13  Y. Misumi, Y. Ishii, M. Hidai, *J. Mol. Catal.* **1993**, *78*, 1.

14  S. H. Han, G. L. Geoffroy, B. D. Dombek, A. L. Rheingold, *Inorg. Chem.* **1988**, *27*, 4355.

**15** K. Kudo, M. Hidai, Y. Uchida, *J. Organomet. Chem.* **1971**, *33*, 393.

**16** Y. Misumi, Y. Ishii, M. Hidai, *Organometallics* **1995**, *14*, 1770.

**17** Y. Ishii, M. Hidai, *Catal. Today* **2001**, *66*, 53.

**18** A. Sisak, F. Ungváry, L. Markó, *Organometallics* **1986**, *5*, 1019.

**19** (a) Y. Misumi, Y. Ishii, M. Hidai, *Chem. Lett.* **1994**, 695; (b) Y. Misumi, Y. Ishii, M. Hidai, *J. Chem. Soc., Dalton Trans.* **1995**, 3489.

**20** A. Fukuoka, T. Sadashima, T. Sugiura, X. Wu, Y. Mizuho, S. Komiya, *J. Organomet. Chem.* **1994**, *473*, 139.

**21** A. Fukuoka, T. Sadashima, I. Endo, N. Ohashi, Y. Kambara, T. Sugiura, K. Miki, N. Kasai, S. Komiya, *Organometallics* **1994**, *13*, 4033.

**22** M. Ferrer, O. Rossell, M. Seco, P. Braunstein, *J. Chem. Soc., Dalton Trans.* **1989** 379.

**23** P. K. Hanna, B. T. Gregg, A. R. Cutler, *Organometallics* **1991**, *10*, 31.

**24** E. J. Crawford, P. K. Hanna, A. R. Cutler, *J. Am. Chem. Soc.* **1989**, *111*, 6891.

**25** M. Akita, O. Mitani, Y. Moro-oka, *J. Chem. Soc., Chem. Commun.* **1989**, 527.

**26** B. T. Gregg, P. K. Hanna, E. J. Crawford, A. R. Cutler, *J. Am. Chem. Soc.* **1991**, *113*, 384.

**27** J. A. Gladysz, *Acc. Chem. Res.* **1984**, *17*, 326.

**28** J. R. Johnson, G. D. Cuny, S. L. Buchwald, *Angew. Chem. Int. Ed. Engl.* **1995**, *34*, 1760.

**29** Y. Ishii, K. Miyashita, K. Kamita, M. Hidai, *J. Am. Chem. Soc.* **1997**, *119*, 6448.

**30** A. Fukuoka, S. Fukugawa, M. Hirano, S. Komiya, *Chem. Lett.* **1997**, 377.

**31** A. Fukuoka, S. Fukugawa, M. Hirano, N. Koga, S. Komiya, *Organometallics* **2001**, *20*, 2065.

**32** (a) M. Hidai, S. Kuwata, Y. Mizobe, *Acc. Chem. Res.* **2000**, *33*, 46; (b) M. Hidai, in: *Perspectives in Organometallic Chemistry* (Eds.: C. G. Screttas, B. R. Steele), The Royal Society of Chemistry, Cambridge, 2003, pp. 62.

**33** M. Hidai, Y. Ishii, S. Kuwata, in *Modern Coordination Chemistry* (Eds.: G. J. Leigh, N. Winterton), The Royal Society of Chemistry, Cambridge, 2002, pp. 208.

**34** (a) M. Hidai, Y. Mizobe, H. Matsuzaka, *J. Organomet. Chem.* **1994**, *473*, 1; (b) M. Hidai, Y. Mizobe, in: *Transition Metal Sulfur Chemistry: Biological and Industrial Significance* (Eds.: E. I. Stiefel, K. Matsumoto), American Chemical Society, Washington, DC, 1996, Chapter 19.

**35** Y. Takagi, H. Matsuzaka, Y. Ishii, M. Hidai, *Organometallics* **1997**, *16*, 4445.

**36** H. Matsuzaka, Y. Takagi, M. Hidai, *Organometallics* **1994**, *13*, 13.

**37** H. Matsuzaka, Y. Takagi, Y. Ishii, M. Nishio, M. Hidai, *Organometallics* **1995**, *14*, 2153.

**38** J.-P. Qü, D. Masui, Y. Ishii, M. Hidai, *Chem. Lett.* **1998**, 1003.

**39** Y. Nishibayashi, M. Yamanashi, I. Wakiji, M. Hidai, *Angew. Chem. Int. Ed.* **2000**, *39*, 2909.

**40** B. M. Trost, S. Matsubara, J. J. Caringi, *J. Am. Chem. Soc.* **1989**, *111*, 8745.

**41** Y. Nishibayashi, I. Wakiji, M. Hidai, *J. Am. Chem. Soc.* **2000**, *122*, 11019.

**42** (a) Y. Nishibayashi, I. Wakiji, Y. Ishii, S. Uemura, M. Hidai, *J. Am. Chem. Soc.* **2001**, *123*, 3393; (b) Y. Nishibayashi, G. Onodera, Y. Inada, M. Hidai, S. Uemura, *Organometallics* **2003**, *22*, 873.

**43** Y. Nishibayashi, Y. Inada, M. Hidai, S. Uemura, *J. Am. Chem. Soc.* **2003**, *125*, 6060.

**44** Y. Nishibayashi, Y. Inada, M. Hidai, S. Uemura, *J. Am. Chem. Soc.* **2002**, *124*, 7900.

**45** Y. Nishibayashi, M. Yoshikawa, Y. Inada, M. Hidai, S. Uemura, *J. Am. Chem. Soc.* **2002**, *124*, 11846.

**46** Y. Nishibayashi, M. Yoshikawa, Y. Inada, M. D. Milton, M. Hidai, S. Uemura, *Angew. Chem. Int. Ed.* **2003**, *42*, 2681.

**47** K. M. Nicholas, *Acc. Chem. Res.* **1987**, *20*, 207.

**48** (a) Y. Inada, Y. Nishibayashi, M. Hidai, S. Uemura, *J. Am. Chem. Soc.* **2002**, *124*, 15172; (b) Y. Nishibayashi, Y. Inada, M. Yoshikawa, M. Hidai, S. Uemura, *Angew. Chem. Int. Ed.* **2003**, *42*, 1495.

**49** Y. Nishibayashi, M. Yamanashi, Y. Takagi, M. Hidai, *Chem. Commun.* **1997**, 859.

**50** H. Shimada, J.-P. Qü, H. Matsuzaka, Y. Ishii, M. Hidai, *Chem. Lett.* **1995**, 671.

**51** S. Kuwata, Y. Mizobe, M. Hidai, *Inorg. Chem.* **1994**, *33*, 3619.

**52** K. D. Demadis, S. M. Malinak, D. Coucouvanis, *Inorg. Chem.* **1996**, *35*, 4038.

**53** Z. Tang, Y. Nomura, Y. Ishii, Y. Mizobe, M. Hidai, *Organometallics* **1997**, *16*, 151.

**54** (a) D. Masui, Y. Ishii, M. Hidai, *Chem. Lett.* **1998**, 717; (b) D. Masui, T. Kochi, Z. Tang, Y. Ishii, Y. Mizobe, M. Hidai, *J. Organomet. Chem.* **2001**, *620*, 69.

**55** T. Shibahara, *Adv. Inorg. Chem.* **1991**, *37*, 143; (b) T. Shibahara, *Coord. Chem. Rev.* **1993**, *123*, 73.

**56** (a) T. Murata, H. Gao, Y. Mizobe, F. Nakano, S. Motomura, T. Tanase, S. Yano, M. Hidai, *J. Am. Chem. Soc.* **1992**, *114*, 8287; (b) T. Murata, Y. Mizobe, H. Gao, Y. Ishii, T. Wakabayashi, F. Nakano, T. Tanase, S. Yano, M. Hidai, I. Echizen, H. Nanikawa, S. Motomura, *J. Am. Chem. Soc.* **1994**, *116*, 3389.

**57** D. Masui, Y. Ishii, M. Hidai, *Bull. Chem. Soc. Jpn.* **2000**, *73*, 931.

**58** T. Wakabayashi, Y. Ishii, T. Murata, Y. Mizobe, M. Hidai, *Tetrahedron Lett.* **1995**, *36*, 5585.

**59** T. Wakabayashi, Y. Ishii, K. Ishikawa, M. Hidai, *Angew. Chem. Int. Ed. Engl.* **1996**, *35*, 2123.

**60** I. Takei, Y. Wakebe, K. Suzuki, Y. Enta, T. Suzuki, Y. Mizobe, M. Hidai, *Organometallics* **2003**, *22*, 4639.

# 10

# Dirhodium Tetraphosphine Catalysts

*George G. Stanley*

## 10.1
### Dirhodium Tetraphosphine Hydroformylation Catalyst

Interest in exploring the effectiveness of multimetallic homogeneous catalysis began in earnest when Muetterties proposed the cluster-surface analogy in 1975 [1]. Of the many areas of catalysis, hydroformylation has had the largest number of reports concerning the use of multimetallic complexes as catalysts [2–3]. Hydroformylation is the largest homogeneous industrial process for converting alkenes, CO, and $H_2$ into aldehyde products (Scheme 10.1) [4].

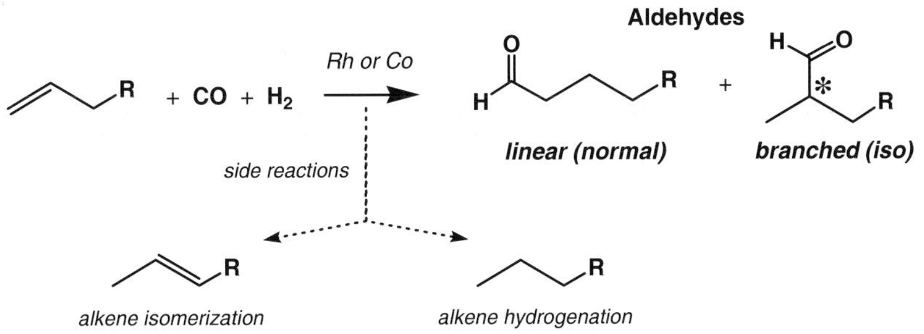

**Scheme 10.1**

Heck proposed in 1961 what is now generally considered to be the correct monometallic mechanism for $HCo(CO)_4$-catalyzed hydroformylation [5]. He also suggested, but did not favor, a bimetallic pathway involving an intermolecular hydride transfer between $HCo(CO)_4$ and $Co(acyl)(CO)_4$ to eliminate the aldehyde product (Scheme 10.2). The monometallic pathway involving the direct reaction of the acyl intermediate with $H_2$, however, has been repeatedly shown to be the dominant catalytic mechanism for 1-alkenes and cyclohexane [6, 7]. Bergman, Halpern, Norton, and Markó [8] have all performed elegant stoichiometric mechanistic studies demonstrating that intermolecular hydride transfers can indeed take place between metal-hydride and metal-acyl species under high concentration conditions to

*Multimetallic Catalysts in Organic Synthesis.* Edited by M. Shibasaki and Y. Yamamoto
Copyright © 2004 WILEY-VCH Verlag GmbH & Co. KGaA, Weinheim
ISBN: 3-527-30828-8

eliminate aldehyde products. Many of the proposals concerning polymetallic cooperativity in hydroformylation have therefore centered around the use of inter- or intramolecular hydride transfers to accelerate the elimination of aldehyde product. The use of appropriately designed ligands to constrain two (or perhaps more) metal centers into proximity to increase the probability of cooperation between them is a common approach by many workers in this area.

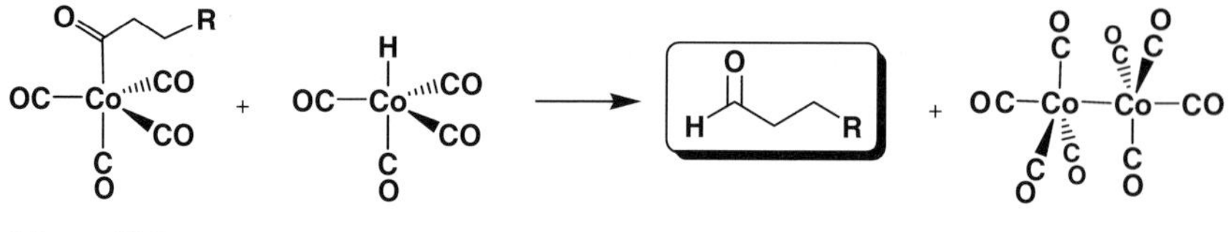

**Scheme 10.2**

The work of the author's group on bimetallic cooperativity in homogeneous catalysis has concentrated on the binucleating tetraphosphine ligands *meso-* and *racemic*-Et,Ph-P4 [9]. These ligands were designed to chelate two metal centers via a single, conformationally flexible, methylene bridge. Both "*open-mode*" bimetallic complexes, in which the metal centers are separated by 5–7 Å [10], and "*closed-mode*" systems, in which the metals are either bonded to one another or are in close contact (M–M < 3 Å), have been characterized [11]. *rac*-Et,Ph-P4, for example, reacts in very high yield with two equivalents of [Rh(nbd)$_2$]BF$_4$ (nbd = norbornadiene) to produce [*rac*-Rh$_2$(nbd)$_2$(Et,Ph-P4)](BF$_4$)$_2$, **1**, which is the precursor to a highly active and highly regioselective bimetallic hydroformylation catalyst [12].

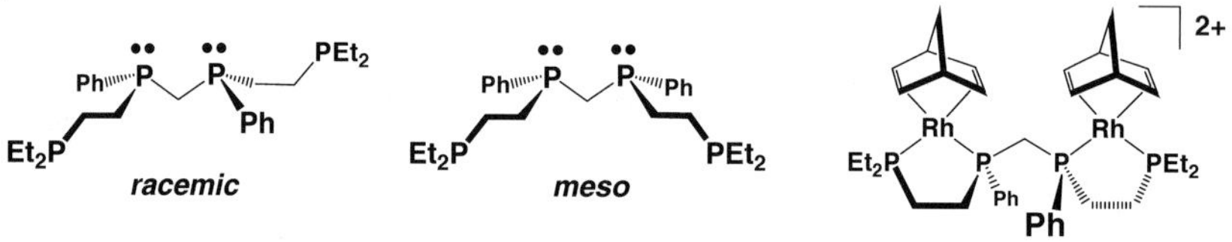

**Figure 10.1** *rac-* and *meso*-et,ph-P4 ligands; catalyst precursor [*rac*-Rh$_2$(nbd)$_2$(Et,Ph-P4)](BF$_4$)$_2$, **1**.

1-Hexene is the standard alkene used, but the results presented are typical for 1-alkenes ranging from propylene to 1-decene. In comparing **1** with the commercial Rh/PPh$_3$ catalyst (Table 10.1), the reaction with **1** is 53% faster and has a considerably higher linear-to-branched (L : B) aldehyde regioselectivity, but is accompanied by more side reactions [13]. The rate and selectivity of the reaction with the monometallic Rh/PPh$_3$ hydroformylation catalyst are quite dependent on the concentration of PPh$_3$. The minimum PPh$_3$ concentration used in industry is around 0.4 M with a 1 mM catalyst loading. The *meso* catalyst precursor generates a very poor overall hydroformylation catalyst (Table 10.1), particularly with respect to the undesirable alkene isomerization and hydrogenation side reactions. The data presented in Table 10.1 are different from those originally reported [12], with the initial turnover frequency (TOF) of the *rac*-dirhodium catalyst being almost twice as high and the process occurring with fewer side reactions. The 1-hexene used in

**Table 10.1** Results of hydroformylations of 1-hexene
(6.2 bar, 1 : 1 H$_2$/CO, 90 °C, acetone solvent, 1 mM catalyst, 1.0 M 1-hexene).

| Catalyst precursor | Initial TO/min[1] | Aldehyde l/b ratio[2] | Alkene isomerization | Alkene hydrogenation |
|---|---|---|---|---|
| rac-Rh$_2$(nbd)$_2$(Et,Ph-P4)](BF$_4$)$_2$ | 20 (1) | 25 : 1 | 2.5% | 3.4% |
| Rh(CO)$_2$(acac) + 0.40 M PPh$_3$ | 13 (1) | 9 : 1 | < 0.5% | < 0.2% |
| meso-Rh$_2$(nbd)$_2$(Et,Ph-P4)](BF$_4$)$_2$ | 0.9 (1) | 14 : 1 | 24% | 10% |

[1] Turnovers per minute (no. of moles product/no. of moles catalyst); initial rate is the initial linear part of the product production curve representing the highest catalytic rate; numbers in parentheses are standard deviations from at least four consistent runs.
[2] Linear-to-branched aldehyde product ratio based on GC analysis.

the initial studies was not purified, and peroxide impurities that it contained deactivated about half of the dirhodium catalyst. There was little effect on the monometallic Rh/PPh$_3$ catalyst (aside from some increased side reactions) due to the large excess of PPh$_3$ present, which scavenges any peroxide (or O$_2$) impurities.

Unlike aryl phosphine- or phosphite-coordinated monometallic rhodium hydroformylation catalysts, **1** does not require any excess phosphine ligand in order to maintain its selectivity or stability. The need for excess phosphine in monometallic rhodium catalysts arises from the relatively weak Rh–PPh$_3$ (or phosphite) bonding. In order to maintain the coordination of two phosphine ligands, which are required for good regioselectivity, a large excess of PPh$_3$ is required to shift the dissociation equilibrium in favor of HRh(CO)(PPh$_3$)$_2$ [14]. In **1**, the chelating and electron-donating Et,Ph-P4 phosphine ligand coordinates the rhodium centers sufficiently strongly that excess phosphine is not needed. Indeed, the addition of excess Et,Ph-P4 ligand strongly inhibits the dirhodium catalyst, as will be discussed later.

The following observations support the proposed bimetallic cooperativity. Model monometallic [Rh(nbd)(P$_2$)](BF$_4$) (P$_2$ = Et$_2$PCH$_2$CH$_2$PEt$_2$, Et$_2$PCH$_2$CH$_2$PMePh, Et$_2$PCH$_2$CH$_2$PPh$_2$, or Ph$_2$PCH$_2$CH$_2$PPh$_2$) catalyst precursors generate very poor hydroformylation catalysts for 1-hexene from both a rate and regioselectivity viewpoint (1–2 turnovers h$^{-1}$, 3 : 1 linear-to-branched aldehyde regioselectivity, 50–70% alkene isomerization and hydrogenation side reactions) [12]. The rac-Et,Ph-P4 ligand, however, links these two *very poor* monometallic catalysts together to produce a highly active and selective bimetallic catalyst.

Further persuasive evidence of bimetallic cooperativity came from bimetallic model systems in which the central methylene group in the Et,Ph-P4 ligand was replaced by p-xylylene or propyl groups, thus limiting the ability of the two rhodium centers to interact with one another. These *"spaced"* bimetallic precursors are also extremely poor hydroformylation catalysts (0.5–6 turnovers h$^{-1}$, 3 : 1 linear-to-branched aldehyde regioselectivity, 50–70% alkene isomerization and hydrogenation side reactions) [12]. The most internally self-consistent check, however, is that the *racemic* bimetallic catalyst is just over 20 times faster for the hydroformylation of 1-hexene than the *meso* catalyst, gives higher product regioselectivity, and far fewer side reactions.

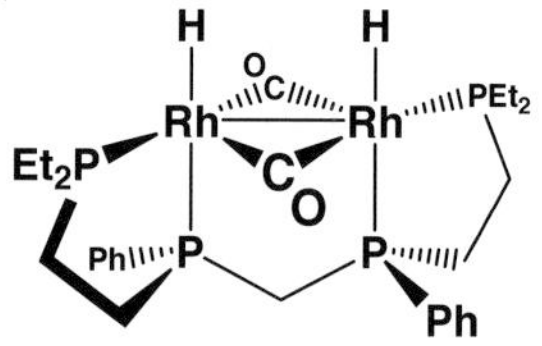

**2**

**Figure 10.2** Proposed dirhodium hydroformylation catalyst [*rac*-Rh$_2$H$_2$($\mu$-CO)$_2$(Et,Ph-P4)]$^{2+}$, **2**.

*In situ* FT-IR and NMR spectroscopic studies have led us to propose that the active catalyst is [*rac*-Rh$_2$H$_2$($\mu$-CO)$_2$(CO)$_x$(Et,Ph-P4)]$^{2+}$ ($x$ = 0, 1, or 2), **2** (Figure 10.2) [15]. The terminal CO ligands are very labile and **2** will generally be drawn without them present. The edge-sharing bioctahedral structure of **2** is unusual for dinuclear Rh$^{II}$ complexes and only a few related systems have been characterized. [Rh$_2$($\mu$-CO)($\mu$-Cl)Cl$_2$(dppm)$_2$(MeOH)]$^+$ has been structurally characterized and needs one additional terminal ligand to reach a full edge-sharing bioctahedral coordination geometry [16]. [Rh$_2$H$_2$($\mu$-H)$_2$(tripod)$_2$] (tripod = MeC(CH$_2$PPh$_2$)$_3$) has also been proposed to have a full edge-sharing bioctahedral structure, but there is scant spectroscopic or structural data on this complex [17]. There are also several examples of Rh$^{II}$ complexes with one or three bridging hydrides [18], but none of these complexes has been demonstrated to be an efficient hydroformylation catalyst.

The change in structural geometry from the far more common $D_{4h}$-like symmetry for Rh$^{II}$-Rh$^{II}$ bimetallic complexes to the unusual edge-sharing bioctahedral structure may have important consequences for the reaction chemistry of these complexes. Figure 10.3 illustrates this $D_{4h}$-like to edge-sharing bioctahedral transformation. In the $D_{4h}$-like motif the two axial ligands (*trans* to the Rh–Rh bond) are very labile. When one transforms this structure to an edge-sharing bioctahedron the axial ligand lability is now electronically distributed between the two new "axial" sites that are *trans* to the bridging ligands. Coupled with the localized cationic charges on each metal center, this electronic lability plays a key role in promoting weak rhodium bonding to the "axial" CO ligands. The dissociation of these axial CO ligands to allow coordination of the alkene substrate appears to be the rate-determining step for this catalyst, as is also the case with many monometallic hydroformylation catalysts. Unfortunately, the terminal phosphine ligands are also imparted with some of this electronic lability, and this leads to the catalyst fragmentation reactions discussed below.

**Figure 10.3** Transformation of $D_{4h}$-like bimetallic structure into an edge-sharing bioctahedral structure. The high lability of the axial L ligands in the $D_{4h}$-like structure is split between the new L and P sites in the bioctahedral structure that are approximately *trans* to the Rh–Rh bond.

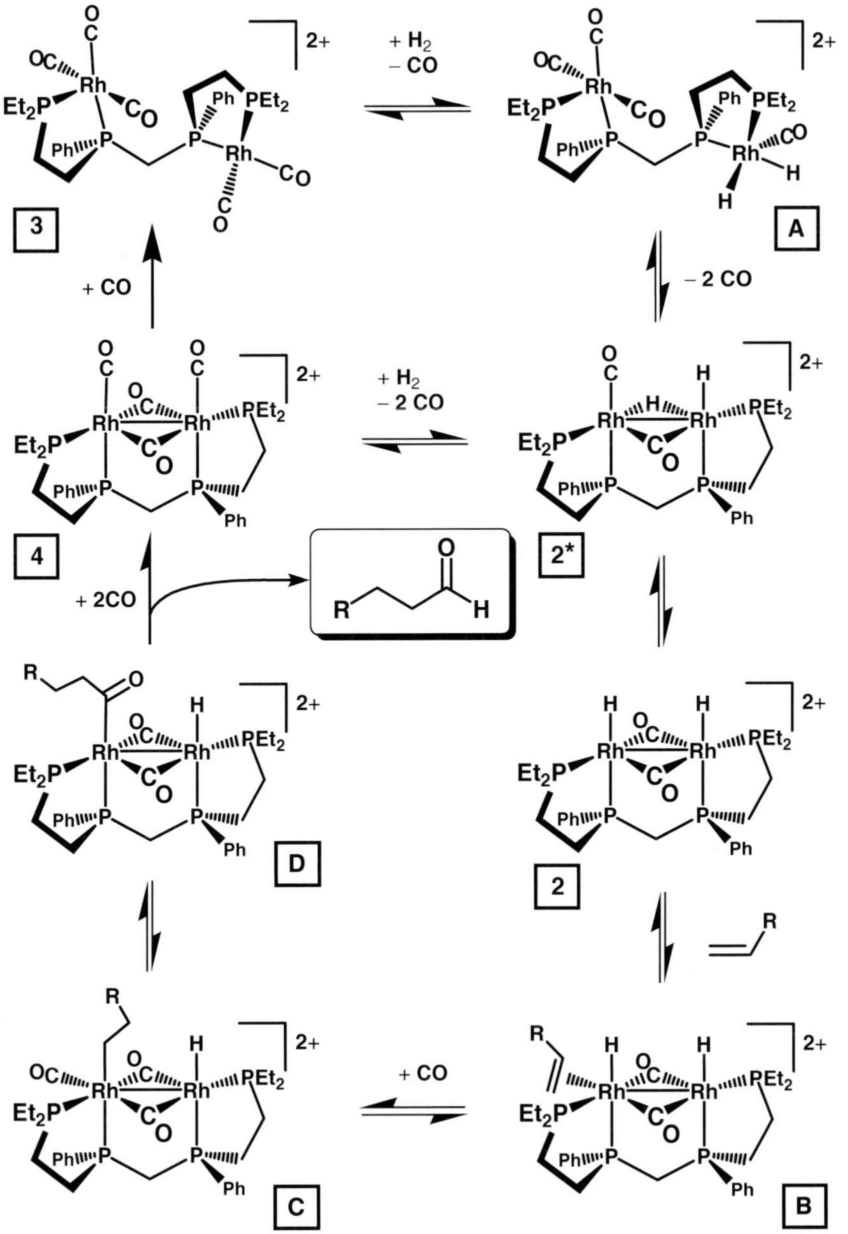

**Figure 10.4** Proposed mechanism for bimetallic hydroformylation.

The proposed mechanism for the dirhodium-catalyzed hydroformylation of alkenes is shown in Figure 10.4. The open-mode complex [$rac$-Rh$_2$(CO)$_5$(Et,Ph-P4)]$^{2+}$, **3**, can be considered as the resting state of the catalyst. We have crystallo-graphically characterized **3**, which has a five-coordinate 18e$^-$ Rh(CO)$_3$P$_2$ unit on one side and a four-coordinate square-planar 16e⁻ $Rh(CO)_2P_2$ unit on the other side. Oxidative addition of $H_2$ to the 16e⁻ Rh center, followed by loss of CO, produces the mixed oxidation state species **A**, which is drawn in Figure 10.4 as a dicationic 16e⁻ $Rh^I$/16e⁻ $Rh^{III}$ dihydride complex. It is at this point that the *first* critical bimetallic cooperativity step is proposed to take place. In monometallic hydroformylation catalysis, a cationic $Rh^{III}$ oxidation state complex such as $[H_2RhP_2(CO)_2]^+$ (P = phosphine ligand) is a very poor hydroformylation catalyst due to the presence of two hydride ligands on the one Rh center. This generally "short-circuits" hydroformylation to simple hydrogenation catalysis [19]. Complex **3** can also be deprotonated by the excess phosphine ligand present to generate the well-known neutral $Rh^I$ hydrido carbonyl complex, $HRhP_2(CO)_2$ [20]. In this bimetallic system, however, another pathway becomes dominant. One can stabilize the dicationic bimetallic dihydride species by performing an *intramolecular* hydride transfer via the bridged intermediate **2*** to form the metal–metal bonded terminal dihydride $Rh^{II}$ dinuclear complex **2**. Now, each Rh center has only a single hydride ligand and is ready to effectively perform hydroformylation.

Complex **2** can readily bind an alkene ligand to form complex **B**. Migratory insertion forms the alkyl species **C**, which can add a CO ligand and perform another migratory insertion to form the acyl complex **D**. As in our originally proposed mechanism [12], we still believe that an *intramolecular* hydride transfer plays a key role at the end of the catalytic cycle to reductively eliminate the final aldehyde product. This generates the transient CO-bridged rhodium complex $[rac-Rh_2(\mu-CO)_2(CO)_2(Et,Ph-P4)]^{2+}$, **4**, which can react with CO to reform **3**, or possibly react directly with $H_2$ to form **2***.

## 10.2
### In Situ Spectroscopic Studies – Fragmentation Problems

In situ FT-IR and NMR spectroscopic studies have provided extremely important information about the nature of the various complexes present under catalytic conditions [15]. FT-IR studies, for example, answered an early question about why the $[rac-Rh_2(nbd)_2(Et,Ph-P4)](BF_4)_2$, **1**, catalyst precursor generates a highly active and selective hydroformylation catalyst, while $rac-Rh_2(\eta^3-allyl)_2(Et,Ph-P4)$ does not [15, 21]. Figure 10.5 shows FT-IR spectra of the catalyst solutions generated from the neutral and dicationic dirhodium catalyst precursors. It is clear that two completely different sets of carbonyl-containing catalyst species are being generated from these precursor complexes. The very poor hydroformylation catalyst generated from $Rh_2(\eta^3-allyl)_2(Et,Ph-P4)$ has carbonyl stretching frequencies that are 100 cm⁻¹ lower in energy relative to those for the highly active and regioselective catalyst generated from dicationic **1** ($D_2$/CO labeling studies confirm that all the bands in the IR spectra shown are due to carbonyls). $rac-Rh_2(\eta^3-allyl)_2(Et,Ph-P4)$ reacts with $H_2$/CO quite quickly to hydroformylate off the allylic groups and produce relatively electron-rich neutral bimetallic species with the rhodium atoms in the +1 or (on average) 0 oxidation states [21]. The poor hydroformylation activity and regio-

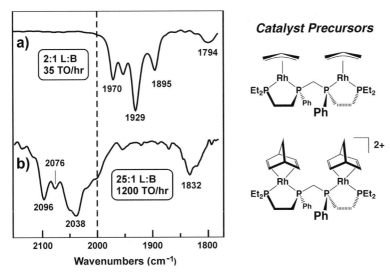

**Figure 10.5** *In situ* FT-IR spectra of the hydroformylation catalyst solutions generated from the indicated precursor species. Conditions: 90 °C, 6.2 bar, 1 : 1 $H_2$/CO.

selectivity of these electron-rich complexes is certainly consistent with previous studies that showed neutral monometallic complexes with electron-donating chelating phosphine ligands to be very poor catalysts [22].

Figure 10.5 clearly demonstrates that the highly active and selective bimetallic complexes generated from the dicationic precursor **1** have considerably lower electron densities on the rhodium atoms, as indicated by the markedly higher $\nu_{CO}$ stretching frequencies relative to the neutral bimetallic complexes generated from $Rh_2(\eta^3$-allyl$)_2$(Et,Ph-P4). The pronounced $\nu_{CO}$ bands of the bridging carbonyls and high-energy terminal carbonyl bands are consistent with the proposed active catalyst being $[rac$-$Rh_2H_2(\mu$-CO$)_2$(CO$)_2$(Et,Ph-P4$)]^{2+}$, **2**, and related complexes such as the open-mode carbonyl $[rac$-$Rh_2$(CO$)_5$(Et,Ph-P4$)]^{2+}$, **3**. The symmetrical $^{31}$P$\{^1$H$\}$ NMR spectrum and single hydride resonance (at $\delta = -7.6$ ppm) seen above 60 °C also support the proposed structure for **2**.

The dicationic charge and unusual $Rh^{II}$ oxidation state offers an ideal explanation for the remarkable hydroformylation activity and regioselectivity seen with **2**. There are, for example, no other examples of active and selective hydroformylation catalysts that have mainly alkylated, strongly donating phosphine ligands (like Et,Ph-P4). The reason for this is generally well understood. The presence of two electron-donating alkylated phosphine ligands increases the electron density on the rhodium atom leading to increased π-back-donation and stronger Rh–CO bonding. This stronger Rh–CO bonding stabilizes the unreactive 18e⁻ five-coordinate complexes $[RhH(CO)_2(P_2)]$ or $[Rh(acyl)(CO)_2(P_2)]$ ($P_2$ = two monodentate phosphines or one chelating bisphosphine). Facile CO (or phosphine) dissociation is needed to generate the catalytically active four-coordinate 16e⁻ complexes needed to allow coordination of alkene or $H_2$ to start and/or finish the hydroformylation catalytic cycle. The fact that our bimetallic catalyst is *dicationic* and has the rhodium centers in the +2 oxidation state compensates for the strongly donating Et,Ph-P4 ligand. This makes the terminal CO ligands sufficiently labile to allow good catalytic activity.

The $^{31}P\{^1H\}$ NMR spectrum of the catalyst solution generated from the dicationic precursor **1** after leaving it to stand at 25 °C under 17–21 bar of $H_2/CO$ for 24–48 h is shown in Figure 10.6. The key point is the production of two new complexes proposed to be [*rac,rac*-$Rh_2H_2(CO)_2(Et,Ph-P4)_2$]($BF_4$)$_2$, **5**, and [*rac*-$RhH_2(\kappa^4$-Et,Ph-P4)]($BF_4$), **6** [13]. Both of these new species strongly point to fragmentation of the initial [$Rh_2$(Et,Ph-P4)]$^{2+}$ catalyst complexes.

Fragmentation has been a major and ongoing problem for multimetallic catalyst systems. Longoni and co-workers reported in 1984 that the $Co_2Rh_2(CO)_{12}$ mixed-metal cluster was more active than either the parent $Co_4(CO)_{12}$ or $Rh_4(CO)_{12}$ cluster species [23]. The higher activity was proposed to be caused by heterobimetallic cooperativity between the Co and Rh centers in the homogeneous cluster. Garland [24], however, showed that the higher activity of the $Co_2Rh_2(CO)_{12}$ mixed-metal cluster was simply due to the more facile fragmentation of this cluster into the highly reactive $HRh(CO)_4$ monometallic catalyst.

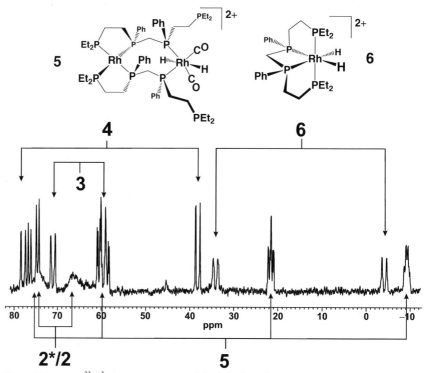

**Figure 10.6** *In situ* $^{31}P\{^1H\}$ NMR spectrum of the hydroformylation catalyst solution generated from **1** after standing for 24–48 h. Proposed structures for complexes **5** and **6** are shown above the spectrum; other structural assignments are discussed in the text. Conditions: 25 °C, 17–21 bar, 1 : 1 $H_2/CO$, acetone solvent.

Kalck and co-workers [25, 26] reported the use of a thiolate-bridged dirhodium catalyst, $Rh_2(\mu\text{-}SR)_2(CO)_4$, for hydroformylation, which was proposed to operate through homobimetallic cooperativity. An unusual aspect of this catalyst was that it showed no hydroformylation activity until $PPh_3$ was added, and then the activity and regioselectivity essentially mirrored that of $HRh(CO)_2(PPh_3)_2$. It has been shown, however, that Kalck's dirhodium complex readily fragments even under mild conditions [27] and that $PPh_3$ simply assists in producing the active $RhH(CO)_2(PPh_3)_2$ monometallic catalyst [28].

The fragmentation of $[rac\text{-}Rh_2H_2(\mu\text{-}CO)_2(CO)_2(Et,Ph\text{-}P4)]^{2+}$, **2**, to ultimately form $[rac,rac\text{-}Rh_2H_2(CO)_2(Et,Ph\text{-}P4)_2](BF_4)_2$, **5**, and $[rac\text{-}RhH_2(\kappa^4\text{-}Et,Ph\text{-}P4)](BF_4)$, **6**, is quite different from Kalck's system in that the fragmented complexes **5** and **6** are very poor hydroformylation catalysts and in fact represent a deactivation pathway for the active and selective bimetallic catalyst system. These can both be generated by dissociation of one of the external phosphine chelate arms in **1** (Scheme 10.3) to form **E**, which leads to loss of the non-chelated rhodium atom via **F**. The resulting transient monometallic complex **G** can either dimerize to form the double-ligand dirhodium complex **5**, or the Et,Ph-P4 ligand can wrap around a single rhodium ultimately forming the $\kappa^4$-coordinated complex **6**. Although we have not, as yet, isolated either of these complexes, the closely related $[rac,rac\text{-}Rh_2(Et,Ph\text{-}P4)_2]^{2+}$ and $[rac\text{-}RhCl_2(\kappa^4\text{-}Et,Ph\text{-}P4)]^+$ have been characterized [29].

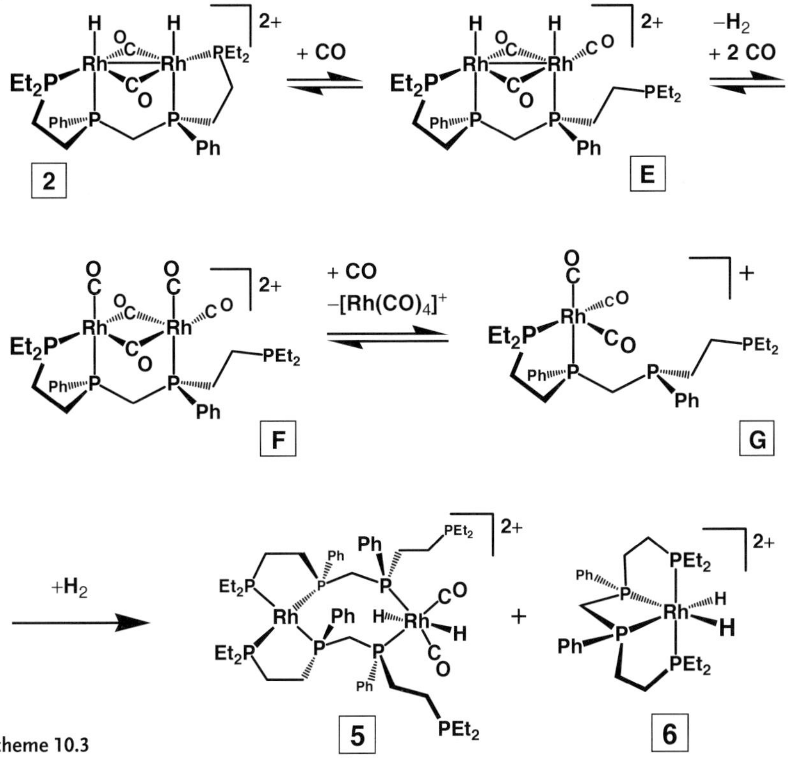

Scheme 10.3

Complex **6** appears to be a remarkably stable 18e$^-$ complex and is the only hydride species that remains after depressurization of the NMR sample followed by N$_2$ purging. Due to its saturated 18e$^-$ electronic configuration it is very unlikely that **6** functions as a catalyst. The double Et,Ph-P4 ligated dirhodium complex **5** is more reactive and we believe that it accounts for almost all of the alkene isomerization and especially the hydrogenation side reactions observed. Complex **5** would certainly be expected to be a very poor hydroformylation catalyst, which is indeed what we find. Evidence for the very poor hydroformylation ability of **5** and **6** comes from the fact that the catalyst solution rapidly deactivates on standing under 6.2 bar of H$_2$/CO at 90 °C in the absence of an alkene substrate. After 50 min under these conditions, **2** shows only about 20% of the activity of the "20 min system", and after 80 min there is essentially no remaining hydroformylation activity. In situ NMR studies show rapid growth in the signals due to **5** and **6** at higher temperatures, almost exactly paralleling the deactivation of the catalyst solution with regard to promoting hydroformylation.

## 10.3
### Effect of Water on Hydroformylation

The dirhodium catalyst **2** is rather sensitive to the nature of the solvent being used. We have studied a number of solvents for hydroformylation and found that polar solvents such as acetone or DMF work best. Given that **2** is insoluble in non-polar solvents, this isn't too surprising, but the polarity of the solvent also appears to play an important role in generating **2** itself. Although the catalyst precursor **1** is extremely soluble in CH$_2$Cl$_2$, we only observe about 25% of the activity, yet the same regioselectivity, in the hydroformylation of 1-hexene compared to the reaction in acetone. The activity of the dirhodium catalyst seems to correlate well with the intensity of the bridging carbonyl bands in the in situ IR spectrum. In CH$_2$Cl$_2$, the bridging CO bands are barely observable, while in acetone they are quite pronounced. The rotation of the open-mode species **A** to closed-mode **2*** (Figure 10.4) has an associated electrostatic barrier due to the presence of two localized cationic charges on the rhodium centers. This implies that the dielectric screening ability of the solvent could play an important role in helping to diffuse this coulombic repulsion, making the rotation to a closed-mode structure easier, thus increasing the concentration of the CO-bridged catalyst **2**.

Although the catalyst (**1** or **2**) is soluble in water, 1-hexene is not. This represents a major limitation of aqueous-phase hydroformylation using the water-soluble sulfonated triphenylphosphine (TPPTS) ligand [30]. To circumvent this problem and to study the effect of a highly polar solvent, we decided to study the effect of water/acetone mixtures on hydroformylation. Much to our surprise, the addition of 30% (*v/v*) water to acetone generated a simple hybrid polar phase solvent system that not only produced a 265% rate increase for **2** (along with considerably improved chemoselectivity in favor of aldehyde products), but also 30–115% rate enhancements for the hydroformylation of 1-hexene with a variety of monometallic rhodium phosphine catalysts [13].

**Table 10.2** Results of hydroformylations of 1-hexene (1.0 M) at 90 °C and 6.2 bar 1 : 1 $H_2/CO$ in acetone using various Rh catalysts (1.0 mM) in the presence and absence of added water.[1]

| Catalyst | % $H_2O$ | Initial TOF $(min^{-1})$ | Aldehyde L : B | % iso |
|---|---|---|---|---|
| 1 | 0 | 20 (1) | 25 : 1 | 2.5 |
| 1 | 30 | 73 (1) | 33 : 1 | < 0.5 |
| Rh/PPh$_3$[2] | 0 | 13 (1) | 9.1 : 1 | < 0.5 |
| Rh/PPh$_3$[2] | 30 | 17 (1) | 14 : 1 | 1.0 |
| Rh/Bisbi[3] | 0 | 25 (2) | 70 : 1 | < 0.5 |
| Rh/Bisbi[3] | 30 | 37 (1) | 80 : 1 | 2.0 |
| Rh/Naphos[3] | 0 | 27 (1) | 120 : 1 | 1.5 |
| Rh/Naphos[3] | 30 | 35(1) | 100 : 1 | 2.2 |
| Rh/Xantphos[3] | 0 | 13 (2) | 80 : 1 | 5.0 |
| Rh/Xantphos[3] | 30 | 28 (1) | 60 : 1 | < 0.5 |

[1] **1** generated from [*rac*-Rh$_2$(nbd)$_2$(Et,Ph-P4)](BF$_4$)$_2$, constant pressure conditions, 1000 rpm, numbers in parentheses for the initial TOF are standard deviations derived from at least four consistent runs; % iso = % alkene isomerization; hydrogenation side reactions less than 0.5% in all cases except for **1** in acetone, where they amount to 3.4%; all reactions convert 98% or more of the starting alkene.
[2] 0.4 M PPh$_3$ (400 equiv.), 1 mM Rh(CO)$_2$(acac).
[3] 5 equivalents of ligand.

The results of hydroformylations of 1-hexene using **2** and some of the best monometallic Rh-ligand combinations in acetone and 30% (*v/v*) water/acetone are presented in Table 10.2. The phosphine ligands used included PPh$_3$, Bisbi [31], Naphos [32], and Xantphos (Figure 10.7) [33]. The very effective bulky bisphosphite ligand developed by Billig and co-workers at Union Carbide (now Dow) was not included due to its sensitivity to fragmentation in the presence of water [34]. Hersh [35] has reported an extensive comparison of these and other Rh catalysts in toluene under somewhat different conditions.

Complex **2** in 30% (*v/v*) water/acetone gave the fastest initial TOF of 73 min$^{-1}$, with a 33 : 1 L : B aldehyde ratio (constant throughout the catalytic run) and virtually no alkene isomerization or hydrogenation side reactions (less than 0.2%). This is in marked contrast to **2** in pure acetone, which gave an initial TOF of 20 min$^{-1}$ and a 25 : 1 L : B aldehyde regioselectivity, along with 2.5% alkene isomerization and 3.4% hydrogenation side reactions.

**Bisbi**  **Naphos**  **Xanthphos**

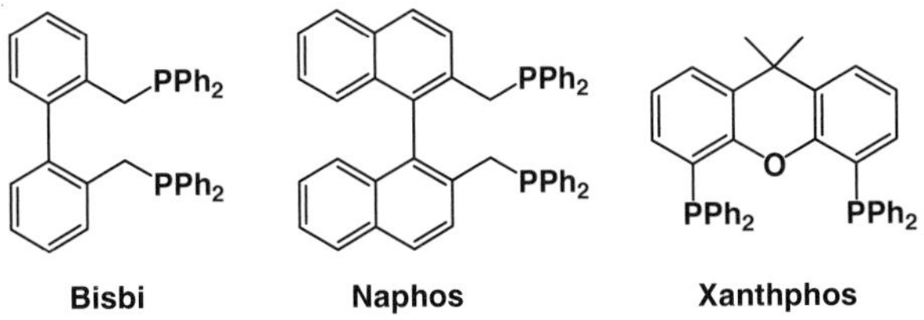

**Figure 10.7** Bisphosphine ligands used in comparison studies.

There is a dramatic decrease in rate for **2** as the water content is increased beyond 30%, with virtually no hydroformylation activity being seen above 60% water content. This is most probably due to the insolubility of 1-hexene in the increasingly polar solvent. The results demonstrate that **2** in 30% ($v/v$) water/acetone gives the highest initial TOF and the fewest side reactions, making it one of the fastest and most selective hydroformylation catalysts known. Although Rh-Naphos gives a better aldehyde L : B ratio, there is increased alkene isomerization, which lowers the overall conversion. Note that the L : B aldehyde ratio tends to exaggerate the higher selectivities: 80 : 1 corresponds to 98.8% linear, while 33 : 1 corresponds to 97.1% linear aldehyde.

The added water also has a generally beneficial effect on the monometallic catalysts. Initial TOF increases of 31, 48, and 30% are found for the $PPh_3$, Bisbi, and Naphos-based catalysts, respectively, upon addition of 30% water to the acetone solvent. The L : B aldehyde regioselectivity increases moderately with $PPh_3$ and slightly with Bisbi, but is somewhat decreased with Naphos. A somewhat higher degree of alkene isomerization is seen with $PPh_3$, Bisbi, and Naphos on addition of water. The catalytic rate is first order in alkene for each of these systems, which means that a local increase in the concentration of the non-polar alkene around the non-polar catalyst enhanced by the very polar solvent could account for the modest rate increases observed.

The increased rate and selectivity seen with the Rh–$PPh_3$ catalyst is consistent with both higher local alkene and $PPh_3$ concentrations. Higher amounts of $PPh_3$, either as a result of reduced ligand dissociation from the rhodium center or from solvent-induced local concentration build-ups, can account for the significant increase in the L : B aldehyde regioselectivity. However, this would normally be expected to decrease the rate of hydroformylation by a similar amount. It is for this reason that we believe there to be some increase in the local alkene concentration that compensates for the increase in that of $PPh_3$. Water is definitely the cause of this effect as the monometallic catalysts show the same activity and selectivity in either acetone or toluene.

Added water, however, causes a rather dramatic 115% increase in the initial TOF for the Rh-Xantphos catalyst and a marked decrease in the alkene isomerization side reaction. Van Leeuwen and co-workers [19] have characterized cationic rhodium complexes in which the central oxygen atom of the Xantphos ligand is coordinated to the metal, and have shown that these are very poor hydroformylation catalysts. We propose that the oxygen atom of the Xantphos ligand can also weakly coordinate to the neutral HRh(CO)(Xantphos) catalyst, leading to partial inhibition through the formation of a saturated five-coordinate complex such as **10** (Scheme 10.4). The added water should engage in hydrogen bonding to the oxygen atom of the coordinated Xantphos ligand, inhibiting the internal Rh–O interaction, thus generating more of the active unsaturated HRh(CO)($\eta^2$-Xantphos) catalyst **8**, which can react with alkene to initiate hydroformylation (Scheme 10.4).

Why does the added water have such a huge effect on the dirhodium catalyst **2**? This is proposed to be mainly due to effective inhibition of the fragmentation of bimetallic **2** into inactive complexes. In situ NMR spectroscopic studies, as discussed

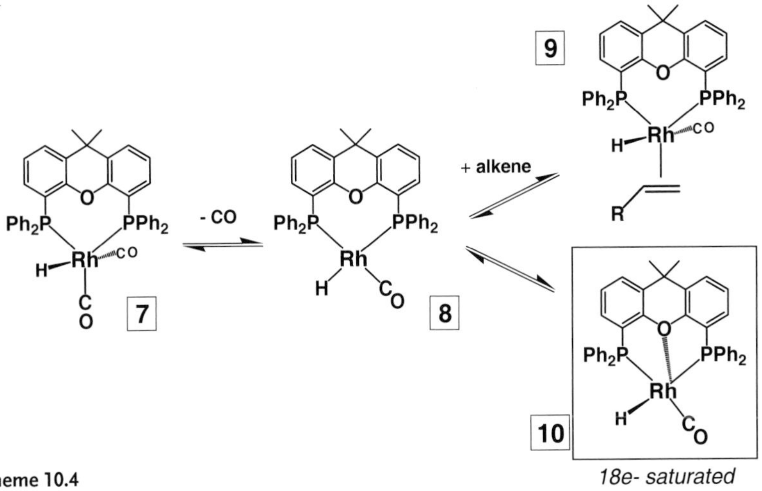

Scheme 10.4

*18e- saturated*

previously, have indicated that when **2** is left to stand under $H_2/CO$, fragmentation occurs to produce [*rac,rac*-$Rh_2H_2(CO)_2(Et,Ph$-$P4)_2](BF_4)_2$, **5**, and [*rac*-$RhH_2(\kappa^4$-$Et,Ph$-$P4)](BF_4)$, **6**, both of which are very poor hydroformylation catalysts. Scheme 10.5 shows the proposed fragmentation reaction sequence that leads to the formation of **5** and **6**. The first step is the dissociation of one of the external phosphine chelate arms to produce complex **E**. This, we believe, is inextricably linked with the $Rh^{II}$ oxidation state and the unusual edge-sharing bioctahedral structure that labilizes the two terminal coordination sites *trans* to the bridging ligands. The dissociation of one of the phosphine ligands reduces the electron density on the Rh centers, promoting the reductive elimination of $H_2$ to produce **F**. The carbonyl ligands at this point can effectively pry loose one of the rhodium centers leading to the monometallic complex **G**, which can either dimerize to form **5**, or have the *rac*-Et,Ph-P4 ligand wrap around the remaining Rh center in a $\kappa^4$-fashion to produce the monometallic complex **6**.

The dramatically enhanced activity and reduced side reactions in the presence of water is consistent with inhibition of the initial phosphine dissociation from **2** that ultimately leads to catalyst fragmentation. $HRh(CO)(TPPTS)_3$, for example, has considerably slower phosphine dissociation equilibria in water relative to $HRh(CO)$ $(PPh_3)_3$ in organic solvents [36]. A similar effect inhibiting the dissociation of the "non-polar" $PEt_3$-like chelate arm into the highly polar water/acetone solvent is proposed here. Inhibiting the first step of the fragmentation reaction sequence shown in Scheme 10.5 is the best way to maintain a high concentration of the active catalyst **2**. Preliminary *in situ* $^{31}P$ NMR studies point to dramatically reduced formation of fragmentation products **5** or **6** in 30% (*v/v*) water/acetone.

As discussed above, prior to alkene addition, the catalyst activity steadily decreases in pure acetone at 90 °C under 5.4–6.1 bar of $H_2/CO$. After 50 min, **2** has only about 20% of the activity of the "20 min system", and after 80 min the system no longer shows any hydroformylation activity. The stabilizing effect of the added water

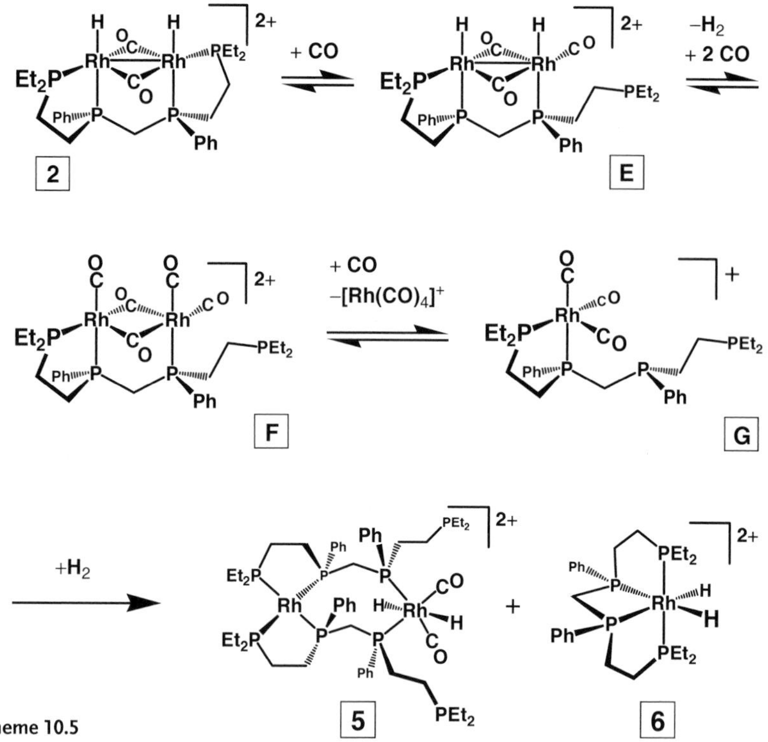

Scheme 10.5

is clearly illustrated with 30% ($v/v$) water/acetone, in the case of which the catalyst can be left to stand under $H_2$/CO at 90 °C for 2 h with only a small loss in activity (10%). In fact, during the initial ~15–20 min required to heat the autoclave to the operating temperature of 90 °C, the catalyst steadily fragments and deactivates in acetone. Thus, different initial heating times can lead to considerable fluctuations in the initial TOF for **2** when using acetone. The improved stability of the catalyst in 30% ($v/v$) water/acetone is further illustrated by the fact that one can easily perform 10,000 turnovers using 0.1 mM catalyst and 1.0 M 1-hexene (initial TOF = 60(3) min$^{-1}$, L : B = 29 : 1, 2% alkene isomerization, > 0.1% alkene hydrogenation).

Phase separation of the product heptaldehyde occurs in the catalytic runs when the water content amounts to 20% or more of the acetone volume, which was one of the reasons why a polar phase solvent system was initially studied. Unfortunately, the dirhodium catalyst is more soluble in the heptaldehyde organic layer than in the water/acetone solvent. Moreover, while the added water considerably stabilizes **2**, it does not completely stop the fragmentation reactions. New, more strongly chelating tetraphosphine ligands are currently being prepared, which, in combination with highly polar solvents, should lead to extremely active and robust dirhodium catalysts.

## 10.4
## Unusual Inhibitory Effect of PPh₃

The PPh₃-induced fragmentation of Kalck's bimetallic $Rh_2(\mu\text{-SR})_2(CO)_4$ complex to produce the active monometallic $HRh(CO)(PPh_3)_2$ hydroformylation catalyst prompted us to study the effect of $PPh_3$ on **2**. The catalytic results for the hydroformylation of 1-hexene starting with the bimetallic precursor **1** and monometallic precursor $Rh(acac)(CO)_2$ (or $[Rh(nbd)_2]^+$) in the presence of varying amounts of $PPh_3$ are shown in Table 10.3. In most runs, minimal alkene hydrogenation side reactions (less than 1%) were observed. An exception is the reference bimetallic catalyst **2** that gives 3.4% alkene hydrogenation, which we believe is mainly caused by the fragmentation product **6**.

The addition of two equivalents of $PPh_3$ to **2** causes a dramatic drop in the L : B aldehyde regioselectivity from 25 to 3.1 and a considerable reduction in the initial TOF to 590 $h^{-1}$. The addition of increasing amounts of $PPh_3$ leads to steadily

**Table 10.3** Hydroformylation of 1-hexene (1 M) in acetone at 90 °C and 6.2 bar 1 : 1 $H_2/CO$ using the bimetallic precursor [rac-$Rh_2(nbd)_2(Et,Ph\text{-}P4)$]$(BF_4)_2$, monometallic $Rh(acac)(CO)_2$, or cationic monometallic [$Rh(nbd)_2$]$(BF_4)$ (all 1 mm).

| Precursor | Equiv. PPh₃[1] | Equiv. PEt₃[1] | Initial TOF[2] (h⁻¹) | Aldehyde L : B | % Alkene isomer[3] |
|---|---|---|---|---|---|
| Bimetallic | 0 | 0 | 1200 | 25 | 2.5 |
| Bimetallic | 0.5 | 0 | 660 | 3.4 | 3.0 |
| Bimetallic | 2 | 0 | 590 | 3.1 | 5.0 |
| Bimetallic | 5 | 0 | 320 | 2.9 | 3.0 |
| Bimetallic | 10 | 0 | 220 | 3.2 | 3.3 |
| Bimetallic | 100 | 0 | 0 | – | 0 |
| Mono | 0 | 0 | 0 | – | 0 |
| Mono | 2 | 0 | 2200 | 2.6 | 6.0 |
| Mono | 5 | 0 | 5300 | 3.2 | 2.9 |
| Mono | 10 | 0 | 6800 | 3.2 | 2.0 |
| Mono | 100 | 0 | 6300 | 5.0 | < 1 |
| Mono | 400 | 0 | 700 | 9.1 | < 1 |
| Mono | 10 | 1.1 | 3100 | 3.1 | 2.0 |
| Mono | 100 | 1.1 | 780 | 4.9 | 1.5 |
| Mono | 10 | 2.2 | 2000 | 3.0 | < 1 |
| Mono | 100 | 2.2 | 450 | 3.6 | < 1 |
| Mono(+)[4] | 10 | 0 | 2 | 1.6 | 80 |
| Mono(+)[4] | 100 | 0 | 300 | 4.2 | 1.8 |
| Mono(+)[4] | 10 | 2.2 | 180 | 3.3 | 1.7 |
| Mono(+)[4] | 100 | 2.2 | 180 | 3.1 | < 1 |

[1] Equivalents of PPh₃ or PEt₃ added relative to the amount of rhodium catalyst precursor.
[2] TOF = turnover frequency, an average of at least three consistent runs; rates are subject to approximately 5% error and have been rounded off to reflect this.
[3] Isomerization (hydrogenation side reactions considerably less than 1% in all cases).
[4] Cationic precursor [Rh(nbd)₂](BF₄) (nbd = norbornadiene).

decreasing catalytic rates until at 100 equivalents essentially no hydroformylation is observed. This is almost completely opposite behavior to that seen with Kalck's bimetallic $Rh_2(\mu\text{-}SR)_2(CO)_4$ system, which is inactive until $PPh_3$ is added, and then the rate and selectivity increase with added $PPh_3$ (up to a point).

Even the addition of 0.5 equivalents of $PPh_3$ causes a dramatic drop in the regioselectivity and catalyst activity, as shown in Table 10.3. The steady deactivation and essentially constant low regioselectivity of the bimetallic system with increasing amounts of $PPh_3$ is not at all consistent with simple fragmentation to a monometallic Wilkinson-like catalyst system. This is illustrated by the parallel runs with the monometallic precursor $Rh(acac)(CO)_2$ and $PPh_3$ (see Table 10.3). At low $PPh_3$ ratios one generates an extremely active but short-lived catalyst with relatively low selectivity. At 100 equivalents of $PPh_3$, the catalyst action starts to slow down and the aldehyde L : B regioselectivity begins to increase. At 400 equivalents (0.4 M) of $PPh_3$, the catalyst initial TOF has decreased to 700 $h^{-1}$, but the L : B regioselectivity has climbed to 9.1. Commercial $Rh/PPh_3$ hydroformylation processes are typically run with at least 0.4 M $PPh_3$ for propylene, but with up to 50% $PPh_3$ by solution weight when the highest L : B aldehyde regioselectivities are needed.

The L : B aldehyde regioselectivity data clearly indicate that the addition of even small amounts of $PPh_3$ strongly inhibits the highly regioselective bimetallic catalyst **2** and generates an alternative catalyst system with much lower regioselectivity. The rate and regioselectivity data, however, indicate that this alternative catalyst is not $HRh(CO)_4$, $HRh(CO)_2(PPh_3)$, or $HRh(CO)(PPh_3)_2$. In Scheme 10.5, we proposed that some species simply referred to as "$[Rh(CO)_4]^+$" is being released into the catalyst solution. If so, this could be intercepted by $PPh_3$, if this ligand is present in high enough concentrations, to produce the standard $HRh(CO)(PPh_3)_2$ catalyst. Once again, however, if this were the case, the rate and regioselectivity would be expected to increase as the $PPh_3$ concentration is increased within the range studied. Instead, in the bimetallic catalytic run with 100 equivalents of $PPh_3$, essentially no hydroformylation catalytic activity was observed, which is completely inconsistent with fragmentation to an $HRh(CO)(PPh_3)_2$-type catalyst.

While the added $PPh_3$ ligand can probably simply inhibit the active bimetallic catalyst, its far more critical role appears to be in transforming the dirhodium catalyst into another structure that has low regioselectivity and moderate activity, but one that is eventually completely inhibited by increasing amounts of $PPh_3$. The most likely entry point for disrupting the formation of the Rh–Rh bonded complex **2** involves the starting carbonyl complex **3** (Figure 10.4). $PPh_3$ coordination could easily disrupt the rotation to **2\*** after $H_2$ oxidative addition. Molecular modeling studies indicate that this open- to closed-mode rotation is strongly influenced by steric effects. For example, the simple tetracarbonyl $[rac\text{-}Rh_2(CO)_4(Et,Ph\text{-}P4)]^{2+}$ (or pentacarbonyl **3**) shows no sign of rotation to give a bridged-carbonyl structure like **4** in the presence of CO. Only when $H_2$ is added are rhodium complexes with bridging carbonyls observed. Indeed, the activity of the catalyst appears to be directly correlated with the presence and intensity of the bridging carbonyl bands in the IR spectrum. Modeling studies clearly demonstrate that it is much easier for the bimetallic complex to rotate into a closed-mode structure when there are two small hydride ligands present on one rhodium center instead of carbonyls. If a PPh$_3$ ligand were present on one or both of the Rh atoms, this could easily sterically block the ability of the complex to form a closed-mode Rh–Rh bonded structure 2* or 2, which we believe to be necessary for high catalyst regioselectivity (and activity).

The coordination of PPh$_3$ is proposed to open up the bimetallic catalyst chelate structure in 3 to form, in essence, a monometallic catalyst center with one alkylated phosphine (PEt$_3$) and a PPh$_3$. This is shown in Scheme 10.6 for one half of 3, although it could, of course, happen to both sides of the bimetallic complex. A proton must be lost from the cationic Rh dihydride in order to generate the neutral Rh–H side of the complex. Cationic monometallic precursors such as [Rh(nbd)$_2$]$^+$ (Table 10.3, entries 17–20) typically generate poor hydroformylation catalysts unless the cationic Rh$^{III}$ dihydride (e.g., [RhH$_2$(CO)$_2$(PPh$_3$)$_2$]$^+$) produced from the oxidative addition of H$_2$ can be deprotonated [19, 20]. PPh$_3$ is just basic enough to deprotonate the [RhH$_2$(CO)$_2$(PPh$_3$)$_2$]$^+$ complex formed from [Rh(nbd)$_2$]$^+$, with 10 equivalents of PPh$_3$ giving a catalyst that is barely active for hydroformylation, but quite good at alkene isomerization. The addition of 100 equivalents of PPh$_3$, however, does shift the deprotonation equilibrium sufficiently to give modest hydroformylation activity (initial TOF = 300 h$^{-1}$) and dramatically reduced alkene isomerization side reactions.

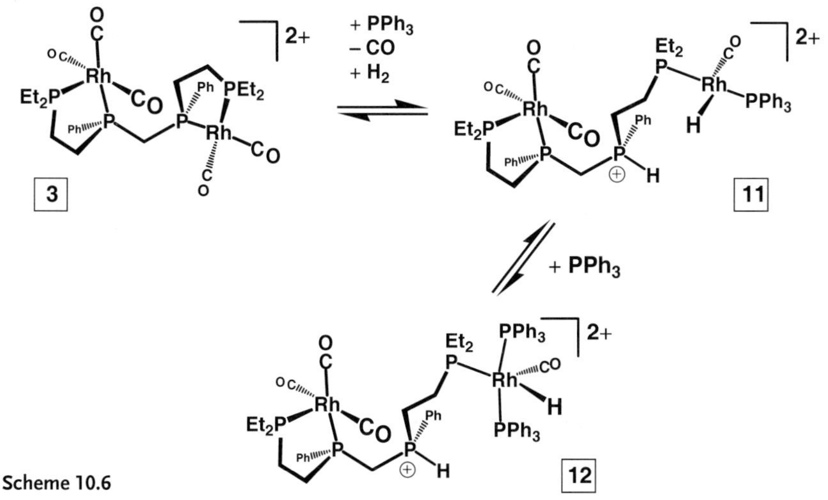

**Scheme 10.6**

The dirhodium catalyst normally avoids the cationic rhodium dihydride "trap" by undergoing an intramolecular hydride transfer from complex A in Figure 10.4 to ultimately form the symmetrical bimetallic dihydride 2 with one hydride per Rh center. While PPh$_3$ can deprotonate [RhH$_2$(CO)$_2$(PPh$_3$)$_2$]$^+$ when present in high enough concentrations, it is probably not as effective for a less acidic cationic dihydride precursor incorporating the more strongly donating Et,Ph-P4 ligand. The considerably more basic Et,Ph-P4 ligand, however, should be able to act as a suitable intramolecular base for this deprotonation. Indeed, the internal free phosphine in 11 (Scheme 10.6) is well situated to act as an intramolecular deprotonating agent, which is how we have drawn it for the proposed "mono-metallic" catalyst **11**.

The validity of **11** as a model for rationalizing the effect of added $PPh_3$ on the dirhodium catalyst was tested by adding 1.1 and 2.2 equivalents of $PEt_3$ to the neutral and cationic monometallic precursor catalytic reactions. In the case of $Rh(CO)_2(acac)$, the addition of $PEt_3$ causes the expected dramatic drop in activity and a considerably smaller (sometimes negligible) drop in regioselectivity (Table 10.3). For example, the initial TOF with 100 equivalents of $PPh_3$ drops from 6300 $h^{-1}$ to 450 $h^{-1}$ when 2.2 equivalents of $PEt_3$ are added, while the L : B aldehyde regioselectivity is reduced from 5.0 to 3.6.

The deprotonating ability of 2.2 equivalents of $PEt_3$ (one equivalent to deprotonate, one equivalent to coordinate to the Rh) was tested on the cationic $[Rh(nbd)_2]^+$ precursor system and it was found to increase the initial rate with 10 equivalents of $PPh_3$ by almost two orders of magnitude, from 2 $h^{-1}$ up to 180 $h^{-1}$. The hydro-formylation run using $Rh(CO)_2(acac)$ with 10 equivalents of $PPh_3$ and 1.1 equivalents of $PEt_3$ showed an initial rate of 3100 $h^{-1}$. This clearly demonstrates that $PEt_3$ only partially deprotonates the cationic rhodium complex and that there is an equilibrium between the cationic and neutral catalyst species. These simple model studies offer reasonable support for the proposed formation of an inefficient monometallic-like catalyst such as **11**.

The essentially complete inhibition of hydroformylation upon addition of 100 equivalents of $PPh_3$ is consistent with the coordination of a second molecule of $PPh_3$ to generate a sterically hindered $HRh(\kappa^1\text{-Et,Ph-P4})(PPh_3)_2$-like complex represented by **12** (Scheme 10.6), which should be essentially inactive for hydro-formylation catalysis. Note that $HRh(CO)(PPh_3)_3$ only forms at considerably higher $PPh_3$ concentrations and is subject to facile ligand dissociation in order to relieve the steric strain of having three bulky phosphines coordinated in a square-planar environment. The smaller "$PEt_2$" arm of our ligand and low CO pressures should make the catalytically inactive $HRh(\kappa^1\text{-Et,Ph-P4})(PPh_3)_2$-like complex considerably more stable at the $PPh_3$ concentrations used in these studies.

The inability of the very simple $PEt_3$ model systems to mimic the zero activity observed when 100 equivalents of $PPh_3$ are added to **1** is reasonable given that $PEt_3$ is certainly not the same as a $\kappa^1$-Et,Ph-P4 ligand having a cationic $Rh(CO)_3$ (or $PPh_3$-substituted) moiety coordinated to the other side, as proposed for **12**. The dramatic regioselectivity lowering effect of even 0.5 equivalents of added $PPh_3$ points to extremely efficient inhibition of the active and selective homobimetallic catalyst, which is probably only present in relatively small amounts compared to the total amount of catalyst precursor being used.

## 10.5
### Catalyst Binding Site Considerations

The high product aldehyde regioselectivity observed for the dirhodium catalyst (33 : 1 linear-to-branched for 1-hexene using water/acetone solvent) is a result of the relatively rigid bimetallic structure of **2** and an appropriate set of steric effects on the Et,Ph-P4 ligand and catalyst. When an alkene coordinates to a typical mono-metallic square-planar hydroformylation catalyst, the other ligands will bend away to form a trigonal bipyramid or square pyramid, which is the least congested coordination geometry (Scheme 10.7). This geometric reorganization results from electronic orbital rehybridization on the metal center and causes the steric directing R-groups on the phosphine ligands to swing away from the incoming alkene substrate. This, in turn, reduces the steric effectiveness of the phosphine for orientating the alkene so that it is inserted into the Rh–H bond in the appropriate manner to give the desired linear (or, in the case of asymmetric hydroformylation, branched) alkyl intermediate species. This geometric reorganization is the basis for the bite-angle hypothesis for monometallic hydroformylation catalysts developed by Casey [37] and extensively explored by van Leeuwen [38]. Larger chelate rings (i.e., larger P–Rh–P bite angles) force the phosphorus atoms in the five-coordinate equatorial-equatorial trigonal-bipyramidal structure shown in Scheme 10.7 to be pushed more towards the coordinated alkene. This causes a larger steric interaction and orientates the alkene in such a way as to favor a migratory hydride insertion to produce a linear alkyl group.

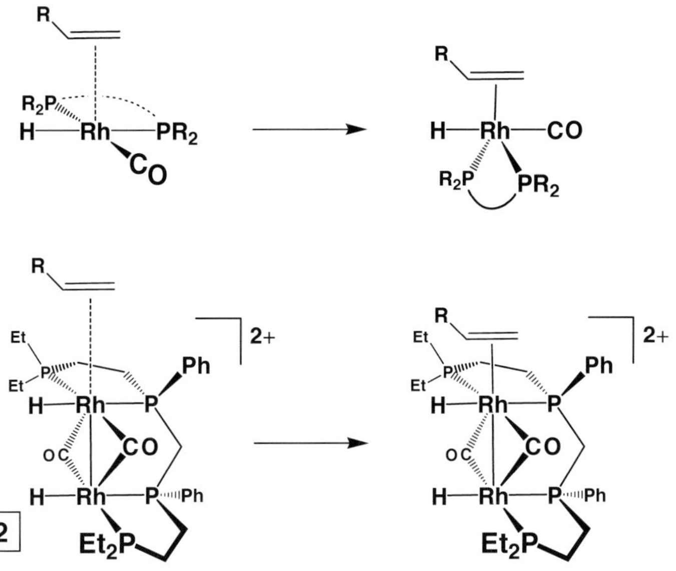

**Scheme 10.7**

The bimetallic catalyst **2**, however, cannot distort like a $d^8$ square-planar complex upon alkene coordination because the Rh–Rh bond and bridging CO ligands prevent any significant movement of the ligand environment away from the alkene. Minimizing the geometric reorganization about the rhodium maximizes the steric effect of the Et,Ph-P4 ligand, directing the alkene insertion into the M–H bond in favor of a linear alkyl group, which ultimately forms the linear aldehyde product.

The Rh–Rh bond and bridging carbonyl ligands play a critical role in defining and enhancing the steric factors present in the alkene binding site on **2**. There are no regioselective monometallic hydroformylation catalysts with phosphine ligands that have the small R groups present in Et,Ph-P4 (ethyl and phenyl). Except for a minor increase in the case of 1-hexene, **2** has essentially constant linear-to-branched regioselectivity across a fairly broad series of alkenes: propylene (20 : 1), 1-butene (20 : 1), 1-pentene (23 : 1), 1-hexene (25 : 1), 1-heptene (21 : 1), and 1-octene (21 : 1). This behavior is quite unusual compared to monometallic hydroformylation catalysts, which tend to show a considerably larger regioselectivity range that increases with longer chain alkene substrates.

These data indicate that the binding site on **2** has little conformational flexibility and is extremely well-defined. This may explain why we get such good agreement between experiment and the molecular modeling calculations, which also demonstrate minimal binding site reorganization on coordination of the alkene substrate. There is one other family of dinuclear complexes that have similarly well-defined binding site properties, namely Doyle's impressive asymmetric cyclopropanation catalysts based on dinuclear Rh$^{II}$ oxidation state complexes with chiral carboxamide and other related ligands [39].

## 10.6
## Future Directions

One of the important ongoing projects associated with this work is the use of Density Functional Theory (DFT, Gaussian 98 program) to calculate essentially all the structures and reaction steps (including transition states) in the proposed bimetallic mechanism shown in Figure 10.4. We have calculated optimized structures for most of these and the DFT results have so far been supportive of the structures of all of the proposed dirhodium species, including the presence of an Rh–Rh single bond in **2**. The relative energies for part of the catalytic cycle are as follows: **A** (30 kcal) → **2\*** (0 kcal) a ⇌ **2** (11 kcal). These energetics are qualitatively consistent with the findings of in situ FT-IR and NMR spectroscopic studies, where we have never seen any sign of the relatively high energy open-mode dihydride complex **A**. There is a rapid dynamic equilibrium, however, between **2\*** and **2**, with the hydridocarbonyl-bridged complex **2\*** being favored at low temperatures. The symmetrical carbonyl-bridged terminal dihydride complex **2** becomes the major species at temperatures > 60 °C.

The preliminary DFT results also support our proposed direct bimetallic reductive elimination of either H$_2$ from **2**, or of aldehyde from species **D** (Figure 10.4). Although these are formally symmetry-forbidden processes, the DFT results are in line with the extended Hückel results on bimetallic reductive eliminations obtained by Hoffmann [40], who found that bridging carbonyls and a lower symmetry of the dimetal unit tends to favor the direct bimetallic reductive elimination process. We have also been able to calculate the CO stretching frequencies for most of these complexes and obtained reasonably good agreement with the experimental IR spectra for both the terminal and bridging Rh–CO bands.

Research with the water/acetone solvent mixture for hydroformylation has led to the remarkable and serendipitous discovery of the new catalytic reaction shown in Scheme 10.8, namely aldehyde-water shift catalysis. This is the reaction of aldehyde and water to produce carboxylic acid and $H_2$. There are no formal reports of this catalytic reaction in the literature, although David Tyler (University of Oregon, USA, personal communication) has observed that this reaction is very slowly catalyzed by a $[(Cp'_2Mo(OH)(H_2O)]^+$ $(Cp' = C_5H_4CH_3)$ complex. It is analogous to the well-known water-gas shift reaction for the conversion of CO and $H_2O$ to make $CO_2$ and $H_2$. The thermodynamics for the aldehyde-water shift catalysis is favorable, with $\Delta G°_{rxn} = -28.4$ kJ mol$^{-1}$ at 90 °C.

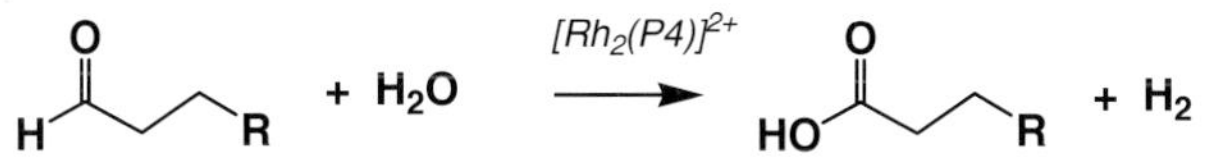

**Scheme 10.8**

The initial TOF for this catalysis in 30% $(v/v)$ water/acetone at 90 °C and under 6.2 bar of CO (small amounts of $H_2$ need to be present) is very fast, around 30 min$^{-1}$, with almost complete selectivity in favor of the carboxylic acid product. This catalysis is inhibited by $H_2$, so the $H_2$ being produced in the catalysis typically prevents complete conversion to the carboxylic acid unless adequate purging is in operation. This reaction can be carried out in tandem with hydroformylation in a single reactor to effectively perform hydrocarboxylation catalysis (Scheme 10.9) under very mild conditions, and gives > 30 : 1 linear-to-branched selectivity in favor of the linear heptanoic acid product when starting with 1-hexene. In a closed autoclave (no purging), we can routinely achieve 75% conversion to carboxylic acid when starting with 1000 equivalents of 1-hexene (1 mM dirhodium catalyst), before the $H_2$ also being produced inhibits the aldehyde-water shift catalysis.

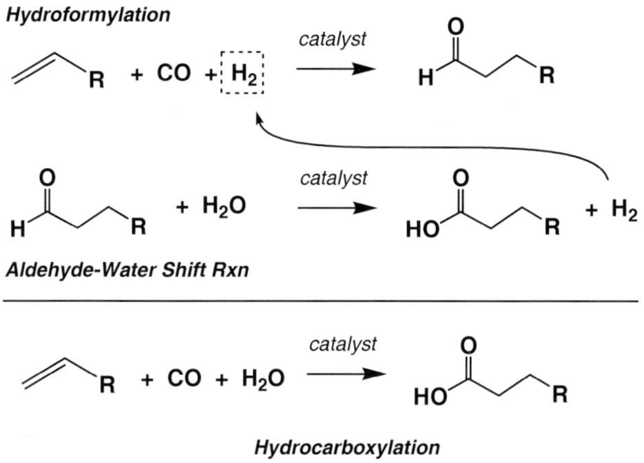

**Scheme 10.9**

Under hydrogen-deficient conditions (initially generated through a fortuitous leak in the autoclave plumbing!), it appears that the dirhodium catalyst system shifts away from the various hydride species back towards the simpler rhodium carbonyl complexes **3** and **4** (Figure 10.4). We propose that the key catalyst species is the CO-bridged dirhodium complex $[rac\text{-}Rh_2(\mu\text{-}CO)_2(CO)_2(Et,Ph\text{-}P4)]^{2+}$, **4**, and that bimetallic cooperativity also plays an important role in the aldehyde-water shift catalysis. DFT calculations on **4** show a remarkable LUMO that has the empty Rh $p_z$ orbitals forming a strong bonding interaction with the terminal and bridging carbonyl $\pi^*$ orbitals, as shown schematically in Figure 10.8. The strong mixing of the bridging carbonyl $\pi^*$ systems with both of the Rh $p_z$ orbitals lowers the energy of this LUMO by 0.8 eV relative to the LUMO in the open-mode (non-CO bridged) isomer $[rac\text{-}Rh_2(CO)_4(Et,Ph\text{-}P4)]^{2+}$. This makes the LUMO in **4** a considerably better acceptor orbital, which, in conjunction with the dicationic charge, can interact with the occupied $\pi$-system of the aldehyde carbonyl group, activating it for nucleophilic attack by water, a key step in the catalytic reaction.

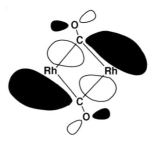

**Figure 10.8** Schematic representation of the LUMO for 4 based on DFT calculations, showing the strong bonding interaction of the Rh $p_z$ orbitals with the carbonyl $\pi^*$ system. Only the in-plane bridging carbonyl interactions are shown for clarity.

The construction of a flow reactor set-up with the aim of achieving 100% conversion in the aldehyde-water shift catalysis, as well as optimization of the reaction conditions, is underway. In situ FT-IR and NMR mechanistic studies are also planned and should be highly useful in helping us understand how this catalysis works. An intriguing application for the aldehyde-water shift catalysis is the conversion of formaldehyde and water to produce formic acid and $H_2$. Formic acid, in turn, can be readily decomposed to $CO_2$ and $H_2$. Formaldehyde/water could, therefore, serve as a high energy density liquid for producing two equivalents of $H_2$ (and one of $CO_2$) for hydrogen fuel cells. We are also studying this reaction using **4** and related variants. This new and unexpected chemistry opens a number of doors for related organic reactions that could be effectively catalyzed by bimetallic complexes.

# References

1. (a) E. L. Muetterties, *Bull. Soc. Chim. Belg.* **1975**, *84*, 959; (b) E. L. Muetterties, *Bull. Soc. Chim. Belg.* **1976**, *85*, 451; (c) E. L. Muetterties, *Angew. Chem. Int. Ed. Engl.* **1978**, *17*, 545; (d) E. L. Muetterties, M. J. Krause, *Angew. Chem. Int. Ed. Engl.* **1983**, *22*, 135.

2. (a) R. C. Ryan, C. U. Pittman, Jr., J. P. O'Connor, *J. Am. Chem. Soc.* **1977**, *99*, 1986; (b) C. U. Pittman, Jr., G. M. Wilemon, W. D. Wilson, R. C. Ryan, *Angew. Chem. Int. Ed. Engl.* **1980**, *19*, 478; (c) M.-J. Don, M. G. Richmond, *J. Mol. Catal.* **1992**, *73*, 181.

3. G. Süss-Fink, G. F. Schmidt, *J. Mol. Catal.* **1987**, *42*, 361.

4. Cf. (a) B. Cornils, in *New Syntheses with Carbon Monoxide* (Ed.: J. Falbe), Springer-Verlag, Berlin, 1980; (b) G. G. Stanley, in *Encyclopedia of Inorganic Chemistry* (Ed.: B. King), Wiley, New York, 1994; (c) P. W. N. M. van Leeuwen, C. Claver (Eds.), *Rhodium-Catalyzed Hydroformylation*, Kluwer Academic, Dordrecht, 2000.

5. R. F. Heck, D. S. Breslow, *J. Am. Chem. Soc.* **1961**, *83*, 4023.

6. M. F. Mirbach, *J. Organomet. Chem.* **1984**, *265*, 205.

7. W. R. Moser, in *ACS Advances in Chemistry Series* (Eds.: W. R. Moser, D. W. Slocum), American Chemical Society, Washington, D.C., Vol. 230, **1992**, chapter 1, p. 14.

8. (a) W. D. Jones, J. M. Huggins, R. G. Bergman, *J. Am. Chem. Soc.* **1981**, *103*, 4415; (b) M. J. Nappa, R. Santi, J. Halpern, *Organometallics* **1985**, *4*, 34; (c) B. D. Martin, K. E. Warner, J. R. Norton, *J. Am. Chem. Soc.* **1986**, *108*, 33; (d) F. Ungváry, L. Markó, *Organometallics* **1983**, *2*, 1608.

9. S. A. Laneman, F. R. Fronczek, G. G. Stanley, *J. Am. Chem. Soc.* **1988**, *110*, 5585.

10. S. A. Laneman, F. R. Fronczek, G. G. Stanley, *Inorg. Chem.* **1989**, *28*, 1872.

11. S. A. Laneman, F. R. Fronczek, G. G. Stanley, *Inorg. Chem.* **1989**, *28*, 1206.

12. M. E. Broussard, B. Juma, S. G. Train, W. J. Peng, S. A. Laneman, G. G. Stanley, *Science* **1993**, *260*, 1784.

13. D. A. Aubry, N. N. Bridges, K. Ezell, G. G. Stanley, *J. Am. Chem. Soc.* **2003**, *125*, 11180–11181.

14. (a) R. L. Pruett, J. A. Smith, *J. Org. Chem.* **1969**, *34*, 327; (b) R. V. Kastrup, J. S. Merola, A. A. Oswald, in *Catalytic Aspects of Metal Phosphine Complexes* (Eds.: E. C. Alyea, D. W. Meek), *Advances in Chemistry Series*, American Chemical Society, Washington, D.C., Vol. 196, chap. 3, **1982**; (c) J. M. Brown, A. G. Kent, *J. Chem. Soc., Perkin Trans.* **1987**, 1597.

15. R. C. Matthews, D. K. Howell, W. J. Peng, S. G. Train, W. D. Treleaven, G. G. Stanley, *Angew. Chem. Int. Ed. Engl.* **1996**, *35*, 2253.

16. (a) F. A. Cotton, C. T. Eagle, A. C. Price, *Inorg. Chem.* **1988**, *27*, 4362; (b) L. J. Tortorelli, P. W. Tinsley, C. Woods, C. J. Janke, *Polyhedron* **1988**, *7*, 315; (c) C. Woods, L. J. Tortorelli, *Polyhedron* **1988**, *7*, 1751.

17. C. Bianchini, A. Meli, F. Laschi, J. A. Ramirez, P. Zanello, A. Vacca, *Inorg. Chem.* **1988**, *27*, 4429.

18. (a) B. R. Sutherland, M. Cowie, *Inorg. Chem.* **1984**, *23*, 1290; (b) C. Bianchini, F. Laschi, D. Masi, C. Mealli, A. Meli, F. M. Ottaviani, D. M. Proserpio, M. Sabat, P. Zanello, *Inorg. Chem.* **1989**, *28*, 2552; (c) S. L. Schiavo, G. Bruno, F. Nicolò, P. Piraino, F. Faraone, *Organometallics* **1985**, *4*, 2091.

19. (a) A. J. Sandee, L. A. van der Veen, J. N. H. Reek, P. C. J. Kamer, M. Lutz, A. L. Spek, P. W. N. M. van Leeuwen, *Angew. Chem. Int. Ed.* **1999**, *38*, 3231; (b) A. J. Sandee, J. N. H. Reek, P. C. J. Kamer, P. W. N. M. van Leeuwen, *J. Am. Chem. Soc.* **2001**, *123*, 8468.

20. R. B. Crabtree, H. Felkin, *J. Mol. Catal.* **1979**, *5*, 75.

21. W. J. Peng, S. G. Train, D. K. Howell, F. R. Fronczek, G. G. Stanley, *Chem. Commun.* **1996**, 2607.

22. G. Consiglio, F. Rama, *J. Mol. Catal.* **1991**, *66*, 1.

23 A. Ceriotti, L. Garlaschilli, G. Lon-
goni, M. C. Malatesta, D. Strumolo,
A. Fumagalli, S. Martinengo, *J. Mol.
Catal.* **1984**, *24*, 309.

24 M. Garland, *Organometallics* **1993**, *12*, 535.

25 P. Kalck, J. M. Frances, P. M. Pfister,
T. G. Southern, A. Thorez, *Chem.
Commun.* **1983**, 510–511.

26 P. Kalck, *Polyhedron* **1988**, *7*, 2441.

27 R. Davis, J. W. Epton, T. G. Southern,
*J. Mol. Catal.* **1992**, *77*, 159.

28 M. Diéguez, C. Claver, A. M. Masdeu-
Bultó, A. Ruiz, P. W. N. M. van
Leeuwen, G. C. Schoemaker,
*Organometallics* **1999**, *18*, 2107–2115.

29 (a) C. Hunt, Jr., B. D. Nelson,
E. G. Harmon, F. P. Fronczek,
S. F. Watkins, D. R. Billodeaux,
G. G. Stanley, *Acta Cryst.* **2000**, *C56*, 546;
(b) C. Hunt, Jr., F. R. Fronczek,
D. R. Billodeaux, G. G. Stanley,
*Inorg. Chem.* **2001**, *40*, 5192.

30 (a) B. Cornils, E. G. Kuntz, *Aqueous-
Phase Organomet. Catal.* **1998**, 271;
(b) B. Cornils, W. A. Herrmann,
R. W. Eckl, *J. Mol. Catal. A: Chem.* **1997**,
*116*, 27.

31 (a) T. J. Devon, G. W. Phillips,
T. A. Puckette, J. L. Stavinoha,
J. J. Vanderbilt, U.S. Patent 4694109,
**1987**; (b) C. P. Casey, E. L. Paulsen,
E. W. Beuttenmueller, B. R. Proft,
L. M. Petrovich, B. A. Matter,
D. R. Powell, *J. Am. Chem. Soc.* **1997**,
*119*, 11817.

32 (a) W. A. Herrmann, C. W. Kohl-
paintner, R. B. Manetsberger,
H. Bahrmann, H. Kottmann, *J. Mol.
Catal. A: Chem.* **1995**, *97*, 65;
(b) D. Gleich, R. Schmid, W. A. Herrmann, *Organometallics* **1998**, *17*, 2141;
(c) H. Klein, R. Jackstell, K.-D. Wiese,
C. Borgmann, M. Beller, *Angew. Chem.
Int. Ed.* **2001**, *40*, 3408.

33 (a) M. Kranenburg, Y. E. M. van der
Burgt, P. C. J. Kamer, P. W. N. M. van
Leeuwen, K. Goubitz, J. Fraanje,
*Organometallics* **1995**, *14*, 3081;
(b) L. A. van der Veen, M. D. K. Boele,
F. R. Bregman, P. C. J. Kamer,

P. W. N. M. van Leeuwen, K. Goubitz,
J. Fraanje, H. Schenk, C. Bo,
*J. Am. Chem. Soc.* **1998**, *120*, 11616.

34 (a) E. Billig, A. G. Abatjoglou,
D. R. Bryant, U.S. Patent 4668651, **1987**;
(b) J. M. Maher, J. E. Babin, E. Billig,
D. E. Bryant, T. W. Leung, U.S. Patent
5288918, **1994**; (b) B. Moasser,
W. L. Gladfelter, D. C. Roe,
*Organometallics* **1995**, *14*, 3832;
(c) A. van Rooy, P. C. J. Kamer,
P. W. N. M. van Leeuwen, K. Goubitz,
J. Fraanje, N. Veldman, A. L. Spek,
*Organometallics* **1996**, *15*, 835.

35 M. P. Magee, W. Luo, W. H. Hersh,
*Organometallics* **2002**, *21*, 362.

36 I. T. Horvath, R. V. Kastrup,
A. A. Oswald, E. J. Mozeleski,
*Catal. Lett.* **1989**, *2*, 85.

37 C. P. Casey, G. T. Whiteker,
M. G. Melville, L. M. Petrovich,
J. A. Gavney, Jr., D. R. Powell,
*J. Am. Chem. Soc.* **1992**, *114*, 5535–5543.

38 (a) L. A. van der Veen, M. D. K. Boele,
F. R. Bregman, P. C. J. Kamer,
P. W. N. M. van Leeuwen, K. Goubitz,
J. Fraanje, H. Schenk, C. Bo, *J. Am.
Chem. Soc.* **1998**, *120*, 11616–11626;
(b) P. Dierkes, P. W. N. M. van
Leeuwen, *J. Chem. Soc., Dalton Trans.*
**1999**, 1519–1530; (c) P. W. N. M. van
Leeuwen, P. C. J. Kamer, J. N. H. Reek,
*Pure Appl. Chem.* **1999**, *71*, 1443–1452.

39 (a) M. P. Doyle, in *Catalysis by Di- and
Polynuclear Metal Cluster Complexes*
(Eds.: R. D. Adams, F. A. Cotton),
Wiley-VCH, New York, p. 249–282, **1998**;
(b) M. P. Doyle, D. C. Forbes, *Chem. Rev.*
**1998**, *98*, 911–935; (c) M. P. Doyle, in
*Comprehensive Organometallic Chemistry
II* (Ed.: L. S. Hegedus), Pergamon Press,
New York, Vol. 2, Chapter 5.2., **1995**;
(d) M. P. Doyle, A. V. Kalinin,
D. G. Ene, *J. Am. Chem. Soc.* **1996**, *118*,
8837; (e) M. P. Doyle,
W. R. Winchester, J. A. A. Hoorn,
V. Lynch, S. H. Simonsen, R. Ghosh,
*J. Am. Chem. Soc.* **1993**, *115*, 9968.

40 G. Trinquier, R. Hoffmann,
*Organometallics* **1984**, *3*, 370–380.

# 11

# Catalysis by Homo- and Heteronuclear Polymetallic Systems

*I. I. Moiseev, A. E. Gekhman, M. V. Tsodikov, V. Ya. Kugel, F. A. Yandieva,*
*L. S. Glebov, G. Yu. Kliger, A. I. Mikaya, V. G. Zaikin, Yu. V. Maksimov,*
*D. I. Kochubey, V. V. Kriventsov, V. P. Mordovin, and J. A. Navio*

## 11.1
## Introduction

Catalysis by metals and alloys is of primary importance in understanding the phenomenon of catalytic action [1]. The field includes catalysis by both solid and colloidal metals. The latter branch borders metal complex catalysis, in which ions and complexes in solutions are the active species of interest. Herein, some results in this field obtained by our group in Moscow are presented.

Our interest in catalysis by polynuclear systems with short metal–metal contacts was prompted by the finding that ethylene can be oxidized to vinyl acetate not only through a stoichiometric reaction with a palladium(II) complex [2] but also through a reaction catalyzed by colloidal low-valence palladium complexes [3]. A typical representative of such cluster complexes is the cluster with the idealized formula $Pd_{561}phen_{60}(OAc)_{180}$, the chemical composition and structure of which have been most extensively studied. A short overview of reactions catalyzed by this cluster in solution is given in Section 11.2.

Section 11.3 deals with hydride complexes formed from the intermetallic $TiFe_{0.95}Zr_{0.03}Mo_{0.02}$ upon its interaction with $H_2$. It was found that these intermetallic hydrides, in combination with the hydrogenating catalyst $Pt/Al_2O_3$, are able to selectively reduce carbon dioxide to carbon monoxide (see Section 11.4). In studying this redox reaction, we found a new catalytic reaction, namely the reductive dehydration of alcohols. The well-known normal dehydration route involves elimination of a water molecule from the alcohol to produce an alkene with the same number of carbon atoms. In contrast, the new reaction affords an alkane rather than an alkene, and the number of carbon atoms in the alkane molecule is at least twice that in the alcohol molecule. This new reaction is considered in Section 11.5.

*Multimetallic Catalysts in Organic Synthesis.* Edited by M. Shibasaki and Y. Yamamoto
Copyright © 2004 WILEY-VCH Verlag GmbH & Co. KGaA, Weinheim
ISBN: 3-527-30828-8

**11.2**
**Palladium Giant Cluster Catalysis**

The first palladium giant cluster [4] was prepared by successive treatment of $Pd(OAc)_2$ with $H_2$ and $O_2$ (1 atm) in an AcOH solution containing 0.5 mol of ligand L (phen or bpy) per Pd atom. The composition and structure of the reaction product were determined by HRTEM, electron diffraction, STM, NMR, EXAFS, and elemental analysis data, along with measurements of molecular mass and magnetic susceptibility [5, 6]. Based on these data, the empirical formula of the substance was found to be $Pd_{570\pm30}L_{63\pm3}(OAc)_{190\pm10}$. Within the framework of the cluster model due to P. Chini [7], this can be approximated by the idealized formula $[Pd_{561}L_{60}](OAc)_{180}$. The experimental data have been rationalized in terms of a scheme in which the molecule of the Pd-561 cluster is supposed to consist of a positively charged metal core ca. 25 Å in diameter containing $570 \pm 30$ densely packed Pd atoms (or 561 atoms in an idealized five-shell icosahedron or cubocta-hedron-shaped core), ~60 neutral L ligands bound to the surface of the metal core, and ~180 outer-sphere $OAc^-$ anions, the latter counterbalancing the positive charge of the metal core (Figure 11.1) [6].

The outer-sphere $OAc^-$ anions can be replaced by other anions. For instance, $O^{2-}$ and $PF_6^-$ anions readily substitute $OAc^-$ anions in an aqueous solution containing $KPF_6$, affording a giant cluster with the idealized formula $[Pd_{561}L_{60}O_{60}](PF_6)_{60}$ [5, 8]. The Pd-561 giant clusters exhibit high catalytic performance in a variety of reactions (see Table 11.1).

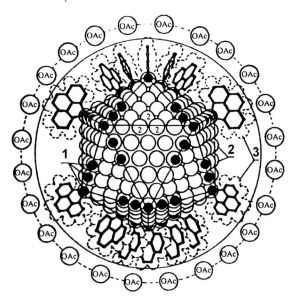

**Figure 11.1** Idealized structure of a Pd-561 cluster:
**1** – Pd atoms coordinated to phen ligands,
**2** – Pd atoms accessible for the coordination of substrate anions or solvent molecules,
**3** – van der Waals' contours of the coordinated phen ligands.

**Table 11.1** Reactions catalyzed by phen-ligated Pd giant clusters.

---

**Oxidation with $O_2$**

$CH_2=CH_2 + \frac{1}{2} O_2 + AcOH \rightarrow CH_2=CHOAc + H_2O$
$CH_2=CHCH_3 + \frac{1}{2} O_2 + AcOH \rightarrow CH_2=CHCH_2OAc + H_2O$
$PhCH_3 + \frac{1}{2} O_2 + AcOH \rightarrow PhCH_2OAc + H_2O$
$CH_2=CHCH_3 + \frac{1}{2} O_2 \rightarrow CH_2=CHCOOH$ or $CH_2=CHCOOEt$
$2 CH_3OH + \frac{1}{2} O_2 \rightarrow HCOOCH_3$
$C_2H_5OH + \frac{1}{2} O_2 \rightarrow CH_3CHO \rightarrow CH_3COOC_2H_5 + (CH_3CO)_2O$

---

**Oxidation with nitrobenzene**

$3 CO + PhNO_2 \rightarrow PhNCO + 2 CO_2$
$3 CO + PhNO_2 + 2 PhOH \rightarrow PhNCO + 2 (PhO)_2CO + CO_2 + H_2O$

---

**Carbonylation reactions**

$PhNO_2 + 3 CO \rightarrow PhNCO + 2 CO_2$
$PhNO_2 + 3 CO + 2 PhOH \rightarrow PhNCO + 2 (PhO)_2CO + CO_2 + H_2O$

---

**Hydrogenation with $H_2$**

$cyclo\text{-}C_6H_{10} + H_2 \rightarrow cyclo\text{-}C_6H_{12}$
$PhNO_2 + 3 H_2 \rightarrow PhNH_2 + 2 H_2O$

---

**Hydrogen-transfer reduction with formic acid**

$CH_3CN + 2 HCOOH \rightarrow C_2H_5NH_2 + 2 CO_2$
$2 CH_3CN + 4 HCOOH \rightarrow (C_2H_5)_2NH + NH_3 + 4 CO_2$
$PhNO_2 + 3 HCOOH \rightarrow PhNH_2 + 3 CO_2 + 2 H_2O$

---

**Acetal (ketal) formation**

$CH_3CHO + 2 EtOH \rightarrow CH_3CH(OEt)_2 + H_2O$
$(CH_3)_2CHO + 2\ ^iPrOH \rightarrow (CH_3)_2CH(O^iPr)_2 + H_2O$

---

**Double-bond migration**

$CH_2=CHCH_2CH_3 \rightarrow cis + trans\text{-}CH_3CH=CHCH_3$
$CH_2=CHCH_2OH \rightarrow CH_3CH_2CHO$

---

**Alkene oligomerization**

$2 CH_2=CH_2 \rightarrow n\text{-butenes}$
$2 CH_2=CHCH_3 \rightarrow n + iso\text{-hexenes}$

---

**Redox disproportionation of benzylic alcohols**

$2 PhCH_2OH \rightarrow PhCHO + PhCH_3 + H_2O$
$2 PhCH(CH_3)OH \rightarrow PhCOCH_3 + PhC_2H_5 + H_2O$

---

**Benzylic alcohol dehydration to form an ether molecule**

$2 PhCH(CH_3)OH \rightarrow PhCH(CH_3)OCH(CH_3)Ph + H_2O$

---

**Norbornadiene allylation**

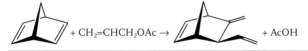

---

Cluster Pd-561 is a catalyst for liquid-phase acetoxylation of ethylene to vinyl acetate under mild conditions (20–60 °C, 1 atm $O_2$) and of propylene to allyl acetate in AcOH solutions; it also catalyzes allylic (but not vinylic!) esterification of higher acyclic alkenes in solutions of carboxylic acids [8]. Oxidative esterification of acyclic alkenes (*viz.*, propylene and 1-hexene) by $O_2$ in AcOH solution catalyzed by the Pd-561 giant cluster produces at least 95–98% yields of allylic esters. No decrease in the selectivity was found even in solutions containing up to 10% water. The only side reaction with the Pd-561 catalysts is the subsequent oxidation of alkenyl esters to form ethylidene and allylidene diacetates. The soluble Pd-561 giant clusters do not induce homoallylic oxidation and give fewer side reactions compared to commercial Pd metal catalysts, which operate at much higher temperatures.

The reaction kinetics for the Pd-561-catalyzed oxidation of ethylene (and propylene) obeys the following equation [5, 6]:

$$r_0 = k\,[\text{Pd-561}]\,\frac{[C_2H_4]\,[O_2]\,[\text{AcOH}]}{(K_1 + [C_2H_4])\,(K_{II} + [O_2])\,(K_{III} + [\text{AcOH}])}$$

The Michaelis–Menten character of the kinetics suggests that the formation of the reaction product is preceded by reversible coordination of the alkene, $O_2$, and AcOH molecules by the cluster.

Observed kinetic isotope effects suggest that the mechanisms of oxidative acetoxylation catalyzed with $Pd^{II}$ and low-valence Pd clusters are different. On the basis of kinetic data, including H/D kinetic isotope effects, the following mechanism has been proposed to rationalize the Pd-561 cluster-catalyzed reaction (Scheme 11.1) [5].

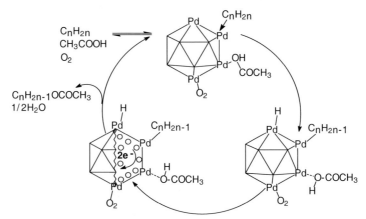

Scheme 11.1 Proposed mechanism of alkene oxidation catalyzed by Pd giant clusters.

The surface of the metal core of a giant cluster is substantially screened by phen or bpy ligands, which restrict access of the reactant and solvent molecules to the cluster core. A study of the effect of additional ligands on the kinetics of the Pd-561 cluster-catalyzed oxidative acetoxylation of $C_2H_4$ and $C_3H_6$ in solution [5, 8] showed that bulky additional ligands, which are capable of strong binding to Pd atoms (e.g., PPh$_3$ and phen), exert only negligible effect on the rate of alkene oxidation. This is due to the fact that these ligands cannot reach the sites suitable for the coordination of the smaller molecules, such as C$_2$H$_4$, O$_2$ or AcOH. Smaller ligands, such as C$_2$H$_5$SH and SCN$^-$, effectively retard oxidation. About 50 "poisonous" ligand molecules per Pd cluster are necessary to inhibit the oxidation of C$_2$H$_4$. However, only ~15 poison molecules per Pd-561 cluster molecule are sufficient for complete inhibition of C$_3$H$_6$ oxidation. This is in line with estimates of the numbers of surface centers accessible for alkene coordination at the cluster metal core based on the idealized model of the giant cluster molecule. Of 252 metal atoms in the outer layer of the Pd metal core, ~20% (50 of 252) of them can participate in the catalytic oxidation of ethylene and ~6% (15 of 252) are implicated in the reaction with the more bulky C$_3$H$_6$ molecule.

The available data show that the Pd-catalyzed oxidative esterification of alkenes can occur through two different pathways: either via a stepwise Pd$^{II}$/Pd$^0$ mechanism or through a route mediated by the low-valence Pd giant clusters, as shown in Scheme 11.2.

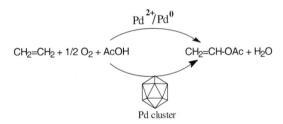

**Scheme 11.2** Formation of vinyl acetate via two catalytic routes.

In the reaction route involving Pd$^{II}$ to Pd$^0$ reduction, an alkene molecule coordinates to a Pd$^{II}$ complex and is subsequently attacked by a nucleophilic species. Because of the enhanced nucleophilicity of H$_2$O molecules or HO$^-$ ions in comparison with AcOH molecules or AcO$^-$ ions, the formation of hydroxyl Pd-organyls of the type X$_3$Pd–CH$_2$(R)OH prevails over the formation of acetates, even at low concentrations of water in aqueous AcOH solution.

The cluster-catalyzed reaction is assumed to involve the insertion of alkene and dioxygen molecules into relatively weak Pd–Pd bonds and subsequent electron transfer from organyl fragments to the oxidant group via the cluster skeleton as a slow step [5]. These two different mechanisms account for the different selectivities of the oxidation reactions [5].

## 11.3
## TiFe$_{0.95}$Zr$_{0.03}$Mo$_{0.02}$ Intermetallic and Its Hydrides

Hydrogen atoms stored in the framework of a metal alloy or metal solid solution possess enhanced mobility and reactivity, as a result of which it is possible to use the hydride material as an effective chemical reagent. Different hydrides, especially binary and ternary hydrides and solid solutions, allow one to engineer materials that not only function as storage and delivery systems, but also as efficient catalysts for various applications [12–14]. The catalytic activities of intermetallic systems based on nickel, zirconium, and rare earth metals in hydrogenation and dehydrogenation have been thoroughly studied [12–16].

Alloys based on the TiFe intermetallic were initially chosen to study $CO_2$ reduction (see Section 11.4). The parent compound TiFe was found to be rather inert with regard to the expected reduction reactions. However, the TiFe intermetallic doped with Mo and Zr atoms, which are known to have a tendency to form rather strong metal–oxygen bonds, was found to catalyze the reduction of $CO_2$ to CO. Moreover, the doped intermetallic hydride was found to convert alcohols into alkanes, with the latter having at least twice the number of carbon atoms in their skeletons as compared to the parent alcohols.

The intermetallic compounds $TiFe_{0.95}Zr_{0.03}Mo_{0.02}$ and TiFe were prepared by the consumable-electrode melting of the starting materials, *viz.*, titanium sponge (TG 100), low-carbon steel (Russian State Standard 11036-75), zirconium iodide, and molybdenum metal. To protect the TiFe alloy from contamination with impurities, the method of scull melting [17–19] was used in this study. Structural information was gleaned from X-ray diffraction, Mössbauer, EXAFS, and XANES spectroscopic data.

The following samples were chosen to study the structural peculiarities of the intermetallics and their hydrides: 1) undoped alloy TiFe (**1**) as a reference material; 2) alloy $TiFe_{0.95}Zr_{0.03}Mo_{0.02}$ (**2**); 3) a hydride of composition $[TiFe_{0.95}Zr_{0.03}Mo_{0.02}]H_2$ (**3**) obtained after absorption of ~1 mol of $H_2$ by alloy **2**; 4) hydride $[TiFe_{0.95}Zr_{0.03}Mo_{0.02}]H$ (**4**) obtained after the removal of ~0.5 mol of $H_2$ from **3** at a temperature not exceeding 185 °C; 5) hydride $[TiFe_{0.95}Zr_{0.03}Mo_{0.02}]H_{0.36}$ (**5**) obtained after the removal of ~0.82 mol of $H_2$ from hydride **3** at 700–920 °C.

The main structural motif of the intermetallic alloy **2**, $TiFe_{0.95}Zr_{0.03}Mo_{0.02}$ (Figure 11.2a), is characterized by reflections with interplanar distances $d = 0.2117$, 0.1493, and 0.1220 nm, which were also observed in the cubic modification of alloy **1**, $\alpha$-TiFe [17, 18]. The X-ray diffraction pattern of **2** at angles $2\Theta = 40.50°$ and $40.75°$ exhibits weak reflections with $d = 0.222$ and 0.216 nm, which distinguishes this alloy from the reference intermetallic **1**. This difference most probably stems from the tetragonal lattice distortion caused by the introduction of the dopant metals Mo and Zr.

The Mössbauer spectra of alloys **1** and **2** exhibit a single line with IS = $-0.16 \pm 0.03$ mm s$^{-1}$ and line width $\Gamma = 0.36 \pm 0.03$ mm s$^{-1}$. Parameters corresponding to the cubic $\alpha$-TiFe structure [19] were used as reference data in analysis of the spectra of the doped alloy samples. The measured spectra also consist of a single line, although its shape and total width differ from the reference values. Computer analysis allowed us to represent the spectrum as a superposition of a single line and a quadruple doublet with the corresponding parameters IS = $-0.17 \pm 0.03$ mm s$^{-1}$, $\Gamma = 0.35 \pm 0.03$ mm s$^{-1}$, relative content $A = 0.74$, and IS = $-0.16 \pm 0.03$ mm s$^{-1}$, quadrupole splitting QS = $0.35 \pm 0.03$ mm s$^{-1}$, $\Gamma = 0.30 \pm 0.03$ mm s$^{-1}$, $A = 0.26$.

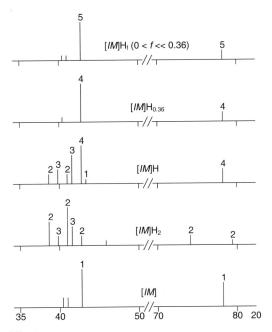

**Figure 11.2** X-ray diffraction patterns:
**1** – TiFe$_{0.95}$Zr$_{0.03}$Mo$_{0.03}$(*IM*), **2** – [*IM*]H$_{1.93}$ (cubic),
**3** – [*IM*]H (orthorhombic), **4** – [*IM*]H$_{0.06}$ (cubic), **5** – [*IM*]H$_{0.06-x}$ (cubic).

As before, the single line corresponds to the cubic α-phase of the alloy, which was predominant in the sample. However, approximately one-third of alloy **2** includes regions in which the local environment of Fe atoms is different from spherically symmetrical (cubic), suggesting pronounced axial distortions. This result is consistent with the X-ray data and suggests that the structure of alloy **2** is considerably distorted compared to that of alloy **1**. A tetragonally distorted structure in which shear deformations of the lattice are caused by the presence of zirconium and molybdenum atoms could be assigned to alloy **2**.

### 11.3.1
### Capacity of the Intermetallic for the Absorption and Thermal Desorption of H$_2$

As can be seen in Figure 11.3, the solubility of H$_2$ in alloy **2** is much higher than that in TiFe at the same hydrogen pressure. A plateau is clearly seen in the absorption isotherm of TiFe (**1**); the saturation pressure $P(1) = 12.5$ atm corresponds to this plateau. According to the data obtained [17–19], the initial region of H$_2$ absorption by the TiFe alloy can be attributed to the formation of a solid solution of hydrogen in the [TiFe]H$_x$ intermetallic, where $x \approx 0.0$–0.1 (the so-called α-phase). The plateau is attributed to a mixture of the α- and β-phases, with transition to the single-phase state of the β-solution. The region of the individual β-phase corresponds to $x = 0.7$–0.8; thereafter, the isotherm exhibits an increase in the hydrogen pressure

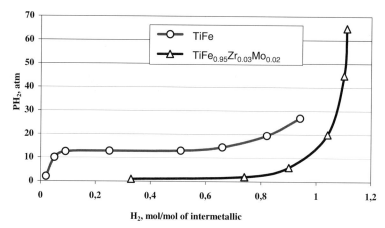

**Figure 11.3** Isotherms of hydrogen absorption by the [TiFe]H$_2$ and [TiFe$_{0.95}$Zr$_{0.03}$Mo$_{0.02}$]H$_2$ systems at 20 °C.

because of additional dissolution of hydrogen in the β-phase. The isotherm of H$_2$ absorption by the alloy **2** also exhibits a plateau corresponding to a saturation pressure of ~2 atm (Figure 11.3). After the plateau region, absorption dramatically increases.

The absorption data show that the interaction of H$_2$ with the TiFe$_{0.95}$Zr$_{0.03}$Mo$_{0.02}$ alloy differs markedly from the interaction of H$_2$ with the undoped TiFe alloy: the hydride phases based on **2** were formed at a considerably lower pressure (~2 atm) than that in the case of **1** (~12.5 atm). Moreover, the saturation of alloy **2** with H$_2$ is characterized by a much shorter activation period (~0.5 h) than that of TiFe (above 24 h). It can be seen in Figure 11.3 that after reaching an H$_2$/(alloy **2**) molar ratio of ~1, the subsequent dissolution of hydrogen requires a dramatic increase in pressure. Thus, the modifying additives of the Zr and Mo atoms in alloy **2** have a considerable effect on the structure and electronic and absorption properties of the TiFe intermetallic compound.

The experiments showed that 1 mol of TiFe$_{0.95}$Zr$_{0.03}$Mo$_{0.02}$ absorbs 1 mol of H$_2$ to form the intermetallic hydride [TiFe$_{0.95}$Zr$_{0.03}$Mo$_{0.02}$]H$_2$ (**3**). The curve for thermal desorption of the absorbed hydrogen from the sample with the stoichiometric composition [TiFe$_{0.95}$Zr$_{0.03}$Mo$_{0.02}$]H$_2$ in an argon atmosphere (Figure 11.4) clearly exhibits three regions. The first of them corresponds to the intense release of ~0.80–0.82 mol of H$_2$ per mol of the sample on heating the hydride from 70 to 185 °C and formation of the hydride [TiFe$_{0.95}$Zr$_{0.03}$Mo$_{0.02}$]H$_{0.36}$ (**4**). The evolved portion of H$_2$ can be conventionally referred to as loosely bound hydrogen (LBH). Practically no H$_2$ was released as the temperature was increased above 185 °C. The next part of the hydrogen bound by the intermetallic (~0.09 mol H$_2$ per mol of the intermetallic compound) was released in the range 700–920 °C. Much higher temperatures are required for thermal desorption of the remaining hydrogen (~0.09 mol/mol).

The intermetallic hydride [TiFe$_{0.95}$Zr$_{0.03}$Mo$_{0.02}$]H$_{0.36}$ formed after heating the hydride [TiFe$_{0.95}$Zr$_{0.03}$Mo$_{0.02}$]H$_2$ up to 700 °C contains so-called strongly bound hydrogen (SBH).

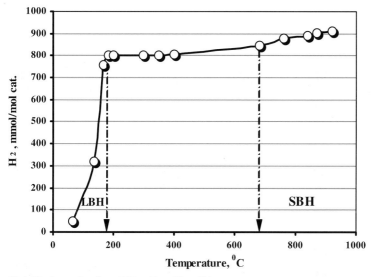

**Figure 11.4** $H_2$ desorption from $[TiFe_{0.95}Zr_{0.03}Mo_{0.02}]H_2$.

It is known that the hydride phases obtained from the hydrogen-accumulating intermetallic compounds are characterized by different strengths of hydrogen binding [17]. The overall picture of $H_2$ thermal desorption from $[TiFe_{0.95}Zr_{0.03}Mo_{0.02}]H_2$ is summarized in Scheme 11.3.

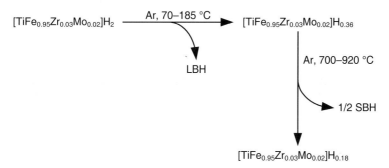

**Scheme 11.3** Conversions of the intermetallic hydride $[TiFe_{0.95}Zr_{0.03}Mo_{0.02}]H_2$.

## 11.3.2
### Structure of Intermetallic Hydrides and Strongly Bound Hydrogen

The X-ray, Mössbauer, and EXAFS spectra of the intermetallic hydrides **3**, **4**, and **5** were studied to elucidate structural changes in intermetallic **2** after hydrogen incorporation into its lattice. We found that the phase composition changes considerably after absorption of the maximum amount of hydrogen (~1 mol $H_2$ per 1 mol of alloy) by alloy **2**. Sample **3** was found to be heterogeneous, with a predominance of a cubic modification isostructural to the $\gamma$-TiFeH$_{1.93}$ hydride [18, 19]

($d$ = 0.232, 0.2196, and 0.209 nm). An orthorhombic modification isostructural to β-TiFeH [20] ($d$ = 0.2260, 0.2185, and 0.2156 nm) is also present in a considerable amount (Figure 11.2b).

The Mössbauer spectrum of sample **3** can be described by two quadrupole doublets with the parameters IS = 0.07 ± 0.03 mm s$^{-1}$, QS = 0.29 ± 0.03 mm s$^{-1}$, $\Gamma$ = 0.29 ± 0.03 mm s$^{-1}$, $A$ = 0.39 and IS = 0.26 ± 0.03 mm s$^{-1}$, QS = 0.26 ± 0.03 mm s$^{-1}$, $\Gamma$ = 0.30 ± 0.03 mm s$^{-1}$, $A$ = 0.61, respectively. The less intense doublet of the former corresponds to the orthorhombic structure of β-TiFeH, whereas the latter can be attributed to the cubic γ-TiFeH$_{1.93}$ hydride [19].

Thus, the X-ray and Mössbauer data consistently indicate that saturation of the initial alloy TiFe$_{0.95}$Zr$_{0.03}$Mo$_{0.02}$ with H$_2$ gives rise to hydride **3**, the lattice of which is completely rearranged, being transformed into a mixture of two modifications of intermetallic hydrides isostructural to the hydrides of the TiFe/H$_2$ system [17–19].

The data corresponding to the intermediate hydride of composition [TiFe$_{0.95}$Zr$_{0.03}$Mo$_{0.02}$]H (**4**) suggest that the material predominantly consists of β-TiFeH (Figure 11.2c) [20]. Along with reflections from this hydride, the X-ray pattern also exhibited reflections typical of TiFeH$_{0.06}$ [21] ($d$ = 0.2106, 0.1485, and 0.1210 nm). Poorly resolved reflections from the cubic γ-TiFeH$_{1.93}$ were also observed.

The Mössbauer spectrum of the hydride [TiFe$_{0.95}$Zr$_{0.03}$Mo$_{0.02}$]H (**4**) can be described by two components: an intense doublet with the parameters IS = –0.05 ± 0.03 mm s$^{-1}$, QS = 0.30 ± 0.03 mm s$^{-1}$, $\Gamma$ = 0.33 ± 0.03 mm s$^{-1}$, and $A$ = 0.65, which correspond to those of the structure of β-TiFeH, and a single line with the parameters IS = –0.10 ± 0.03 mm s$^{-1}$, $\Gamma$ = 0.30 ± 0.03 mm s$^{-1}$, and $A$ = 0.35, which belongs to the α-TiFe$_{0.95}$Zr$_{0.03}$Mo$_{0.02}$ alloy.

The X-ray diffraction pattern of hydride **5** [TiFe$_{0.95}$Zr$_{0.03}$Mo$_{0.02}$]H$_{0.36}$ containing ~0.18 mol of H$_2$ (Figure 11.2d) consists of a set of lines with interplanar distances mainly related to cubic α-TiFe. It is likely that a reflection with $d$ = 0.224 nm is also indicative of tetragonally distorted structures.

Analysis of the [200] reflection profiles from the parent TiFe$_{0.95}$Zr$_{0.03}$Mo$_{0.02}$ alloy and the intermetallic [TiFe$_{0.95}$Zr$_{0.03}$Mo$_{0.02}$]H$_{0.36}$ showed that, after the removal of ~0.82 mol of H$_2$ from hydride **3**, not only a line shift by ~$\Delta 2\Theta$ = 0.02°, but also a considerable line-broadening compared to the initial alloy occur. It should be noted that these changes are irreversible [22].

The Mössbauer spectrum of hydride **5** [TiFe$_{0.95}$Zr$_{0.03}$Mo$_{0.02}$]H$_{0.36}$ is very similar to that of the parent alloy **2**. Both spectra feature an intense single line and a doublet with the respective parameters IS = –0.14 ± 0.03 mm s$^{-1}$, $\Gamma$ = 0.36 ± 0.03 mm s$^{-1}$, $A$ = 0.74 and IS = –0.14 ± 0.03 mm s$^{-1}$, QS = 0.27 ± 0.03 mm s$^{-1}$, $\Gamma$ = 0.28 ± 0.03 mm s$^{-1}$, $A$ = 0.26. As in the case of alloy **2**, the single line characterizes a cubic α-phase, while the doublet characterizes the spatial regions of the alloy in which the local Fe environment differs from a spherically symmetrical one.

The desorption data obtained in this work clearly indicate that at least two types of lattice hydrogen are dissolved in the hydrides under study. The first is the "loosely bound" hydrogen contained in the hydrides [γ-TiFe$_{0.95}$Zr$_{0.03}$Mo$_{0.02}$]H$_{1.93}$/[β-TiFe$_{0.95}$Zr$_{0.03}$Mo$_{0.02}$]H. The loosely bound hydrogen can be removed from the lattice in the temperature range 70–185 °C. The second form is the "strongly bound" hydrogen dissolved in the lattice of the [TiFe$_{0.95}$Zr$_{0.03}$Mo$_{0.02}$]H$_{0.36}$ hydride **5**. This form can only be removed in the temperature range 700–920 °C.

The above-mentioned data suggest that the presence and distribution of strongly bound hydrogen in the intermetallic lattice cannot be probed by the X-ray and Mössbauer techniques. In this context, EXAFS measurements seem to be informative for elucidating the arrangement of atomic hydrogen in the lattice.

The RAD curves of the titanium and iron environments obtained from the corresponding EXAFS spectra are shown in Figures 11.5 and 11.6, respectively. The results suggest that all the samples exhibit a well-resolved fine structure up to 8 Å. It is known [23] that the structure of the TiFe intermetallic alloy is a primitive cubic lattice in which each metal atom is surrounded by eight atoms of another metal, which are located on the cube vertices with a distance of 2.56 Å, and by six atoms of the same metal, which constitute an octahedron with a distance of 2.98 Å. Therefore, the polyhedral environments of both metals are identical.

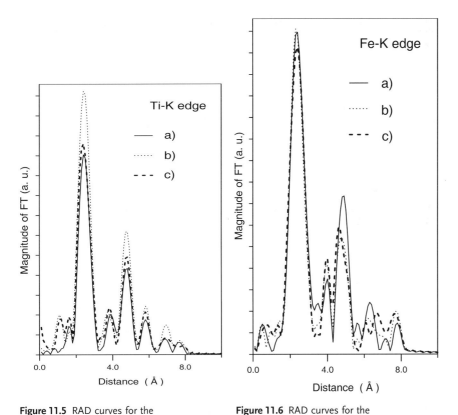

**Figure 11.5** RAD curves for the local environment of titanium atoms:
**a** – initial TiFe$_{0.95}$Zr$_{0.03}$Mo$_{0.02}$ intermetallic,
**b** – [TiFe$_{0.95}$Zr$_{0.03}$Mo$_{0.02}$]H$_{0.36}$,
**c** – [TiFe$_{0.95}$Zr$_{0.03}$Mo$_{0.02}$]H$_{<0.1}$ after CO$_2$ reduction.

**Figure 11.6** RAD curves for the local environment of iron atoms:
**a** – initial TiFe$_{0.95}$Zr$_{0.03}$Mo$_{0.02}$ intermetallic,
**b** – [TiFe$_{0.95}$Zr$_{0.03}$Mo$_{0.02}$]H$_{0.36}$,
**c** – [TiFe$_{0.95}$Zr$_{0.03}$Mo$_{0.02}$]H$_f$ after CO$_2$ reduction, where $f > 0.1$.

A distinctive feature of alloy **2** ($TiFe_{0.95}Zr_{0.03}Mo_{0.02}$) is the replacement of a portion of the iron atoms by zirconium and molybdenum atoms. Detailed analysis of the EXAFS data for **1** and **2** suggests that the set of interatomic distances for the closest environments of the Ti and Fe atoms is similar in both intermetallic compounds. However, in order to reach coordination numbers close to the observed values, very high Debye–Waller factors should be assumed. This seems to be a consequence of tetragonal distortion of the TiFe structure, as was suggested for the parent alloy **2**. Due to this structural distortion, a set of close interatomic distances appeared. It is possible that the Debye approximation does not hold in this case. Nevertheless, for the parent alloy **2** the RAD curves for the titanium and iron environments are similar.

The situation changed after $H_2$ treatment. Figure 11.6 demonstrates that the environment of Fe at a hydrogen content of ~0.18 mol $H_2$ (sample **5**, $[TiFe_{0.95}Zr_{0.03}Mo_{0.02}]H_{0.36}$) does not differ markedly from that observed for the initial alloy **2**. A considerable change in the RAD curves for the Fe environment (decrease in the intensity of reflections and the splitting of reflections) was observed only at 1 and 0.5 mol $H_2$ per mol unit of $[TiFe_{0.95}Zr_{0.03}Mo_{0.02}]H_2$ (**3**) and $[TiFe_{0.95}Zr_{0.03}Mo_{0.02}]H$ (**4**). The intensity of the peak for the Ti environment (for sample **5**) increased at the expense of that for the Fe environment. All the other RAD curves characterizing the Ti environment changed dramatically on going from one hydride to another (Figure 11.5). The RAD curves for both Ti and Fe became identical to that observed for the parent alloy **2** after the complete, or almost complete, removal of hydrogen from the intermetallic.

Two reasons for the changes in peak intensities in the RAD of titanium can be considered. First, the degree of structural distortion may increase in the series TiFe $\rightarrow$ $TiFe_{0.95}Zr_{0.03}Mo_{0.02}$ $\rightarrow$ $[TiFe_{0.95}Zr_{0.03}Mo_{0.02}]H_{0.36}$. However, this factor should result in analogous changes in the environments of both Ti and Fe, which was not observed.

The second and most probable reason may be an asymmetrical arrangement of the hydrogen atoms. According to neutron diffraction data, H atoms in the α- and δ-phases of intermetallic solid solutions of hydrogen in the TiFe alloy are located at the octahedral positions between four titanium atoms (with Ti–H distance 2.0 Å) and two iron atoms (with Fe–H distance 1.5 Å) (Figure 11.7a) [23]. The Ti–H distance is typical of titanium hydrides. Since iron does not form bulk hydrides, it is believed that the H atoms are primarily bound to titanium, although the Ti–H interatomic distances are longer than the Fe–H distances. However, the situation with alloy **2** is different inasmuch as both zirconium and molybdenum are known to form hydrides. Therefore, a shift of hydrogen atoms from the positions that they occupy in the TiFe alloy with a symmetrical octahedral arrangement of the Ti and Fe atoms towards the dopant atoms would be expected. The location of hydrogen atoms at tetrahedral positions cannot be excluded (Figure 11.7b). Moreover, these positions should be occupied first because they are more energetically favorable, whereas the positions at iron atoms will be occupied only at higher hydrogen concentrations. In particular, the observed anomalous behavior of the hydride $[TiFe_{0.95}Zr_{0.03}Mo_{0.02}]H_{0.36}$ could be rationalized by assuming coordination of four H atoms around

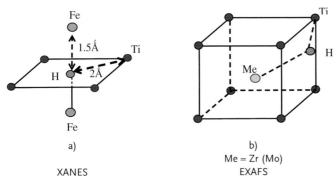

**Figure 11.7** Local structures of [TiFe]H$_{1.93}$ polyhedron (a) and [TiFe$_{0.95}$Zr$_{0.03}$Mo$_{0.02}$]H$_{0.36}$ (b).

each Zr or Mo of the alloy. In this case, the self-focusing effect of secondary photoelectrons, which is well known in EXAFS spectroscopy [24], can be expected. The effect can be observed when the third atom is situated on a straight line between the central and scattering atoms. The observed increase in the peak intensity may be a manifestation of this effect. Hence, the EXAFS data can be considered as an indication that the atoms of "strongly bound" hydrogen in the hydride [TiFe$_{0.95}$Zr$_{0.03}$ Mo$_{0.02}$]H$_{0.36}$ are mainly bound to the titanium atoms. In terms of this scheme, the observed increase in the RAD peak intensities for Ti in the hydride [TiFe$_{0.95}$Zr$_{0.03}$ Mo$_{0.02}$]H$_{0.36}$ may be considered as an indication that the H atoms are located in positions close to a straight line between the titanium and dopant atoms (Figure 11.7b). Note that no such binding was observed in the H-saturated alloy TiFe, for which no strong H-bonding was found by neutron diffraction, indicating that the hydrogen atom localization differs from that proposed for [TiFe$_{0.95}$Zr$_{0.03}$Mo$_{0.02}$]H$_{0.36}$ [23].

The XANES spectra are in good agreement with the EXAFS data (Figures 11.8 and 11.9). Changes in the XANES spectra for the iron atoms were observed only at a hydrogen content in excess of 0.18 mol H$_2$ per mol unit of [TiFe$_{0.95}$Zr$_{0.03}$Mo$_{0.02}$]H$_2$ or [TiFe$_{0.95}$Zr$_{0.03}$Mo$_{0.02}$]H of the alloy (Figure 11.9). Changes in the XANES spectra for titanium were observed at any H concentration (Figure 11.8).

Moreover, after hydrogen dissolution, the titanium absorption edge in the XANES spectrum of [TiFe$_{0.95}$Zr$_{0.03}$Mo$_{0.02}$]H$_{0.36}$ was shifted to higher energies. This shift of the absorption edge to higher energies can be interpreted in terms of a decrease in the electron density at the metal atom due to electron density transfer to the H atoms of "strongly bound" hydrogen (primarily binding with titanium, hydrogen atoms in the test alloy **1** are negatively charged, which is consistent with the results of quantum-chemical calculations for titanium hydride [25]) or a change in the distribution of the density states of titanium p-electrons over the Fermi level due to a redistribution of electrons between levels. On detecting this effect, one should also take into account the fact that the saturation of an intermetallic compound with hydrogen results in its mechanical crushing. It is possible that this material degradation, which results in the transformation of granules into dust-like particles

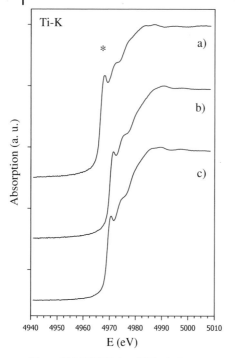

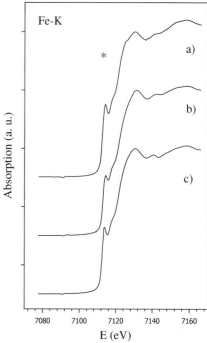

**Figure 11.8** XANES data (Ti-K edge):
**a** – initial TiFe$_{0.95}$Zr$_{0.03}$Mo$_{0.02}$ dintermetallic,
**b** – [TiFe$_{0.95}$Zr$_{0.03}$Mo$_{0.02}$]H$_{0.36}$,
**c** – [TiFe$_{0.95}$Zr$_{0.03}$Mo$_{0.02}$]H$_f$
after CO$_2$ reduction, where $f < 0.1$.

**Figure 11.9** XANES spectra (K-Fe edge):
**a** – initial TiFe$_{0.95}$Zr$_{0.03}$Mo$_{0.02}$ intermetallic,
**b** – [TiFe$_{0.95}$Zr$_{0.03}$Mo$_{0.02}$]H$_{0.36}$,
**c** – [TiFe$_{0.95}$Zr$_{0.03}$Mo$_{0.02}$]H$_f$
after CO$_2$ reduction, where $f < 0.1$.

and which can be observed visually, is based on deep structural changes. The crystallite dispersity in the course of this process increased by almost 30%. The broadening of a coherent scattering region, along with a noticeable increase in the specific surface area [26], is indicative of the formation of structural defects.

Hence, the data obtained by X-ray diffraction and Mössbauer spectroscopy indicate that the introduction of a small amount of Zr and Mo atoms into the α-TiFe alloy with the formation of TiFe$_{0.95}$Zr$_{0.03}$Mo$_{0.02}$ results in a distortion of the highly symmetrical cubic structure of the binary intermetallic alloy. The heterogeneous hydride [γ-TiFe$_{0.95}$Zr$_{0.03}$Mo$_{0.02}$]H$_{1.93}$/[β-TiFe$_{0.95}$Zr$_{0.03}$Mo$_{0.02}$]H$_{1.1}$ (sample **3**) contains the "weakly bound" lattice hydrogen that corresponds to ~0.82 mol H$_2$ per mol of the alloy. The hydride [TiFe$_{0.95}$Zr$_{0.03}$Mo$_{0.02}$]H$_{0.36}$ (sample **5**), obtained after the release of the "weakly bound" hydrogen at temperatures not exceeding 185 °C, includes the "strongly bound" lattice hydrogen, which can only be desorbed at temperatures of 700–920 °C. Data from Mössbauer spectroscopy, EXAFS, and XANES, as well as the proximity of the decomposition temperatures of hydrides [TiFe$_{0.95}$Zr$_{0.03}$Mo$_{0.02}$]H$_{0.36}$ and TiH$_2$, allowed us to conclude that the "strongly bound" lattice hydrogen is coordinated around the titanium. Because of the high affinity of titanium for hydrogen, the transition metal binds atomic H more tightly, producing lattice strain, deformation of the immediate surroundings, and crystal embrittlement to form a nanostructure.

## 11.4
## Stoichiometric CO$_2$ Reduction

The world's reserves of fossil fuels (natural gas, oil, and coal) amount to $10^{13}$ t of carbon. The global amount of CO$_2$ and its derivatives on Earth, including the atmosphere and the world's oceans, is at least ten times larger, amounting to $10^{14}$ t of carbon. Carbon dioxide is both an industrially renewable raw material and an atmospheric pollutant, one of the major contributors to the greenhouse effect. Gas-works processing of natural gas produces CO$_2$ as a final product. Many electric power stations produce exhaust gases with a high concentration of CO$_2$ suitable for chemical use. Utilization of CO$_2$ is thus both an important environmental concern and a challenging goal of so-called C$_1$ chemistry [27].

In the presence of metal-containing catalysts, hydrogenation of CO$_2$ produces methanol, gaseous C$_1$–C$_4$ hydrocarbons, and CO [28–32]. Carbon dioxide also takes part in the reverse water-gas shift reaction (RWGS), giving rise to CO and light hydrocarbons, mainly methane [28, 33]. A more selective CO$_2$ to CO transformation occurs in the presence of metal oxide and sulfide catalysts [33, 34]. The selectivity of CO$_2$ to CO transformation in the presence of a pre-reduced iron/alumina catalyst can reach 97%; however, the degree of CO$_2$ conversion remains low [35–37]. Selective CO formation is also achieved in the hydrogenation of CO$_2$ on a Pd/Ru membrane modified by a nickel coating [36]. In the presence of cobalt and iron filamentary crystals with special purity quartz particles, direct transformation of CO$_2$ into ethylene and propylene has been attained; the content of these products in the resulting gas mixture was up to 10% [37].

Among CO$_2$ reductions with dihydrogen, only methanation is allowed thermodynamically under standard conditions (see Table 11.2, reaction 5).

**Table 11.2** Thermodynamics of reactions involving CO$_2$ (kJ mol$^{-1}$) [75].

| Reactions | $\Delta H^0$ | $-T\Delta S^0$ | $\Delta G^0$ |
|---|---|---|---|
| $H_{2(g)} + CO_{2(g)} \rightarrow CO_{(g)} + H_2O_{(g)}$ | 41.20 | 22.6 | 18.60 |
| $H_{2(g)} + CO_{2(g)} \rightarrow CO_{(g)} + H_2O_{(l)}$ | −2.80 | 22.8 | 20.00 |
| $H_{2(g)} + CO_{2(g)} \rightarrow HCOOH_{(l)}$ | −31.20 | 64.2 | 33.00 |
| $2 H_{2(g)} + CO_{2(g)} \rightarrow CH_2O_{(g)} + H_2O_{(l)}$ | −9.00 | 55.0 | 44.00 |
| $3 H_{2(g)} + CO_{2(g)} \rightarrow CH_3OH_{(l)} + H_2O_{(l)}$ | −131.30 | 122.1 | −9.20 |
| $4 H_{2(g)} + CO_{2(g)} \rightarrow CH_{4(g)} + 2 H_2O_{(l)}$ | −252.90 | 122.1 | −130.80 |
| $CO_{2(g)} + CH_{4(g)} \rightarrow 2 CO_{(g)} + 2 H_{2(g)}$ | 191.80 | – | 288.00 |

The hydride form of the $[TiFe_{0.95}Zr_{0.03}Mo_{0.02}]H_x$ intermetallic compound ($x \leq 2$) reacts with $CO_2$ to give CO and a small amount of methane at 20 °C [38, 39]. The selectivity of $CO_2$ hydrogenation depends on the means by which hydrogen is supplied to the reaction vessel. It is known that hydride phases prepared from the hydrogen-accumulating intermetallic compounds are characterized by different strengths of hydrogen binding [17].

Our catalytic experiments were performed in a laboratory high-pressure flow-circulation set-up made of stainless steel. A functional diagram of the set-up is shown in Figure 11.10, along with descriptions of its constituent units. The $[TiFe_{0.95}Zr_{0.03}Mo_{0.02}]H_{0.36}$ intermetallic (**5**, denoted here as **I**) [26, 38], an industrial platinum/alumina catalyst (AP-64) (**II**), and a mixture of these (**III**) were used as catalysts.

Prior to loading in the reactor (see Figure 11.10, block 4), catalyst **I** was crushed to ca. 2–3 mm pellets using a ball mill with corundum balls.

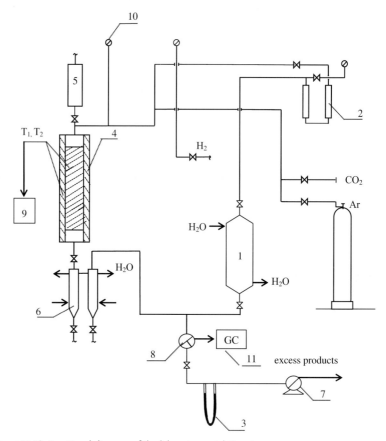

**Figure 11.10** Functional diagram of the laboratory catalytic setup:
(1) gas circulation pump, (2) flow meters, (3) pressure gauge, (4) reactor,
(5) dosing device, (6) cooled gas separators, (7) gas meter, (8) three-way valve,
(9) control unit, (10) pressure gauge, (11) LKhM-80MD.

Catalyst **III** was a thoroughly mixed blend consisting of pellets of intermetallic **I** and small cylinders of the industrial $Pt/Al_2O_3$ catalyst **II**. In reference experiments, intermetallic compound **I** and catalyst **II** were loaded into the reactor in such a way as to prevent their mixing: the upper layer of the reaction vessel contained catalyst **II**, and the lower layer was intermetallic compound **I**. The finely ground intermetallic compound **I** (60 g) and catalyst **II** (6.3 g) were loaded into the reactor and the resulting heterogeneous system was activated by treatment with a hydrogen flow at 100 °C (1 atm) for 10–14 h. After activation, the system was cooled to 25–30 °C, and high-pressure (120–135 atm) $H_2$ was rapidly introduced in circulation mode. After $H_2$ absorption by the intermetallic and formation of the hydride $[TiFe_{0.95}Zr_{0.03}Mo_{0.02}]H_2$, the weakly bound hydrogen (WBH) was eliminated by heating the system to 350–430 °C; thereafter, the Ar atmosphere was replaced by $CO_2$ and the reduction of $CO_2$ was studied at 350 °C and 430 °C at an initial pressure of 10–12 atm. When the degree of hydrogenation showed a marked decrease, the reaction products, together with the remaining $CO_2$, were replaced by fresh $CO_2$. This operation was repeated several times until no hydrogenation of $CO_2$ could be detected.

After completion of $CO_2$ reduction, the gas phase was replaced by fresh $CO_2$ and, under conditions of its circulation, cyclohexane was added to the catalyst layer from a dosing device with a flow rate of ~0.08–0.1 $h^{-1}$ for 2–3 h. Thereafter, the catalytic cycle was considered complete. The catalytic system was then subjected to activation once more by treating the catalyst with $H_2$ at 100 °C for 10–14 h followed by repeated absorption of $H_2$ and conduction of the initial cycle under the above-described conditions. In experiments with platinum/alumina catalyst **II**, hydrogen was fed to the system together with $CO_2$. The total operation times for the catalytic systems **I**, **II**, and **III** were 227 h, 67 h, and 277.4 h, respectively.

The reaction products were analyzed by GC in the on-line mode using an LKhM 80-MD chromatograph (see Figure 11.10, block 11). The gas components were identified by adding reference compounds; their concentrations were calculated by the absolute calibration method. The $H_2$, CO, $CH_4$, and $CO_2$ contents of the samples were analyzed using an analytical packed column (5 m × 4 mm) (block 1) with SKT carbon sorbent (0.25–0.5 mm), a heat conductivity detector, and special purity grade Ar as the carrier gas. The $C_1$–$C_4$ hydrocarbon gases were analyzed on an analytical packed column (2 m × 4 mm) composed of modified $\alpha$-$Al_2O_3$ (0.25–0.50 mm) containing 2% squalane, using a flame-ionization detector.

The results of GC analyses were used to calculate the amount of $CO_2$ in the reaction vessel after completion of the hydrogenation and the amount of $H_2$ consumed in the formation of CO and $C_1$–$C_4$ hydrocarbons. On the basis of these data, the full balance for $CO_2$ and $H_2$ could be deduced and, hence, the degree of $CO_2$ conversion and the total amount of $H_2$ consumed in $CO_2$ transformation and evolved from the intermetallic compound to the reaction bulk could be calculated. The procedure for the calculation of the degree of $CO_2$ conversion and the amounts of the reaction products formed has been described elsewhere [33, 45].

The intermetallic hydride $[TiFe_{0.95}Zr_{0.03}Mo_{0.02}]H_2$ was found to react with $CO_2$ even at room temperature. However, the reaction was not selective. Much more

**Table 11.3** Selectivity of $CO_2$ hydrogenation; influence of the nature of the reagent (430 °C, 10–12 atm).

| Reagent | $\tau_{45\%}$ (h) | $W_0$ mmol/$g_{cat}$ h | $S_{CO}$ (%) | Product composition (mol %) | | |
|---|---|---|---|---|---|---|
| | | | | CO | $CH_4$ | $C_2$–$C_5$ |
| $[TiFe_{0.95}Zr_{0.03}Mo_{0.02}]H_{0.36}$ | 0.25 | 3.3 | 85 | 85 | 8.5 | 6 |
| $Pt/\gamma$-$Al_2O_3$ ($CO_2$:$H_2$ = 1 : 1) | 23 | 0.34 | 42 | 42 | 28 | 10 |
| $[TiFe_{0.95}Zr_{0.03}Mo_{0.02}]H_{0.36}$ + $Pt/\gamma$-$Al_2O_3$ | 0.25 | 3.7 | 99 | 99 | 0.1 | 0.5[a] |

[a] +0.4% formaldehyde.

selective reduction was observed when $CO_2$ reacted with the strongly bound hydrogen (SBH).

Hydrogenation of $CO_2$ with the SBH ($[TiFe_{0.95}Zr_{0.03}Mo_{0.02}]H_{0.36}$) was found to occur most readily at 430 °C to produce CO and $C_1$–$C_5$ hydrocarbons: 60% of the carbon dioxide was converted into CO with a selectivity of 80% during the first hour. Virtually complete $CO_2$ conversion was observed in the second hour of circulation; however, the selectivity of CO formation decreased to 50% and the yield of hydrocarbons increased. A much lower degree of $CO_2$ conversion (22%) with high selectivity in favor of $CO_2$ reduction to CO (98–100%) was observed after replacement of the hydrogenation products with a fresh portion of $CO_2$. These data indicate that CO hydrogenation can produce hydrocarbons.

The selectivity in favor of CO formation did not exceed 23% at a 40% conversion of $CO_2$ in the reaction of an $H_2$/$CO_2$ mixture in the presence of the $Pt/Al_2O_3$ catalyst (430 °C, 10 atm). The major reaction products are $C_1$–$C_4$ hydrocarbons, mainly methane (see Table 11.3).

In the presence of a $[TiFe_{0.95}Zr_{0.03}Mo_{0.02}]H_{0.36}$–$Pt/Al_2O_3$ mixture, the conversion of $CO_2$ reached 58% with 98–100% selectivity in favor of CO formation at 430 °C.

When the intermetallic catalyst or its mixture with $Pt/Al_2O_3$ was used, the $CO_2$ conversion rapidly decreased during the course of the experiment. However, after removal of the hydrogenation products by Ar circulation and subsequent replacement of the Ar by a fresh portion of $CO_2$, the catalytic activity in $CO_2$ hydrogenation was restored until the SBH was exhausted. Slow evolution of SBH into the reaction vessel was observed during experiments on $CO_2$ reduction. Accumulation of $H_2$ causes a decrease in the CO selectivity due to more extensive hydrogenation of carbon oxides with dihydrogen from the gas phase.

The correlation between the selectivity of CO formation and SBH content in the reaction system suggests the participation of SBH in the reactions giving rise to CO formation (Figure 11.11). Both the reduction of $CO_2$ to hydrocarbons and, possibly, carbon deposition on the surface of the heterogeneous system occur at the expense of the gas-phase $H_2$.

The amount of carbon deposited on the surface of the intermetallic catalyst after 11 h of the gas-phase reaction was found to be 0.14 g and corresponded to the

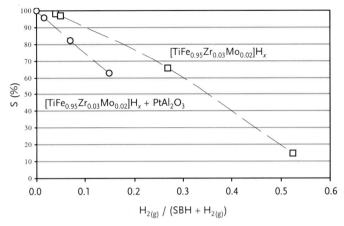

**Figure 11.11** Selectivity of CO$_2$ to CO conversion as a function of H$_2$ content in the gas phase ($T = 430\,°C$, $P = 10$ atm).

transformation of 0.012 mol of CO$_2$. The total amount of CO$_2$ converted was 0.079 mol, of which 84.8% was converted into CO and C$_1$–C$_5$ hydrocarbons, while 15.2% was converted into carbon.

Apparently, the selective CO$_2$ to CO reduction is a result of stoichiometric reaction between CO$_2$ and SBH hydride bound to the metal atoms of the intermetallic lattice, whereas the hydrocarbon formation involves hydrogenation of CO with H$_2$ from the gas phase.

Typical kinetic curves for the transformation of CO$_2$ in the presence of the mixed intermetallic and Pt/Al$_2$O$_3$ catalyst (Figure 11.12) show that the rate of CO$_2$ transformation per operation cycle decreases as the formation of CO proceeds. The curve for CO accumulation shows a maximum. A poorly defined induction period is typical of plots of the yield of hydrocarbons *vs.* the reaction time. All the above data, and the increase in the hydrocarbon yield as the CO content decreases, are consistent with the consecutive nature of CO$_2$ hydrogenation, i.e. CO$_2 \rightarrow$ CO $\rightarrow$ C$_1$–C$_4$.

One of the major components of the intermetallic catalyst is iron. The formation of stable iron carbonyl complexes can prevent CO$_2$ chemisorption at the active sites of [TiFe$_{0.95}$Zr$_{0.03}$Mo$_{0.02}$]H$_x$. Diffusion of the SBH hydride through the lattice of the intermetallic [TiFe$_{0.95}$Zr$_{0.03}$Mo$_{0.02}$]H$_x$ towards the surface might be an additional factor leading to the decrease in the CO$_2$ conversion during the reaction (see Figure 11.12).

The initial rate of CO$_2$ reduction with H$_2$ in the presence of the platinum/alumina catalyst is an order of magnitude lower than the initial rate of stoichiometric reactions of CO$_2$ with [TiFe$_{0.95}$Zr$_{0.03}$Mo$_{0.02}$]H$_x$ or the mixture of [TiFe$_{0.95}$Zr$_{0.03}$Mo$_{0.02}$]H$_x$ and Pt/Al$_2$O$_3$ (see Table 11.3).

Experiments on cyclohexane dehydrogenation coupled with CO$_2$ reduction were carried out with the catalysts [TiFe$_{0.95}$Zr$_{0.03}$Mo$_{0.02}$]H$_x$ and its mixture with Pt/Al$_2$O$_3$, in which SBH was consumed almost completely in the first cycle of the experiment.

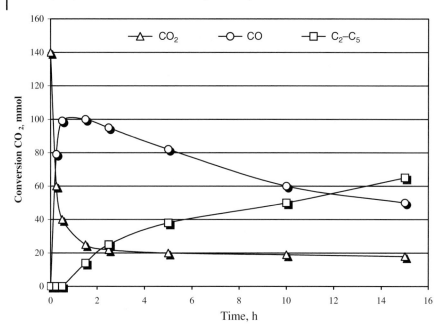

**Figure 11.12** Typical kinetic curves for $CO_2$ hydrogenation.

## 11.4.1
### Reduction of $CO_2$ with Hydrocarbons

The reduction of $CO_2$ with hydrocarbons is thermodynamically restricted under standard conditions.

The $CO_2/CH_4$ reaction requires much higher temperature than the $CO_2/C_6H_{12}$ reaction because of the entropy factor. Dehydrogenation of cyclohexane catalyzed by a nickel–zirconium intermetallic compound combined with the platinum/alumina catalyst produces benzene in a yield substantially exceeding that achieved in the presence of traditional catalysts [40]. Dehydrogenation of cyclohexane in a $CO_2$ atmosphere is catalyzed by the $[TiFe_{0.95}Zr_{0.03}Mo_{0.02}]H_x$–$Pt/Al_2O_3$ mixture at 430 °C. Both the catalyst components are also active in this reaction. The activity of the pure $[TiFe_{0.95}Zr_{0.03}Mo_{0.02}]H_x$ catalyst is much lower than that of the mixed or industrial $Pt/Al_2O_3$ catalyst. The selectivity in favor of benzene production decreases in the series $[TiFe_{0.95}Zr_{0.03}Mo_{0.02}]H_x + Pt/Al_2O_3 > Pt/Al_2O_3 > [TiFe_{0.95}Zr_{0.03}Mo_{0.02}]H_x$.

During dehydrogenation of cyclohexane, the concentration of molecular hydrogen in the reaction vessel markedly increases and $CO_2$ hydrogenation substantially accelerates.

$$cyclo\text{-}C_6H_{12} \rightarrow C_6H_6 + 3\ H_2$$
$$\underline{3\ CO_2 + 3\ H_2 \rightarrow 3\ CO + 3\ H_2O}$$
$$cyclo\text{-}C_6H_{12} + 3\ CO_2 \rightarrow C_6H_6 + 3\ CO + 3\ H_2O$$

In parallel to the coupled reaction, dehydrogenation of cyclohexane and a number of side reactions occur. For instance, an increase in the concentration of molecular hydrogen decreases the selectivity in favor of CO formation due to extensive hydrogenation of $CO_2$ with molecular hydrogen to give hydrocarbons. Even after termination of the cyclohexane supply to the reaction zone, a prolonged circulation of the gaseous products over the catalysts $[TiFe_{0.95}Zr_{0.03}Mo_{0.02}]H_x$ or $[TiFe_{0.95}Zr_{0.03}Mo_{0.02}]H_x$–$Pt/Al_2O_3$ results in the transformation of $CO_2$ and CO into hydrocarbons. However, a certain amount of hydrogen formed through cyclohexane dehydrogenation seems to be absorbed by the intermetallic compound to form SBH. As a consequence of the SBH formation, the selectivity of CO formation again reaches 99–100% at a carbon dioxide conversion of ~20% after the replacement of hydrogenation products by a fresh portion of $CO_2$. Besides the above reactions, cyclohexane hydrogenolysis occurs to some extent.

Primary and secondary alcohols can also be used as hydrogen donors. A number of reactions involving alcohol participation should be taken into account in the $CO_2/H_2$/alcohol system.

The intermetallic compound $[TiFe_{0.95}Zr_{0.03}Mo_{0.02}]H_x$ contains two types of absorbed hydrogen, one being strongly bound and the other being weakly bound to the intermetallic lattice. The reactivities of these hydrogen species in $CO_2$ hydrogenation are different. Selective $CO_2$ to CO reduction, to degrees of conversion of up to 60%, is attained only when SBH participates in the process. The gaseous hydrogen is predominantly involved in CO reduction to produce hydrocarbons.

## 11.4.2
### Mechanistic Aspects: Role of SBH in Selective CO₂ Reduction

In most hydrogen-accumulating systems of the $ABH_x$ type, including TiFe, $H_2$ absorption is accompanied by the formation of a boundary layer of interstitial solid solutions and hydride phases characterized by a constant composition [19]. However, $H_2$ absorption by such intermetallic compounds only commences after the formation of the iron metallic phase. This is due to a low catalytic activity of the Ti–Fe binary system with respect to hydrogen dissociation on its surface.

Modification of the intermetallic compound with small amounts of Zr and Mo leads to marked acceleration in the process of $H_2$ absorption with retention of the cubic crystal structure and relatively high absorption capacity [17, 41].

A study by Mössbauer spectroscopy showed that the incorporation of hydrogen into the TiFe structure leads to a substantial decrease in the occupancy of the valence level of iron by electrons and, therefore, a partial transfer of the electron density to the hydrogen atoms linked to the lattice [19].

Evidently, the high rate of $CO_2$ reduction and the enhanced selectivity of CO formation are due to chemisorption of $CO_2$ at the sites bearing hydride hydrogen. The subsequent steps may involve destruction of the surface formyl groups (designated by subscript 's') to form CO and water, which is typical of mechanisms postulated in homogeneous catalysis [42] (see Scheme 11.4).

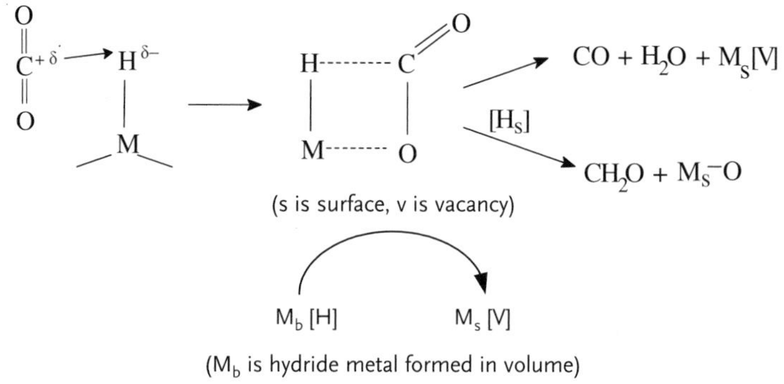

Scheme 11.4 Proposed mechanism of $CO_2$ activation.

When hydrogen attached to the intermetallic surface has been completely consumed for $CO_2$ reduction, a vacancy appears in the coordination sphere of the hydride-forming metal atom; apparently, hydrogen migrates by way of this vacancy to the surface of the system. This may account for the fact that during $CO_2$ reduction virtually all of the SBH is involved in the reduction.

A decrease in the rate of $CO_2$ reduction in the reverse water-gas shift reaction that occurs on the surface of nickel catalysts is due to the strong adsorption of CO, which blocks the active sites [28, 37]. The accumulation of CO on the surface of the present system might result in the formation of stable iron carbonyl complexes, which hamper the access of $CO_2$ to the active sites. Treatment of the surface with Ar or fresh $CO_2$ restores the ability of I to reduce $CO_2$.

In the presence of the catalytic composition III, consisting of the intermetallic compound and $Pt/Al_2O_3$ catalyst, the selectivity of CO formation upon $CO_2$ hydrogenation at 430 °C is markedly higher than that in the presence of either of these individual catalysts at the same temperature. Among other reasons, this synergistic effect may be due to a spillover of hydride hydrogen occurring at the surface of the intermetallic compound to the surface of the platinum/alumina catalyst and its participation in $CO_2$ reduction. It might also be suggested that the phase transfer of hydrogen is favored by the contact of highly dispersed intermetallic particles arising as a result of embrittlement of I with the pellets of the alumino-platinum catalyst. In the presence of the mixed catalyst (III), the absorption of $H_2$ starts at a lower temperature than in the presence of either catalysts I and II loaded separately. This points to a possible role of phase transfer of $H_2$.

$H_2$ phase transfer is observed in hydrogenation reactions over mixed catalytic compositions, in particular, a mixture of nickel–zirconium intermetallic compound and the platinum/alumina catalyst [43, 44]. One of the most widely accepted views on the spillover mechanism implies the heterophase interaction of heterogeneous systems, resulting in tight contact of the surface layers of the pellets of different catalysts [45, 46].

The synergism in $CO_2$ hydrogenation coupled with cyclohexane dehydrogenation in the presence of catalyst III is evidently due to different functions of the system components. The first stage of this process, that is, the dehydrogenation of cyclohexane, is efficiently catalyzed by catalyst **II** and produces molecular hydrogen in a relatively high concentration; as a consequence, extensive hydrogenation of $CO_2$ to CO and light $C_1$–$C_5$ hydrocarbons ensues. However, a certain amount of the $H_2$ formed is absorbed from the reaction zone by the intermetallic compound and consumed almost entirely for $CO_2$ hydrogenation. The reversibility of absorption at elevated temperatures is also indicated by the data for the thermal desorption of $H_2$ (see Figure 11.4).

The formation of active sites containing SBH leads to a higher selectivity in the reduction of $CO_2$ to give CO after the gaseous products containing a high concentration of molecular hydrogen, obtained through cyclohexane dehydrogenation, have been replaced by a fresh portion of $CO_2$.

## 11.5
## Reductive Dehydration of Alcohols

The current production of synthetic methanol is close to 30 million tons per year. The estimated ethanol producing capacity at chemical and biochemical plants is nearly 20 million tons per year. In addition, millions of tons of low-molecular-weight linear $C_3$–$C_5$ alcohols are manufactured annually from alkenes [47, 48].

Interest in the chemistry of alcohols started at the time of Butlerov and Sabatier, and related studies have held a central position in the development of fundamental aspects of catalysis, as well as in industrial applications.

Oxidation and oxidative dehydrogenation processes, as well as oxidative condensation to esters, are widely employed in industry for the production of ethyl acetate and methyl formate, formaldehyde, and acetaldehyde [47]. Reduction of alcohols to alkanes with retention of the length and geometry of carbon skeleton is used in fine organic synthesis [49].

$$ROH + H_2 \rightarrow R\text{–}H + H_2O$$

Acid-catalyzed dehydration of ethanol was widely used for the production of alkenes (in particular of ethylene) in the era prior to petrochemistry, and the method is still used today in countries such as Brazil, where inexpensive ethanol is available from the fermentation of biomass [47]:

$$C_2H_5OH \xrightarrow{\ H^+\ } CH_2{=}CH_2 + H_2O$$

The carbon chain of an alcohol may be lengthened by carbonylation [49]:

$$CH_3OH + CO \xrightarrow[\text{180 °C, 10–20 atm}]{\text{Rh/CH}_3\text{J}} CH_3COOH$$

The number of carbons may also be increased by the synthesis of diene hydrocarbons, as performed industrially by the methods of Lebedev [50] and Ostromislensky [51]:

$$2 \ C_2H_5OH \rightarrow CH_2=CH^-CH=CH_2 + 2 \ H_2O + H_2$$

$$CH_3CHO + CH_3CH_2CH_2OH \xrightarrow[Al_2O_3]{} CH_2=CH^-CH=CH^-CH_3 + 2 \ H_2O \quad (11.1)$$

Another known condensation reaction is Guerber's reaction discovered in 1899 [52, 53]. This reaction, which involves cross-condensation between alcohols, is currently attracting attention as a route for the production of precursors of quality-improving additives for engine fuels [54, 55]:

$$CH_3CH_2CH_2OH + CH_3OH \xrightarrow[180-220\ ^\circ C]{} (CH_3)_2CHCH_2OH \quad (11.2)$$

The Mobil company has developed a method for the conversion of methanol into a mixture of hydrocarbons with a boiling range corresponding to that of engine fuels [47, 56]:

$$CH_3OH \xrightarrow{H^+} [H_2O + (CH_3)_2O] \xrightarrow[400-420\ ^\circ C,\ 2-15\ atm]{ZSM\text{-}5}$$

$$C_nH_{2n+2} + ArH + C_nH_{2n} + H_2O \quad (11.3)$$

In 1986, this method was successfully commercialized in New Zealand to produce about 600 thousand tons of engine fuel with octane number 95 that was free of N- and S-containing compounds [47].

Our studies revealed a new reaction, so-called reductive dehydration [57]. The new reaction allows one to convert alcohols to alkanes with at least twice the number of carbons in the hydrocarbon chain as compared to the original alcohol:

$$2 \ ROH + H_2 \rightarrow R\text{-}R + 2 \ H_2O$$

Earlier studies [57–61] of the mechanism of the Fischer–Tropsch reaction over fused iron catalysts at 1 atm have established that the alcohol molecules are not only incorporated into the growing chain but also interact with the catalyst and participate in the creation of an initiating center responsible for the formation of the reaction products. It is especially important to note that alkenes related to alcohols are incapable of behaving similarly.

It is interesting that an alcohol is unable to play the role of an initiating agent and to become incorporated into the growing chain at elevated pressures of CO and $H_2$. Apparently, these molecules can compete with alcohols for the coordination sites at the active center [62–64].

These facts impelled us to study the transformation of various alcohols under reduction conditions in the presence of a series of modified iron catalysts. The promoted and unpromoted reduced molten iron catalysts (PRMIC and RMIC, respectively) were investigated in the first set of experiments. These catalysts are well known in the synthesis of alcohols from CO and $H_2$ as well as in the Fischer–Tropsch synthesis. Intermetallic catalytic systems have been used in the later experiments.

## 11.5.1
## Cycloalkanones and Cycloalkanols

Using cyclopentanol and cyclohexanol as starting compounds, it was found that, over PRMIC in the temperature range 250–250 °C and at a $H_2$ pressure of 0.1–1 MPa, these cycloalkanols react with hydrogen with the elimination of a water molecule and give rise to di- and polycyclic hydrocarbons. The carbon skeletons of the latter were constructed from the cycloalkyl/cycloalkenyl groups of the starting alcohols. Among the products derived from cyclopentanol, various hydrocarbons and their functionalized derivatives (I–XI) were identified (Table 11.4 and Scheme 11.5).

**Table 11.4** Typical products of cyclopentanol and cyclopentanone conversions over PRMIC.

| No. | Products | No. | Products |
|-----|----------|-----|----------|
| I | | VII | |
| II | | VIII | |
| III | | IX | |
| IV | with one C=C bond | X | |
| V | | XI | |
| VI | with one C=C bond | | |

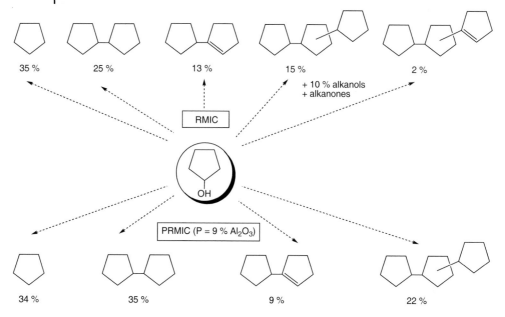

**Scheme 11.5** Reductive conversion of cyclopentanol 250 °C,
$P_{H2} = 0.7$ MPa, $V_{H2} = 10^3$ h$^{-1}$, $V_{alcohol} = 160$–$180$ g [h kg]$^{-1}$.

**Table 11.5** Di- and polycycloalkanes formed from cyclopentanol over PRMIC
(250 °C, $P_{H2} = 0.7$ mPa, space velocity of H$_2$ $10^3$ h$^{-1}$, 160–180 g (h kg)$^{-1}$).

| No. | PRMIC composition (%) | Conversion of alcohols (%) | Selectivity[1] (%) | | | | | | | |
|-----|------------------------|------------------------------|------|------|------|------|------|------|------|------------------|
| | | | I | II | III | IV | V | VI | VIII | Σ VIII, IX, X, XI |
| 1 | Dopant-free | 98 | 35 | 0 | 25 | 13 | 15 | 2 | 0 | 10 |
| 2 | 9% Al$_2$O$_3$ | 99.5 | 34 | 0 | 35 | 9 | 22 | 0 | 0 | 0 |
| 3 | 4.7% V$_2$O$_5$ | 99.5 | 38 | 1 | 40 | 1 | 19 | 1 | 0 | 0 |
| 4 | 1% Al$_2$O$_3$ + 1.5% BaO | 99.5 | 27 | 0 | 12 | 7 | 36 | 6 | 6 | 6 |
| 5[2] | 3.1% Al$_2$O$_3$ + 2.2% CaO + 0.7% K$_2$O | 96 | 23 | 0 | 13 | 13 | 6 | 1 | 4 | 41 |
| 6 | 3.8% BaO + 4.5% V$_2$O$_5$ | 99.5 | 35 | 3 | 23 | 5 | 31 | 1 | 0 | 2 |
| 7 | 4.7% V$_2$O$_5$ + 1% Cu | 99.5 | 18 | 0 | 32 | 2 | 40 | 1 | 7 | 0 |

[1] Designations of the reaction products are the same as in Table 11.4.
[2] 240 g (h kg)$^{-1}$.

For the reductive dehydration of alcohols: $P_{H2} = 0.7$ MPa, volume rate of hydrogen $10^3$ h$^{-1}$, space rate of alcohol 160–170 g h$^{-1}$ (kg$_{cat}$)$^{-1}$; temperatures in the range 205–270 °C are optimal for cyclopentanone and 300–350 °C for cyclohexanol. In the case of cyclopentanol, the best results were achieved in the presence of the catalyst promoted with a mixture of vanadium pentoxide (4.7%) and copper oxide (1%) (Tables 11.4 and 11.5). The cyclopentanol conversion exceeded 99.5% and the sum selectivity for di-, tri-, and tetracyclic hydrocarbons (III + V + VII) was about 79% with this catalyst. The high conversion of cyclohexanol (~99%) gave rise to cyclohexylcyclohexane (4–12%) and dicyclohexylcyclohexane (up to 4%).

Thermodynamic calculations on the system cycloalkanol + hydrogen indicated that at the temperatures and reagent ratios used in the experiments an equilibrium mixture of cycloalkanol and cycloalkanone should contain up to 10% of the latter, despite the presence of hydrogen. As we did not find any noticeable amounts of cycloalkanones in the reaction products, it may be suggested that one of the ways in which alcohols are converted into alkanes involves acid-catalyzed reactions involving cycloalkanones. It should be noted that the formation of hydrocarbons from cycloalkanones in the presence of Ni/kieselguhr has been described previously [65].

Therefore, we also studied reactions of cyclopentanone and cyclohexanone with hydrogen over catalysts of various compositions. The reactions were performed under optimal conditions used for the conversion of the corresponding alcohols into di- and polycyclic hydrocarbons (Tables 11.6 and 11.7).

The data in Table 11.7 illustrate that the total yield of di-, tri-, and tetracyclic hydrocarbons was somewhat higher when $V_2O_5$ and CuO were employed as promoters. Variation of the ratio between the promoters has a small effect on the sum yield of the products of reductive dehydration. In the case of cyclohexanone, the total yield of the $C_{12}$ and $C_{18}$ reaction products (cyclohexylcyclohexane, dicyclohexylcyclohexane, cyclohexylcyclohexene, phenylcyclohexane, cyclohexyl-cyclohexanone, cyclohexylcyclohexanol and dicyclohexylcyclohexanone) varied between 30 and 60% depending on the nature of the promoter (Table 11.7).

Moreover, one can see from Table 11.5 that, as a whole, the composition and distribution of the products obtained from cyclopentanol changed only slightly even when significant amounts of such an acidic promoter as $Al_2O_3$ were incorporated into the catalyst (see also Scheme 11.5).

Similarly to cyclopentanol, the reaction of cyclohexanone does not show a strong dependence on the nature of the promoter (Scheme 11.6).

Comparison of the data presented in Tables 11.5–11.7 indicates that the yields of hydrocarbons are essentially the same starting from either cycloalkanols or cycloalkanones in the presence of catalysts of the same composition. This suggests that the transformation of an alcohol into di- or polycycloalkanes may involve an intermediate dehydrogenation of the alcohol to a ketone. Conversely, the conversion of a cycloalkanone to a hydrocarbon may be considered to involve the reduction of the ketone to the cycloalkanol as an intermediate step.

**Table 11.6** Di- and polycyclic hydrocarbons formed from cyclopentanone over PRMIC (250 °C, $P_{H2}$ = 0.8 MPa, 160 g (h kg)$^{-1}$).

| No. | PRMIC composition[1] (%) | Conversion (%) | Selectivity[2] (%) | | | | | |
|---|---|---|---|---|---|---|---|---|
| | | | I | III | IV | V | VII | Σ VIII, IX, X, XI |
| 1 | Dopant-free | 99.5 | 18 | 7 | 6 | 6 | 5 | 48 |
| 2 | 9% $Al_2O_3$ | 99.5 | 30 | 35 | 8 | 22 | 0 | 13 |
| 3 | 4.7% $V_2O_5$ | 99.5 | 14 | 29 | 8 | 42 | 0 | 5 |
| 4 | 1% $Al_2O_3$ 1.5%BaO | 99.5 | 10 | 14 | 10 | 36 | 8 | 12 |
| 5 | 3.8% BaO 4.5% $V_2O_5$ | 99.5 | 19 | 30 | 5 | 33 | 0 | 13 |
| 6 | 3.1% $Al_2O_3$ + 2.2% CaO + 0.7% $K_2O$ | 98 | 18 | 7 | 6 | 6 | 5 | 48 |
| 7 | 4.7% $V_2O_5$ 0.5% CuO | 99.5 | 5 | 42 | 2 | 46 | 5 | 0 |
| 8 | 4.7% $V_2O_5$ 1%CuO | 99.5 | 7 | 35 | 3 | 50 | 4 | 1 |
| 9 | 4.7% $V_2O_5$ 2%CuO | 99.5 | 5 | 40 | 2 | 47 | 6 | 0 |
| 10 | 4.7% $V_2O_5$ 2.5%CuO | 99.5 | 6 | 3 | 4 | 48 | 6 | 0 |
| 11 | 6% $V_2O_5$ 2%CuO | 99.5 | 5 | 41 | 2 | 46 | 6 | 0 |

[1] PRMIC composition relates to the non-reduced sample.
[2] Designations of the reaction products are the same as in Table 11.4.

**Table 11.7** Di- and tricyclic fused hydrocarbons formed from cyclohexanone over PRMIC ($P_{H2}$ (0.8–1) mPa, 160 g (h kg)$^{-1}$).

| No. | Catalyst composition[1] (%) | T (°C) | Conversion (%) | Selectivity[2] (%) | | | | | | | | | | | |
|---|---|---|---|---|---|---|---|---|---|---|---|---|---|---|---|
| | | | | XII | XIII | XIV | XV | XVI | XVII | XVIII | XIX | XX | XXI | XXII | XXIII |
| 1 | 4.7% $V_2O_5$ | 300 | 99.5 | 20 | 29 | 18 | 24 | 0 | 0 | 7 | 0 | 2 | 0 | 0 | 0 |
| 2 | 4.7% $V_2O_5$ | 350 | 99.5 | 25 | 34 | 9 | 19 | 5 | 3 | 5 | 0 | 0 | 0 | 0 | 0 |
| 3 | 4.7% $V_2O_5$ 1% CuO | 300 | 99.5 | 23 | 22 | 17 | 29 | 0 | 0 | 9 | 0 | 0 | 0 | 0 | 0 |
| 4 | 3.1% $Al_2O_3$ + 2.2% CaO + 0.7% $K_2O$ | 300 | 93 | 20 | 6 | 1 | 13 | 13 | 1 | 2 | 3 | 24 | 5 | 2 | 4 |

[1] PRMIC composition relates to the non-reduced sample.
[2] XII – cyclohexane; XIII – cyclohexene; XIV – benzene; XV – cyclohexylcyclohexane (**CC**); XVI – **CC** with one C=C bond; XVII – phenylcyclohexane; XVIII – dicyclohexylcyclohexane (**DCC**); XIX – **DCC** with one C=C bond; XX – cyclohexanol; XXI – cyclohexylcyclohexanone; XXII – cyclohexylcyclohexanol; XXIII – dicyclohexylcyclohexanone.

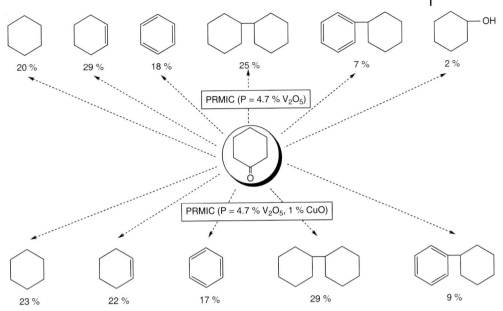

**Scheme 11.6** Reductive conversion of cyclohexanone 300 °C, $P_{H2} = 0.8–1.0$ MPa, $V_{alcohol} = 160$ g [h kg]$^{-1}$.

## 11.5.2
## Benzyl Alcohol and Benzaldehyde

In contrast to cycloalkanones and cycloalkanols, benzaldehyde and benzyl alcohol possess no hydrogen atoms in the α-position to the functional group. Consequently, reactions such as crotonic and aldol condensations, which are in principle possible for cycloalkanones (see previous section), cannot be expected for these substrates. At the same time, the reduction of benzaldehyde over an iron catalyst and tungsten sulfide at a hydrogen pressure of 10 MPa and at 280–330 °C is well known [66, 67].

In the presence of typical methathesis catalysts (WCl$_6$/BuLi, WCl$_6$/LiAlH$_4$, WCl$_6$/s-BuLi [68, 69]), benzaldehyde is transformed into cis- and trans-stilbenes, i.e. possible precursors of diphenylethane. It is important to take into account that, in contrast to the aforementioned systems, PRMIC does not catalyze the methathesis of alkenes.

Our experiments showed that both benzaldehyde and benzyl alcohol react with hydrogen ($P_{H2} = 0.1$ atm) at 200–250 °C in the presence of PRMIC (containing ~4.7% V$_2$O$_5$ as promoter), giving rise to products containing double arylalkyl residues. 1,2-Diphenylethane predominates among such products, but small amounts of trans-stilbenes and even hexaphenylpropane were also observed (cis-stilbenes were not found, see Table 11.8).

Comparison of the reactivities of benzaldehyde and benzyl alcohol under the reaction conditions indicates that the product composition changes only slightly on going from the alcohol to the aldehyde (Scheme 11.7).

**Table 11.8** Products of the conversion of benzaldehyde and its derivatives over PRMIC.

| No. | Reagents | T (°C) | Conv. (%) | Specific velocity of reagent loading, g [h L]$^{-1}$ | Selectivity[1] (%) | | | | | | | |
|---|---|---|---|---|---|---|---|---|---|---|---|---|
| | | | | | XXIV | XXV | XXVI | XXVII | XXVIII | XXIX | XXX | XXXI |
| 1 | PhCHO | 220 | 72 | 240 | 0 | 60 | 32 | 6 | 0.6 | – | 1 | 0.4 |
| 2 | PhCHO | 250 | 97 | 240 | 1.5 | 82 | 15 | 0.1 | 0.1 | – | 0.1 | 0.5 |
| 3 | PhCH$_2$OH | 250 | 96 | 520 | 0.1 | 57 | 29 | 0.1 | 0.5 | 12 | – | 0.8 |
| 4 | (PhCH$_2$)$_2$O | 220 | 38 | 120 | 0 | 93 | 5 | 0.1 | 0 | 1 | 0.1 | – |
| 5[2] | PhCHO | 220 | 47 | 210 | 0 | 65 | 4 | 0.1 | 0 | – | 31 | 0 |
| 6 | PhCH(OH)COPh | 220 | 95 | 90 | 0 | 0.1 | 97 | 0 | 2 | 0 | 0 | 0 |

[1] XXIV – PhH; XXV – PhCH$_3$; XXVI – (PhCH$_2$)$_2$; XXVII – PhCH=CHPh; XXVIII – PhCH$_2$CH(Ph)CH$_2$Ph; XXIX – PhCHO; XXX – PhCH$_2$OH; XXXI – (PhCH$_2$)$_2$O;
[2] D$_2$O was used.

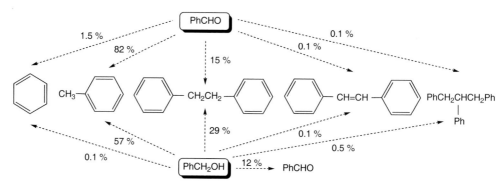

**Scheme 11.7** Reductive conversions of benzaldehyde and benzyl alcohol
PhCHO $\xrightarrow{H_2}$ PhCH$_2$OH $\xrightarrow{H_2}$ (PhCH$_2$)$_2$ + H$_2$O
250 °C, PRMIC (P = 4.7 V$_2$O$_5$).

## 11.5.3
## 3-Methylbutan-1-ol

The above results show that not only the reductive dehydration of cycloalkanols and benzyl alcohol, but also their reduction to cycloalkanes/phenylalkanes with retention of their carbon skeletons, occurs in the presence of PRMIC (see Schemes 11.5–11.7):

$$c\text{-}C_nH_{2n-1}OH + H_2 \rightarrow c\text{-}C_nH_{2n} + H_2O \qquad (11.4)$$

$$C_6H_5CH_2OH + H_2 \rightarrow C_6H_5CH_3 + H_2O$$

Reduction to alkanes appeared to become the major process in the case of alkanols $C_nH_{2n+1}OH$ under the conditions used for the reductive dehydration of cyclo-alkanols. This process has been avoided by using a composite catalytic system comprising the intermetallic compound $[TiFe_{0.95}Zr_{0.03}Mo_{0.02}]H_{0.36}$, which served simultaneously as a catalyst and as a source (reservoir) of hydrogen, in combination with a small amount of the well-known hydrogenating alumino-platinum catalyst $Pt/\gamma\text{-}Al_2O_3$ [70].

Our experiments revealed that the yield of $C_{10}$ isoalkanes does not exceed 2.5% in the presence of the alumino-platinum catalyst when the intermetallic compound is absent. The yield of isoalkanes reached almost 3.2% in the presence of $[TiFe_{0.95}Zr_{0.03}Mo_{0.02}]H_{0.36}$ alone, i.e. in the absence of $Pt/\gamma\text{-}Al_2O_3$. When using the combination $Pt/\gamma\text{-}Al_2O_3 + [TiFe_{0.95}Zr_{0.03}Mo_{0.02}]H_{0.36}$, the yield of isoalkanes was at least 10% (Table 11.9).

Table 11.9 Conversion of 3-methylbutan-l-ol in the presence of various catalytic systems (Ar, $(Ar : H_2)* = 1 : 1$, 350 °C, 50 atm., space velocity 0.01–0.02 h$^{-1}$).

| Catalysts | *$Pt/\gamma\text{-}Al_2O_3$ | $[TiFe_{0.95}Zr_{0.03}Mo_{0.02}]H_{0.36}$ | $[TiFe_{0.95}Zr_{0.03}Mo_{0.02}]H_{0.36}$ + $Pt/\gamma\text{-}Al_2O_3$ |
|---|---|---|---|
| Conversion of iso-pentanol (% mass) | 69.6 | 43.8 | 79.5 |
| Yield of gases based on converted alcohol (% mass) | 52.1 | 31.1 | 64.6 |
| **Gas composition (% mass)** | | | |
| CO | 0.4 | – | 16.8 |
| $CO_2$ | 16.9 | – | 7.7 |
| $\Sigma C_1$–$C_5$ | 82.7 | 100 | 75.5 |
| Yield of liquid products (% mass) | 47.9 | 68.9 | 35.4 |
| **Composition of liquid products (% mass)** | | | |
| *Hydrocarbons* | 41.6 | 16.8 | 84.7 |
| i-$C_6$ (iso-methylpentanes) | – | – | 9.1 |
| i-$C_7$ (dimethylpentanes) | 29.6 | – | 7.1 |
| i-$C_8$ (methylheptanes) | 4.5 | 11.1 | 9.1 |
| i-$C_9$ (methyloctanes) | 7.5 | – | 12.9 |
| i-$C_{10}$ (dimethyloctanes) | – | 5.7 | 42.3 |
| i-$C_{11}$ (methyldecanes) | – | – | 2.1 |
| i-$C_{14}$ (tetramethyldecanes) | – | – | 2.1 |
| *Oxygen-containing* | 58.4 | 83.2 | 15.3 |
| $C_5$ (dimethylpropanal) | 36.7 | – | – |
| $C_5$ (3-methylbutanal) | – | 55.9 | – |
| $C_5$ (methylbutanal) | – | – | 5.7 |
| R-O-R (di-isoamyl ether) | 2.9 | 14 | 2.0 |
| R-COO-R (isoamyl valerate) | 18.8 | 13.3 | 7.6 |

From the data presented in Table 11.9, it can be deduced that the activity of the catalysts under study increases in the following order: $[TiFe_{0.95}Zr_{0.03}Mo_{0.02}]H_{0.36}$ < $Pt/\gamma\text{-}Al_2O_3$ < $(Pt/\gamma\text{-}Al_2O_3 + [TiFe_{0.95}Zr_{0.03}Mo_{0.02}]H_{0.36})$.

The yield of gaseous products (mainly $C_5$ alkanes) is lowest in the presence of $[TiFe_{0.95}Zr_{0.03}Mo_{0.02}]H_{0.36}$ but is considerably higher over $Pt/\gamma\text{-}Al_2O_3$ and the $(Pt/\gamma\text{-}Al_2O_3 + [TiFe_{0.95}Zr_{0.03}Mo_{0.02}]H_{0.36})$ mixture. The yield of liquid products is higher when using the intermetallic compound alone than in the cases of the alumino-platinum catalyst or the mixture of both catalysts.

The composition of the liquid fraction in the case of IMC and the mixture of catalysts, however, differs significantly. Almost one-third of the alcohol does not undergo reduction, and instead is converted to dimethyloctane through reductive dehydration with coupling of its alkyl residues.

$$2\ \underset{H_3C}{\overset{H_3C}{>}}CHCH_2CH_2OH + 2\ H_2 \rightarrow \underset{H_3C}{\overset{H_3C}{>}}CHCH_2CH_2\text{--}CH_2CH_2CH\overset{CH_3}{\underset{CH_3}{<}} + 2\ H_2O$$

In addition to the major product, 2,7-dimethyloctane, other isoalkanes such as methylpentane, dimethyloctanes, dimethylhexanes, dimethylheptanes, methyl-decanes, and hexamethyldecane are produced. The formation of these minor reaction products gives evidence of the occurrence of cracking processes under the vigorous experimental conditions that accompany the main reaction.

**Table 11.10** Products of 2-methylpropan-1-ol conversion in the presence of the $TiFe_{0.095}Zr_{0.03}Mo_{0.02} + Pt/\gamma\text{-}Al_2O_3$ catalytic system ($T = 350\ °C$; $P = 50$ atm., space velocity $0.01–0.02\ h^{-1}$).

| | |
|---|---|
| Conversion of 3-methylpropan-1-ol (% mass) | 67 |
| Yield of gaseous products (% mass) | 12 |
| **Composition of gaseous products (% mass)** | |
| $CO_2$ | 35.8 |
| $\Sigma C_1\text{--}C_5$ | 64.2 |
| Yield of liquid products (% mass) | 88 |
| **Composition (% mass)** | |
| *Hydrocarbons* | 84.6 |
| i-$C_7$ (dimethylpentanes) | 7.9 |
| i-$C_8$ (methylheptanes) | 52.6 |
| i-$C_{10}$ (dimethyloctanes) | 4.6 |
| i-$C_{17}$ (iso-methylheptadecanes) | 8.7 |
| i-$C_{18}$ (methyloctadecanes) | 5.7 |
| *Aromatics* | |
| Xylenes | 5.1 |
| *Oxygen-containing products* | 15.4 |
| $C_4$ (2-methylpropanal) | 8.9 |
| R-COO-R (isobutyl isobutyrate) | 6.5 |

11.5.4
**2-Methylpropan-1-ol**

Experiments performed in the presence of the combined $Pt/\gamma\text{-}Al_2O_3$ + $[TiFe_{0.95}Zr_{0.03}Mo_{0.02}]H_{0.36}$ catalyst showed that of the total hydrocarbon fraction obtained from the conversion of 2-methylpropan-1-ol, $C_8$ hydrocarbons accounted for at least 60% by weight. The yield of gaseous products from the conversion of 2-methylpropan-1-ol was several times lower than those for similar reactions of isoamyl alcohol, as revealed by the data presented in Table 11.10. The yield of the liquid products increased noticeably on going from 3-methylbutan-1-ol to 2-methylpropan-1-ol. As for the reaction of 3-methylbutan-1-ol, products originating from the coupling of two alkyl groups are predominant in the hydrocarbon fraction (Figures 11.13 and 11.14).

Besides the formation of isoalkanes, aromatization processes occur in the system as well. This is evidenced by the presence of aromatic hydrocarbons, specifically xylenes.

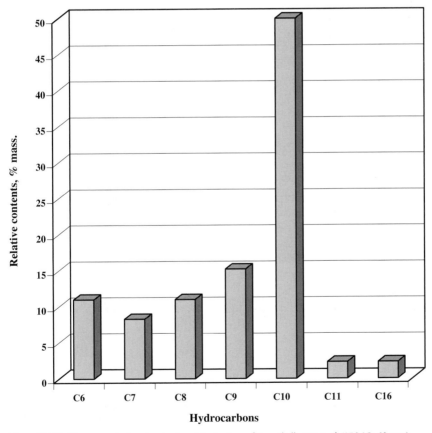

**Figure 11.13** Alkane products of reductive dehydration of 3-methylbutan-1-ol (350 °C; 60 atm).

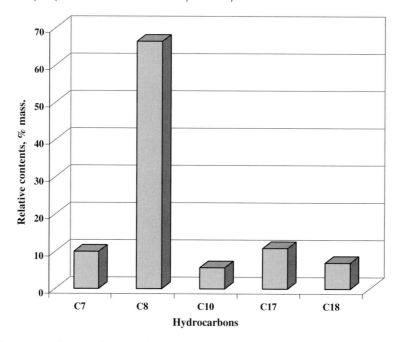

**Figure 11.14** Alkane products of reductive dehydration of 2-methylpropan-1-ol (350 °C; 60 atm).

## 11.5.5
### Ethanol

In contrast to isobutanol and isopentanol, which form mainly the products of hydrocarbon skeleton dimerization, ethanol is converted to $C_6$–$C_{11+}$ alkanes over a catalyst system composed of the intermetallic [$TiFe_{0.95}Zr_{0.03}Mo_{0.02}$]$H_{0.36}$ and commercial platinum/alumina catalyst. The total activity and selectivity in favor of alkane formation depends strongly on the experimental conditions. The yield of alkanes obtained by alcohol dehydration can reach up to 50%.

The reactivity of alcohols appeared to be rather sensitive to the replacement of inert argon by carbon dioxide. While an argon atmosphere most likely favors the attachment of alcohol molecules to the catalyst surface under the experimental conditions (50 atm, 350 °C), the role of $CO_2$ is more complicated. We found that $CO_2$ can be reduced to CO in the presence of the catalytic system under study. Table 11.11 gives details of the alkene content of the products of ethanol conversion obtained under Ar and $CO_2$ atmospheres. As can be seen, under Ar the oxygen-containing products predominated. When the gas phase was changed from Ar to $CO_2$, alkanes were formed as reaction products.

The carbon monoxide formed in this reaction may change the properties of the active centers at the transition metal atoms. It may also react with the alcohol molecules or with intermediates (Table 11.11).

**Table 11.11** Products of ethanol conversion over the $[TiFe_{0.095}Zr_{0.03}Mo_{0.02}]H_{0.36} + Pt/\gamma\text{-}Al_2O_3$ catalytic system and the same system after its pre-treatment with $CO_2$ ($T = 350\,°C$; $P_{Ar} = 50$ atm., space velocity 0.01–0.02 $h^{-1}$).

| Pre-treatment of catalyst | – | $CO_2$ |
|---|---|---|
| Conversion of ethanol (% mass) | 41.6 | 96 |
| Yield of gaseous products (% mass) | 87.7 | 37.1 |
| **Composition of gaseous products (% mass)** | | |
| $CO_2$ | – | 11.2 |
| CO | 49.5 | 45.6 |
| $\Sigma C_1\text{–}C_4$ | 50.6 | 43.2 |
| Yield of liquid products (% mass) | 12.3 | 62.9 |
| **Composition (% mass)** | | |
| *Hydrocarbons* | 0.54 | 87.5 |
| $C_6$ (*n*-hexane) | – | 21.5 |
| $C_6$ (iso-hexane) | 0.54 | 48.1 |
| i-$C_7$ (dimethylpentanes) | – | 2.9 |
| i-$C_8$ (methylheptanes) | – | 2.2 |
| *n*-$C_9$ (*n*-nonane) | – | 1.6 |
| i-$C_9$ (iso-nonane) | – | 8.4 |
| i-$C_{10}$ (iso-decanes) | – | 2.8 |
| *Oxygen-containing products* | 99.5 | 12.5 |
| acetaldehyde | 62.7 | – |
| dimethyl ether | 16.7 | 1.1 |
| butanal | 3.6 | – |
| methyl acetate | 1.5 | – |
| butanol | 8.7 | 0.7 |
| diethyl ether | 0.9 | – |
| ethyl butyl ether | 5.4 | 6.9 |
| dibutoxyethane | – | 0.5 |
| diethoxypropane | – | 2.6 |
| ethyl butyl ether | – | 0.7 |

## 11.5.6
## Mechanistic Speculations

The expected primary products of the conversion of alcohols in the presence of metal catalysts are aldehydes and ketones arising through dehydrogenation. As noted above, thermodynamic estimates predict that up to 10% of the alcohol can be dehydrogenated to form carbonyl compounds even at $P_{H2} = 1$ MPa. Meanwhile, ketones and aldehydes can undergo crotonic-type condensation on the acidic centers of the catalyst. Indeed, Ostromislensky's reaction (Eq. 11.1) [51] may possibly proceed through dehydration of the alcohol with the formation of an alkene, followed by interaction of the alkene with the protonated form of an aldehyde:

$$CH_3CH_2CH_2OH \xrightarrow[Al_2O_3]{} CH_2{=}CH{-}CH_3 + H_2O$$

$$CH_3CH{=}O + H^+ \rightarrow CH_3C^+H{-}OH \xrightarrow{CH_3CH{=}CH_2} CH_3C^+H{-}CH_2{-}CHOH{-}CH_3$$

$$CH_3C^+H{-}CH_2{-}CHOH{-}CH_3 \xrightarrow{-H^+,\,-H_2O} CH_3CH{=}CH{-}CH{=}CH_2$$

Similarly, in the case of cyclohexanol, one could expect the formation of the enol, the enone, and the corresponding diene, etc., as intermediate compounds:

The ultimate hydrogenation product from all of these intermediates should be dicyclohexyl. At first sight, this scheme might account for the appearance of the above series of hydrocarbons among the observed reaction products. This is the mechanism that was proposed [65] to explain the formation of dicyclohexyl from cyclohexanone in the presence of Ni/keiselguhr catalyst at 250–300 °C and at atmospheric pressure. The acidic carrier used in the present work may accelerate the condensation of cyclohexanone to form the corresponding enone. However, the following facts that have emerged from our experiments seemingly contradict the above suppositions:

a) In experiments with cycloalkanols and cyclohexanone, the content of acidic promoters was quite low. In addition, the incorporation of basic additives, such as BaO, CaO, and $K_2O$, reducing the acidity of the catalyst, did not have a significant effect on its activity (see Table 11.6, exp. 6 and Table 11.7, exp. 4).

b) Aldol condensation is impossible in the case of benzaldehyde. However, a considerable amount of 1,2-diphenylethane was found among the products obtained from the reactions of both benzaldehyde and benzyl alcohol (Table 11.8, expts. 1, 2, and 3). The origin of this compound might be the intermediate formation of benzoin, which is almost completely converted to 1,2-diphenylethane under even milder conditions than benzaldehyde or benzyl alcohol (Table 11.8, exp. 6). Benzoin condensation, however, is specifically catalyzed only by strong bases ($CN^-$; carbanions) [69]. Such anionic centers are most likely absent from the catalysts under study. A scheme invoking the simultaneous action of the same catalyst as both a strong acid and as a strong base would seem to be highly unlikely.

c) Acid-catalyzed condensation of aldehydes in the case of 2-methylpropan-1-ol should result in the formation of the thermodynamically more stable 2,2,4-trimethylpentane. However, dimethylhexanes dominate among the $C_8$ hydrocarbons obtained from the conversion of 2-methylpropan-1-ol.

The foregoing discussion seems wholly inconsistent with mechanisms invoking the predominant dehydrogenation of alcohols and conversion of carbonyl compounds with the catalysts under study acting as strong acids or bases.

It is known [71] that alcohols react stoichiometrically with compounds containing low-valence Ti and Nb with coupling of their alkyl residues. Similarly, metallic La reacts with aliphatic alcohols and arylalkanols in the presence of chlorotrimethylsilane and catalytic amounts of iodine [72]:

$$R-OH + M \rightarrow R-R + M(OH)_2 \tag{11.5}$$

$$\text{(structure: Ph–C(OH)(CH_3)–Ph)} + M \longrightarrow \text{(structure: Ph–C(CH_3)(Ph)–C(Ph)(Ph))} + M(OH)_2 \tag{11.6}$$

It may be proposed that the metal catalyst in our experiments serves as a one-electron reducing agent that is regenerated by hydrogen. Additional data are, however, required in order to make firm conclusions about the possible mechanism of the observed reactions. The most useful could be kinetic data for such supposed reactions as the reduction of oxides and hydroxides under the conditions of reductive dehydration of cycloalkanols and other alcohols. Moreover, it would be difficult to rationalize the formation of products composed of three or four alkyl residues, as observed in the case of ethanol, in the framework of a scheme including reactions according to Eqs. (11.4) and (11.5).

In addition to the above hypotheses, another possible scheme for the conversion of alcohols should be considered. The first catalytic step of the reaction could be expected to be oxidative addition of the alcohol or carbonyl compound molecule to the M–M bond. In the case of an alcohol, such a process could give rise to an intermediate with a vicinal disposition of the alkyl and hydroxy groups:

$$M-M + R-CH_2-OH \rightarrow \underset{\underset{M}{\overset{|}{\underset{}{}}}{\overset{CH_2R}{\overset{|}{}}} - \underset{\underset{M}{\overset{|}{}}}{\overset{OH}{\overset{|}{}}}$$

Note that the metal alkyl derivative ($M-CH_2-R$) may take part in a side reaction, namely in alcohol reduction to an alkane with retention of the number of carbon atoms.

The reductive elimination of two $RCH_2$ groups could give rise to the formation of alkane $RCH_2CH_2R$. However, to complete the catalytic cycle it is necessary to reduce the MOH groups arising at the catalyst surface with hydrogen. As noted above, such a reduction of hydroxides of Fe, Ti and of other components of the catalytic system needs additional confirmation.

It cannot be ruled out that alkyl $RCH_2$ is capable of transferring one of the hydrogen atoms from the coordinated alkyl group to the neighboring hydroxyl group:

$$\begin{array}{c} \underset{\underset{M}{\overset{|}{}}}{\overset{R}{\overset{H}{\underset{C}{}}}} H \quad \underset{\underset{M}{\overset{|}{}}}{\overset{H}{\overset{|}{O}}} \quad \longrightarrow \quad \underset{\underset{M}{\overset{|}{}}}{\overset{R}{\underset{C}{}}} H \quad H \underset{\underset{M}{\overset{|}{}}}{\overset{H}{O}} \\ \mathbf{1} \end{array} \tag{11.7}$$

The carbene complex (1) generated in this step can be a source of both hydrocarbons with the same number of carbon atoms as in the starting alcohol and of alkanes with a larger number of carbon atoms. The idea of formation of carbenes in the form of hydrates and adducts with dimethyl ether, etc., was proposed by Olah [71] to explain the mechanism of reaction according to Eq. (11.2). Scheme 11.7 was supposed [73] to explain the formation of trans-stilbene from benzyl alcohol through the reaction with the Pd/Mo cluster. Note that thermodynamic estimates of the data in Table 11.12 indicate that the energy needed for the formation of methylene carbene from methanol is close to that required for the formation of synthesis gas and even slightly lower than that for decomposition of ethylene molecule (Table 11.12) [73, 74].

A simultaneous occurrence of acid-base catalyzed processes (aldol condensation) and redox reactions (hydrogenation–dehydrogenation) is traditionally supposed to explain the Guerber' reaction accomplished at 150–220 °C in the liquid phase (solution of strong base $NaOCH_3$ in absolute alcohol) in the presence of the Raney Cu and related catalysts [54, 55].

Obviously, the presence of such a strong base increases the probability of such reactions as aldol condensation involving aldehydes arising from the dehydrogenation of alcohols. It was found, however, that the addition of propanal to a mixture of methanol and propanol decreased the yield of isobutanol [54].

Another feature that is contradictory to the traditional mechanism of the Guerber' reaction is a high selectivity in favor of isobutanol formation. It seems logical that the intermediate formation of propanal and its cross-condensation with formaldehyde should ultimately give rise to 2-methylpropanediol. However, neither this product nor formaldehyde was found.

One of the inexplicable features of the Guerber' reaction is the appearance of a methyl group at the β-carbon atom of the long-chain alcohol. To explain this fact, the authors suggested the participation of the coordinated methylene ligands as reaction intermediates [54]. Naturally, under our experimental conditions (temperature above 250 °C), all the discussed mechanisms can compete with one another, and, in addition, cracking, isomerization reactions, etc., can also occur.

**Table 11.12** Thermodynamic constants of methylene formation.

| | |
|---|---|
| $CH_{4\,(g)} \rightarrow CH_{2\,(g)} + H_{2(g)}$ <br> $\Delta H°_{298} = 111.28$ kcal mol$^{-1}$ <br> $\Delta G°_{298} = 101.4$ kcal mol$^{-1}$ | $CO_{(g)} + 2\,H_{2\,(g)} \rightarrow CH_{2\,(g)} + H_2O_{(g)}$ <br> $\Delta H°_{298} = 62.02$ kcal mol$^{-1}$ <br> $\Delta G°_{298} = 66.2$ kcal mol$^{-1}$ |
| $CH_3F_{(g)} \rightarrow CH_{2\,(g)} + HF_{(g)}$ <br> $\Delta H°_{298} = 87.7$ kcal mol$^{-1}$ <br> $\Delta G°_{298} = 79.52$ kcal mol$^{-1}$ | $\frac{1}{2}\,C_2H_{4\,(g)} \rightarrow CH_{2\,(g)}$ <br> $\Delta H°_{298} = 80.85$ kcal mol$^{-1}$ <br> $\Delta G°_{298} = 74.77$ kcal mol$^{-1}$ |
| $CH_3OH_{(g)} \rightarrow CH_{2\,(g)} + H_2O_{(g)}$ <br> $\Delta H°_{298} = 83.65$ kcal mol$^{-1}$ <br> $\Delta G°_{298} = 73.39$ kcal mol$^{-1}$ | |

Note that the composite catalytic system $Pt/\gamma\text{-}Al_2O_3/intermetallic$ was found to perform a number of unusual reactions. In the catalysis products, especially of ethanol and of its higher homologues, esters and acids were found (see Tables 11.9–11.11). The formation of the latter products seems to be rather unexpected under reductive conditions.

Hence, this work represents the first instance of the conversion of cycloalkanols and methanol homologues into cycloalkanes and alkanes. It seems that the reaction can be employed as a bridge between renewable raw materials (alcohols are the products of fermentation of any biomass) and alkanes, including components of engine fuels. Further studies of these catalytic systems, including efforts to elucidate the mechanisms of their action, are in progress.

## 11.6
## Conclusion

Most theories of catalysis by metals and alloys are based on the theoretical concepts of solid-state physics. Catalysis by metal complexes and ions in solution is mainly focused on the oxidation state of a metal atom and changes in a substrate molecule when it enters into the coordination sphere of the catalyst ion. As a metal particle approaches molecular dimensions, the band structure, which is a decisive factor in bulk metal catalysis, starts to lose meaning. For instance, the above-mentioned palladium giant cluster, the molecule of which contains about 600 metal atoms, exhibits a noticeable distinction between the metal complex-like sites and bulk metal [76]. Electrostatic effects result in such a charge distribution in this cationic particle that the effective charge of the metal atoms on the vertices of the metal polyhedron differs noticeably from that on the faces and inside the polyhedron [77]. All of the above-mentioned features complicate theoretical analysis of the observed catalytic phenomena.

The structures of intermetallics and their hydride complexes are less well understood than those of the palladium giant clusters, the structural features of the latter being based mainly on indirect findings. The observed embrittlement and crushing of the intermetallics upon their saturation with hydrogen suggests that nanosize species are again involved. Hence, band theory is of little value in relation to these intermetallic hydrides. Indirect findings give evidence for the formation of Ti–H, Mo–H, and Zr–H bonds. Some features of these species (viz., the hydride nature of the H atom and a high affinity of the metal atom for the oxygen atom of the $CO_2$ molecule) may be important in delineating the mechanism of $CO_2$ reduction. The reactivity of the metal–metal bonds can change on going from homometallic to heterometallic catalysts, similar to the changes seen in the series C–X where X = C, Cl, Br, etc.

In spite of the significant progress outlined in the above, studies of these complicated but intriguing objects are still at a relatively early stage.

## Acknowledgements

This work was financially supported by the Russian Foundation for Basic Research (project No. 02-03-32853), the Program of the President of the Russian Federation "Leading scientific schools of Russia" (grant No. 1764.2003.03), the Program of Presidium of the Russian Academy of Sciences "Purposeful synthesis of substances with predicted properties and creation of functionalized materials based thereon", and the Program of the Ministry of Industry and Science of the Russian Federation "Chemical transformations involving nano- and supramolecular structures", NATO grants (ENVIR.LG 971292, EST.CLG.977957).

## References

1  V. PONEC, G. C. BOND, *Catalysis by Metals and Alloys*, Elsevier, Amsterdam, **1995**, p. 734.

2  I. I. MOISEEV, M. N. VARGAFTIK, YA. K. SYRKIN, *Doklady Akad. Nauk SSSR*, **1960**, *133*, 377 (in Russian).

3  M. N. VARGAFTIK, V. P. ZAGORODNIKOV, I. P. STOLAROV, I. I. MOISEEV, D. I. KOCHUBEY, V. A. LIKHOLOBOV, A. L. CHUVILIN, K. I. ZAMARAEV, *J. Mol. Catal.* **1989**, *53*, 315.

4  M. N. VARGAFTIK, V. P. ZAGORODNIKOV, I. P. STOLAROV, I. I. MOISEEV, V. A. LIKHOLOBOV, D. I. KOCHUBEY, A. L. CHUVILIN, V. I. ZAIKOVSKY, K. I. ZAMARAEV, G. I. TIMOFEEVA, *J. Chem. Soc., Chem. Commun.* **1985**, 937.

5  M. N. VARGAFTIK, V. P. ZAGORODNIKOV, I. P. STOLAROV, I. I. MOISEEV, D. I. KOCHUBEY, V. A. LIKHOLOBOV, A. L. CHUVILIN, K. I. ZAMARAEV, *J. Mol. Catal.* **1989**, *53*, 315.

6  M. N. VARGAFTIK, I. I. MOISEEV, D. I. KOCHUBEY, K. I. ZAMARAEV, *Faraday Discuss.* **1991**, *92*, 13.

7  P. CHINI, *J. Organometal. Chem.* **1980**, *200*, 37.

8  I. I. MOISEEV, M. N. VARGAFTIK, *Catalysis by Di- and Polynuclear Metal Complexes* (Eds.: F. A. COTTON, R. ADAMS), Wiley-VCH, New York, 1998, p. 395.

9  K. J. BUSCHOW, P. C. P. BOUTEN, A. R. MIEDEMA, *Rep. Prog. Phys.* **1982**, *45*, 1039.

10  V. V. LUNIN, O. V. KRYUKOV, *Kataliz: Fundamental'nye prikladnye issledovaniya* (Catalysis: Basic and Applied Studies) (Eds.: O. A. PETRICH, V. V. LUNIN), Mos. Gos. Univ., Moscow, 1987, p. 86.

11  B. A. KOLACHEV, R. E. SHALIN, A. A. YLIYN, *Splavy – nakopiteli vodoroda* (*Alloys for Hydrogen Storage*), Metallurgy, Moscow, **1995**, p. 384.

12  O. V. CHETINA, O. V. LUNIN, G. V. ISAGULYANTS, *Neftekhimiya* **1988**, *28*, 757.

13  V. V. LUNIN, O. V. CHETINA, *Zh. Fiz. Khim.* **1990**, *64*, 3019.

14  K. A. MANOVYAN, P. V. AFANAS'EV, V. V. LUNIN, *Kinet. Katal.* **1992**, *33*, 566.

15  E. V. EVDOKIMOVA, V. V. LUNIN, P. V. AFANAS'EV, I. I. MOISEEV, *Mendeleev Commun.* **1993**, *1*, 1.

16  I. I. MOISEEV, E. N. EVDOKIMOVA, V. V. LUNIN, P. V. AFANAS'EV, A. E. GEKHMAN, A. R. GROMOV, *Dokl. Ross. Akad. Nauk* **1993**, *332*, 195.

17  G. ALEFELD, J. VOLKL (Eds.), *Hydrogen in Metals: Topics and Applied Physics; II: Application-Oriented Properties*, Springer, Berlin, **1978**, *29*, p. 430.

18  J. J. REILLY, R. H. WISWALL, *Inorg. Chem.* **1974**, *13*, 218.

19  O. K. SHENOY, B. D. DUNLAB, P. J. VICCARO, D. NIAREHOS, *Mössbauer Spectroscopy and Its Chemical Application* (Eds.: J. G. STEVENS, O. K. SHENOY), **1981**, p. 501.

**20** P. Fischer, *Mater. Res. Bull.* **1978**, *13*, 931.

**21** P. Thompson, *Phys. F.* **1980**, *10*, 57.

**22** D. I. Kochubey, V. V. Kriventsov, Yu. V. Maksimov, M. V. Tsodikov, F. A. Yandieva, V. P. Mordovin, O. A. Navio, I. I. Moiseev, *Kinetika I Katalyz* **2003**, *44*, 1 (in Russian).

**23** S. Sakaki, K. Ohkubo, *Inorg. Chem.* **1989**, *28*, 2583.

**24** D. I. Kochubey, *EXAFS Spectroscopy of Catalysts*, Nauka, Novosibirsk, **1992**.

**25** A. I. Ivanovsky, V. A. Gubanov, E. Z. Kurmaev, *Phys. Chem. Solids* **1985**, *46*, 823.

**26** M. V. Tsodikov, V. Ya. Kugel, E. V. Slivinskii, V. P. Mordovin, *Izv. Akad. Nauk, Ser. Khim.* **1995**, *10*, 2066.

**27** V. D. Rusanov, *Hydrogen Energy and Technology: Research and Development in Russia, Hydrogen Energy Progress* (Eds.: T. N. Veziroglu et al.), Frankfurt/Main, **1996**, p. 31.

**28** R. P. A. Sneeden, *J. Mol. Catal.* **1982**, *17*, 349.

**29** V. V. Lunin, O. V. Kryukov, in *Kataliz: Fundamental´nye i prikladnye issledovaniya Catalysis: Fundamental and Applied Studies* (Eds.: O. A. Petrich, V. V. Lunin), Izd-vo MGU, Moscow, **1987**, 86 (in Russian).

**30** P. Braunstein, D. Matt, D. Mobel, *Chem. Rev.* **1988**, *88*, 747.

**31** P. G. Jessop, T. Ikariga, R. Noyori, *Chem. Rev.* **1995**, *95*, 259.

**32** O. V. Krylov, A. Kh. Mamedov, *Usp. Khim.* **1995**, *64*, 935 *Russ. Chem. Rev.* **1995**, *64*, 877 (Engl. Transl.).

**33** H. Ando, Y. Matsumura, Y. Souma, *J. Mol. Catal., A: Chem.* **2000**, *154*, 23.

**34** M. Pijolat, V. Perrichon, M. Primet, P. Bussiere, *J. Mol. Catal.* **1982**, *17*, 367.

**35** T. Osaki, H. Toada, T. Horiuchi, H. Yamakita, *React. Kinet. Catal. Lett.* **1993**, *51*, 39.

**36** V. M. Gryaznov, S. G. Gul'yanova, Yu. M. Serov, V. D. Yagodovskii, *Zh. Fiz. Khim.* **1981**, *55*, 1306 *J. Phys. Chem. USSR* **1981**, *55* (Engl. Transl.).

**37** V. M. Gryaznov, Yu. M. Serov, N. B. Polyanskii, *Dokl. Akad. Nauk* **1998**, *359*, 647 *Dokl. Chem.* **1998** (Engl. Transl.).

**38** M. V. Tsodikov, V. Ya. Kugel, E. V. Slivinskii, V. P. Mordovin, *Izv. Akad. Nauk, Ser. Khim.* **1995**, 2066 *Russ. Chem. Bull.* **1995**, *44*, 1983 (Engl. Transl.).

**39** M. V. Tsodikov, V. Ya. Kugel, E. V. Slivinskii, V. G. Zaikin, V. P. Mordovin, G. Colon, M. C. Hidalgo, J. A. Navio, *Langmuir* **1999**, *15*, 6601.

**40** K. N. Semenenko, V. V. Burnashova, N. A. Yakovleva, E. A. Ganich, *Izv. Akad. Nauk, Ser. Khim.* **1998**, *47*, 214 *Russ. Chem. Bull.* **1998**, *47*, 209 (Engl. Transl.).

**41** The Netherlands Pat. 7513159 (1977).

**42** C. H. Cheng, D. E. Mendriksen, R. Eisenberg, *J. Am. Chem. Soc.* **1977**, *99*, 2791.

**43** O. V. Chetina, V. V. Lunin, G. V. Isagulyants, *Izv. Akad. Nauk SSSR, Ser. Khim.* **1988**, 2405 *Bull. Acad. Sci. USSR, Div. Chem. Sci.* **1988**, *37*, 2168 (Engl. Transl.).

**44** W. C. Conner, G. M. Pajonk, S. J. Teichner, in *"Spillover of Sorbed Species", Advances in Catalysis* **1984**, *34*, 1.

**45** S. J. Tauster, S. C. Fung, R. L. Garten, *J. Am. Chem. Soc.* **1978**, *100*, 170.

**46** K. Fujimoto, S. Toyoshi, in *Proc. 7th Int. Congr. Catal.* **1981**, 235.

**47** K. Weissermel, H. J. Arpe, *Industrial Organic Chemistry*, VCH, Weinheim, **1993**.

**48** A. Ya. Rozovsky, G. N. Lin, *Theoretic Basis of the Methanol Synthesis Process*, Khimiya, Moscow, **1990** (in Russian).

**49** J. March, *Advanced Organic Chemistry. Reaction, Mechanisms and Structures*, Wiley, New York, **1992**, p. 969–970.

**50** S. V. Lebedev, *Zh. Obsh. Khimii* **1933**, *3*, 698.

**51** I. I. Ostromislensky, *Zhurn. Russ. Phys.-Chem. Obsh.* **1915**, *47*, 1494.

**52** K. V. Vatsuro, G. L. Mishchenko, *Named Reactions in Organic Chemistry*, Khimiya, Moscow, **1976**.

**53** M. Guerber, *Compt. rend.* **1899**, *128*, 511; **1912**, *155*, 1156.

**54** C. Carlini, M. D. Girolamo, M. Marchionna, A. M. R. Galleti, G. Sbrana, *J. Mol. Catal., A: Chem.* **2002**, *84*, 273.

**55** C. Carlini, M. D. Girolamo, A. Maciani, M. Marchionna, M. Noviello, A. M. R. Galleti, G. Sbrana, *J. Mol. Catal., A: Chem.* **2003**, *200*, 137.

56 B. Cornils, W. A. Herrmann, R. Schlögl, C. H. Wong (Eds.), *Catalysis from A to Z – A Concise Encyclopedia*, Wiley-VCH, Weinheim, 2000.

57 M. J. Astle, *The Chemistry of Petrochemicals*, Reinhold, New York, 1956, p. 30.

58 J. T. Kummer, H. H. Padurski, W. B. Spencer, P. H. Emmett, *J. Am. Chem. Soc.* 1951, *73*, 564.

59 J. T. Kummer, P. H. Emmett, *J. Am. Chem. Soc.* 1953, *75*, 5177.

60 W. K. Hall, R. J. Kokes, P. H. Emmett, *J. Am. Chem. Soc.* 1957, *79*, 2984.

61 H. H. Podgurski, P. H. Emmett, *J. Phys. Chem.* 1953, *57*, 159.

62 Ju. B. Kryukov, A. N. Bashkirov, R. A. Friedman, L. B. Liberov, R. M. Smirnova, G. T. Lyssenko, A. A. Pegov, *Petrochemistry* 1964, *6*, 868 (in Russian).

63 Ju. B. Kryukov, A. N. Bashkirov, R. A. Friedman, L. B. Liberov, R. M. Smirnova, G. T. Lyssenko, A. A. Pegov, *Petrochemistry* 1965, *1*, 62 (in Russian).

64 Yu. B. Kogan, Ju. B. Kryukov, E. V. Kamzolkina, A. N. Bashkirov, *Izv. Acad. Nauk SSSR. Ser. Khim.* 1952, 649 (in Russian).

65 S. D. Mekhtiev, A. F. Aliev, S. M. Imamova, *Doklady Akad. Nauk SSSR* 1954, *99*, 773 (in Russian).

66 W. Ipatiev, *Katalytische Reaktionen bei hohen Temperaturen und Drucken. XIV. Hydrogenisation des Benzylaldehydes und des Benzylalkohols in Gegenwart von Eisen*, Ber. 1908, *41*, 993.

67 Yu. Kryukov, A. N. Shuikin, Z. T. Shumaeva, A. N. Bashkirov, *Petrochemistry* 1970, *10*, 83 (in Russian).

68 K. B. Sharpless, M. A. Umbreit, M. T. Nich, T. C. Flood, *J. Am. Chem. Soc.* 1972, *94*, 6538.

69 I. A. Kop'eva, I. A. Oreshkin, B. A. Dolgoplosk, *Izv. Akad. Nauk SSSR. Ser. Khim.* 1988, 1912 (in Russian).

70 M. V. Tsodikov, V. Ya. Kugel, F. A. Yandieva, G. A. Kliger, L. S. Glebov, A. I. Mikaia, V. G. Zaikin, E. V. Slivinsky, N. A. Plate, A. E. Gekhman, I. I. Moiseev, *Kinet. Catal.* 2004 (in press).

71 G. A. Olah, A. Molnar, *Hydrocarbon Chemistry*, Wiley, New York, 1995, pp. 16, 85, 88, 527.

72 T. Nishino et al., *Tetrahedron Lett.* 2002, *43*, 3689.

73 T. A. Stromnova, I. N. Busygina, S. B. Katser, A. S. Antsyshkina, I. I. Moiseev, *Doklady Chemistry* 1987, *295*, 890 (in Russian).

74 O. M. Nefedov, A. I. Ioffe, L. G. Men'shikov, *Chemistry of Carbenes*, Khimiya, Moscow, 1990 (in Russian).

75 D. R. Stull, E. F. Westrum, G. C. Sinke, *The Chemical Thermodynamics of Organic Compounds*, Wiley, New York, 1969.

76 Y. Volokitin, J. Sinzig, L. J. (d)e Jongh, G. Schmidt, M. N. Vargaftik, I. I. Moiseev, *Nature* 1996, *384*, 621.

77 L. A. Abramova, S. P. Baranov, A. A. Dulov, M. N. Vargaftik, I. I. Moiseev, *Doklady Akad. Nauk* 2001, *377*, 344 (in Russian).

# Subject Index

*Multimetallic Catalysts in Organic Synthesis*. Edited by M. Shibasaki and Y. Yamamoto
Copyright © 2004 WILEY-VCH Verlag GmbH & Co. KGaA, Weinheim
ISBN: 3-527-30828-8